OXFORD MONOGRAPHS ON GEOLOGY AND GEOPHYSICS NO. 14

OXFORD MONOGRAPHS ON GEOLOGY AND GEOPHYSICS

1. De Verle P. Harris: *Mineral resources appraisal: mineral endowment, resources, and potential supply: concepts, methods, and cases*
2. J. J. Veevers (ed.): *Phanerozoic earth history of Australia*
3. Yang Zunyi, Cheng Yuqi, and Wang Hongzhen: *The geology of China*
4. Lin-gun Liu and William A. Bassett: *Elements, oxides, and silicates: High-pressure phases with implications for the Earth's interior*
5. Antoni Hoffman and Matthew H. Nitecki (eds.): *Problematic fossil taxa*
6. S. Mahmood Naqvi and John J. W. Rogers: *Precambrian geology of India*
7. Chih-Pei Chang and T. N. Krishnamurti (eds.): *Monsoon meteorology*
8. Zvi Ben-Avraham (ed.): *The evolution of the Pacific Ocean margins*
9. Ian McDougall and T. Mark Harrison: *Geochronology and thermochronology by the* $^{40}AR/^{39}AR$ *method*
10. Walter C. Sweet: *The conodonta: morphology, taxonomy, paleoecology, and evolutionary history of a long-extinct animal phylum*
11. H. J. Melosh: *Impact cratering: a geologic process*
12. J. W. Cowie and M. D. Brasier (eds.): *The Precambrian-Cambrian boundary*
13. C. S. Hutchison: *Geological evolution of south-east Asia*
14. Anthony J. Naldrett: *Magmatic sulfide deposits*

Magmatic Sulfide Deposits

ANTHONY J. NALDRETT
University of Toronto

New York · Oxford
CLARENDON PRESS · OXFORD UNIVERSITY PRESS
1989

Oxford University Press

Oxford New York Toronto
Delhi Bombay Calcutta Madras Karachi
Petaling Jaya Singapore Hong Kong Tokyo
Nairobi Dar es Salaam Cape Town
Melbourne Auckland

and associated companies in
Berlin Ibadan

Published by Oxford University Press, Inc.,
200 Madison Avenue, New York, New York 10016

Oxford is a registered trademark of Oxford University Press

Library of Congress Cataloging-in-Publication Data

Naldrett, A. J.
Magmatic sulfide deposits / Anthony J. Naldrett
p. cm.—(Oxford monographs on geology and geophysics ; no. 14)
Bibliography: p.
Includes index.
ISBN 0-19-505119-X
1. Sulfides. 2. Rocks, Igneous. I. Title. II. Series.
QE389.2.N35 1989
552′.1—dc20 89-3409
CIP

9 8 7 6 5 4 3 2 1

Printed in the United States of America

PREFACE

What we have to do is to be for ever curiously testing new opinions and courting new impressions, never acquiescing in a facile orthodoxy. . . . Philosophical theories or ideas, as points of view, instruments of criticism, may help us to gather up what might otherwise pass unregarded by us "Philosophy is the microscope of thought."

From the "Conclusion" to Pater's *The Renaissance: Studies in Art and Poetry*

New concepts in relation to magmatic sulfide ores have appeared particularly rapidly over the last 25 years. Twenty-five years ago ultramafic lavas were thought impossible and Kambalda meant gold not nickel; the meteorite impact origin of the Sudbury structure was an idea that had just arrived in one person's mind, and was not a viable hypothesis; the knowledge of the distribution, let alone laws governing the geochemical behavior of the platinum group elements, was minimal; the external controls governing the segregation of massive Ni–Cu sulfide deposits from their ultramafic or mafic source magmas was only just being appreciated, most workers were still seeking for internal controls; and our understanding of internal processes in layered intrusions was still firmly rooted in Skaergaard-based, closed-system crystal settling models because the application of fluid dynamics, except for Stoke's Law, lay far in the future. As readers continue to turn the pages of this volume, they will discover that it is impossible to think about magmatic ores today without also considering one or more of these concepts. They will also discover that some of the concepts are mutually exclusive.

It is great inconvenience for a man, in a haste, that there are three distinct roads between Calais and Paris; in behalf of which, there is so much to be said by the several deputies from the towns which lie along them that half a day is easily spent in settling which you'll take.

From Volume 7, Chapter III of Laurence Sterne's *Life and opinions of Tristam Shandy, Gent.*

This book is an attempt to freeze and preserve, at this instant in time, a description of the roads that lead to an understanding of magmatic sulfide ores. I make no excuses that it is a personal view in which certain roads are favored over others. Hopefully, this may be helpful to the "man in a haste." My view draws heavily on the experiences that I have been fortunate enough to have had. In this respect I am grateful to Falconbridge Ltd., particularly to Mr. G. P. Mitchell and to Mr. A. M. Clarke for the opportunity to first become acquainted with the subject as a mine geologist in Sudbury in 1957. Both Dr. A. R. Graham, research geologist for Falconbridge, and Mr. T. Podolsky, then research geologist for INCO Ltd., encouraged my early work, assisting me with access to deposits and, from time to time, with company information. The late Professor J. E. Hawley of Queen's University taught me how to approach research on ore deposits, and Dr. G. Kullerud gave me the opportunity to work with him and enjoy the stimulating environment of the Geophysical Laboratory and of Geoscience in general in Washington in the mid-1960s. A succession of very understanding chairmen in the Department of Geology of the University of Toronto have since provided an almost ideal environment in which to work. I am also most grateful to Mr. W. Ewers and to the late Dr. A. Gaskin of the CSIRO Division of Mineralogy, Australia; to Professor G. von Gruenewaldt of the University of Pretoria's Institute for Geological Research on the Bushveld Complex; and to Dr. Z. Johan of the Centre de Recherches sur la Synthese et Chimie des Mineraux, CNRS, France for making it possible for me to spend sabbatical years at their institutions. The benefits that I gained from these years will be immediately apparent to the readers of this book. I owe particular thanks to Dr. Johan and the CNRS since much of the revision of the book was undertaken during my 12 months with them. I also thank Dr. C. Coats of Falconbridge Ltd., Dr. J. M. Duke of the Geological Survey of Canada, Professor A. C. Brown of the University of Montreal, and Dr. M. Zientek of the U.S. Geological Survey for their comprehensive reviews of early versions of the manuscript.

The true success of every academic is best assessed by the success of his students. I have been very fortunate and am most proud of mine. Their success is written in the bibliography of this book.

Last, but most important, I thank my wife, Sylvia, who has not only provided verbal encouragement on the many occasions it was needed, but who for one year demonstrated this by leaving the house at 3 A.M. several mornings per month to keep the neutron activation analysis going at a critical stage in the studies. Her contribution was neither rewarded financially nor is it adequately acknowledged in the bibliography.

Toronto A.J.N.
February 1989

CONTENTS

Magmatic Sulfide Deposits

1. INTRODUCTION

1.1 DISTRIBUTION OF THE TRANSITION METALS

This book is concerned with magmatic sulfide deposits, which are one of the principle concentrations of the group VIII transition metals: iron (Fe), cobalt (Co), nickel (Ni), palladium (Pd), platinum (Pt), rhodium (Rh), ruthenium (Ru), iridium (Ir), and osmium (Os), that are available to humanity. These metals are most heavily concentrated in the core of the earth, becoming progressively less abundant in the successive outer shells of the mantle and the crust. Table 1.1 lists the probable concentrations of certain of these metals in carbonaceous chondrites, meteorites whose composition is best believed to reflect the composition of our planetary system apart from hydrogen and helium, and in the earth as a whole. These two estimates are quite similar, the earth being perhaps two to three times richer in these metals than the planetary system.

Table 1.1 also gives the average concentrations of certain group VIII metals in the earth's crust. With the exception of Fe, these metals are rarely found in the crust, being present at concentrations of only 1/100 to 1/1,000 of that of the bulk earth. The outstanding chemical characteristic of the group VIII metals is their siderophile nature—their tendency to alloy with Fe. At a very early stage in the earth's evolution they were dissolved in native Fe, moving with it as it sank to the center of the earth to form the core. Although Fe was the main instrument of the distribution of these metals, its general abundance and its ability to combine with oxygen more easily than other transition metals has resulted in significant amounts of Fe remaining in the oxidized outer portions of the earth.

Most of the other transition metals are available in workable concentrations because, first, they are 5 to 30 times more concentrated in the mantle than in the crust (Table 1.1). Second, there has been a constant, if slow, interchange of material between the mantle and the crust throughout the earth's history and, under certain circumstances, this interchange has given rise to economic concentrations of metals.

In the case of Ni, a concentration of about 10,000 ppm or greater is required for an ore body, 4 times that of typical mantle material. In the case of Pt, most deposits are mined at concentrations of 5 ppm or greater, a factor of 70 times the level in the mantle. Of course, a combination of Ni and Pt, along with other valuable metals, such as copper (Cu) and gold (Au), allows a deposit to be worked at concentrations lower than those given for the individual metals.

1.2 CONCENTRATION OF NI AND OTHER TRANSITION METALS IN THE CRUST

Nickel and the other group VIII transition metals are transferred from the mantle and concentrated in the crust in three principal ways:

1.2.1 Laterite Formation

This normally involves the thrusting of large slabs of mantle into the crust along zones of weakness such as subduction zones and major lateral faults. The slabs are not mineable in themselves, but when they are stripped of overlying rocks and exposed to the combined effects of rain water, oxygen, and carbon dioxide from the atmosphere, the constituent minerals break down and elements, such as silicon (Si) and magnesium (Mg), are leached and carried away to the oceans. This chemical erosion leaves a residue enriched in Fe and Ni. Under ideal conditions, usually those found in tropical or equatorial areas, Ni is also leached but redeposited toward the bottom of the weathered zone to form a mineable concentration of Ni laterite, assaying generally from 15,000 to 30,000 ppm Ni.

Where erosion involves both the physical disintegration and the chemical dissolution of rocks, the minute, very dense particles of Pt and other platinum group elements (PGE) originally occurring as inert native metals in mantle slabs are sometimes concentrated in stream bed gravels.

1.2.2 Segregation of Magmatic Sulfides

The formation of a magmatic sulfide ore is illustrated schematically in Figure 1.1. Portions of the mantle melt to give rise to ultramafic or basaltic magma, which then rises upward because of its low density relative to the mantle to intrude the crust or to extrude as lava. This magma is usually not particularly concentrated in transition metals. Either before or while it is cooling and crystallizing, however, certain processes (discussed in

TABLE 1.1. *Abundance of Some Transitional Metals in Earth and Planetary Materials (ppm)*

	Fe	Ni	Pd	Ir	Pt
Carbonaceous chondrites (Planetary system)	270,000	16,000	0.545	0.540	1.02
Total earth (32.4% of core + 67.6% of mantle)	334,000	30,000	1.25	1.14	2.79
Earth's crust	50,000	75	0.001	0.001	0.005
Earth's mantle (Alpine peridotite)	60,000	2,500	0.018	0.006	0.073
Earth's core (Iron meteorites)	907,000	88,000	3.82	3.53	8.45

Chapter 4) may cause droplets of liquid sulfide to form within it. In addition to being siderophile, the group VIII transition metals are also chalcophile; that is, they tend to form bonds with sulfur rather than with the oxygen, to which they are forced to bond when dissolved in a silicate magma. Thus the sulfides concentrate the transition metals by factors ranging from 100 to 100,000 over their initial levels in the host magma. The droplets are much denser than the enclosing magma and settle to form concentrations at the base or at other specific horizons within the cooling body of magma. The sulfide concentrations cool and solidify, becoming important targets for mining, as they often contain 20,000 to 50,000 ppm Ni.

Platinum group elements (PGE) occur in and are recovered from most magmatic sulfide deposits, but certain ores, which contain particularly high concentrations of PGEs, are economic primarily because of their contents of these metals. Because of the very low concentrations of PGE in natural magmas, ores with high PGE concentrations form only under special conditions, in which sulfides have the opportunity to concentrate metals from a much larger relative proportion of magma than is necessary for economic Ni deposits. These conditions appear to develop in layered intrusions, especially the large stratiform complexes. At present, there is considerable debate over whether the sulfides are brought into contact with a large mass of magma, thus scavenging the PGE, or whether the PGE are scavenged by aqueous fluids, which bring them to the sites where the sulfides accumulate. This is discussed in Chapter 5. Concentrations of PGE characterizing the ores noted primarily for their Ni and Cu range from 0.2 to 2 ppm, whereas ores noted primarily for their PGE generally contain 4 ppm Pt or more.

1.2.3 Formation of Deep Sea Fe–Mn Nodules and Crusts

In the deep oceans, particularly adjacent to the ocean ridges where basaltic magma is constantly rising from the mantle to form the ocean floor, seawater is drawn down into the basalts and the underlying mantle. It is heated from contact with the recently molten lavas, leaching metals from them. The heated water rises, emerging at the ocean floor as hot springs and contributing small

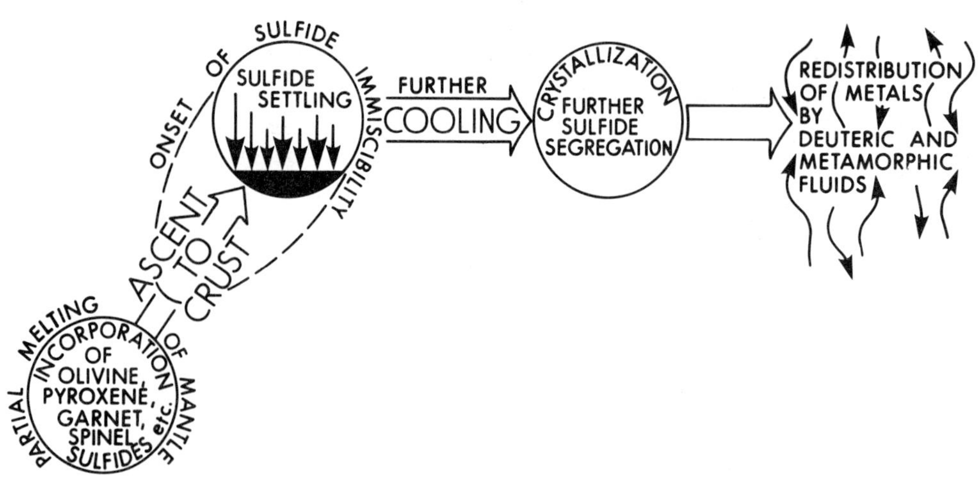

Fig. 1.1. Stages in the formation of a typical magmatic ore.

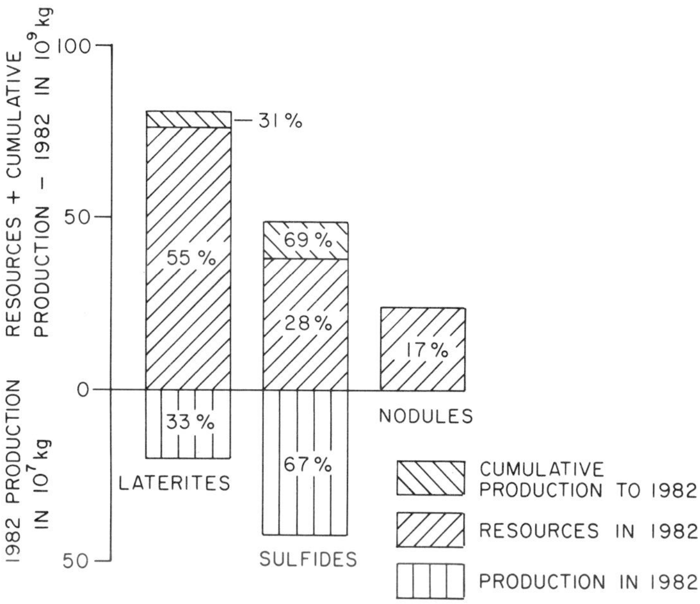

Fig. 1.2. Resources and typical annual production of Ni in the three principal deposit-types: magmatic sulfides, Ni-laterites and deep-sea Mn nodules.

quantities of dissolved metals to those already present. Some of these metals, together with contributions from the weathering of landmasses, are extracted from the seawater in small, cauliflower-shaped nodules, rich in Fe and manganese (Mn), which grow on the ocean bottoms. Typical metal-rich Mn nodules contain 240,000 ppm Mn, 16,000 ppm Ni, 14,000 ppm Cu, and 2,100 ppm Co. Mn crusts form at shallower depths than the nodules (500–1500 m), are poorer in Ni (8,000 ppm), richer in Co (15,000 ppm), and contain up to 1 ppm Pt.

Other generally more complicated mechanisms have also given rise to concentrations of certain group VIII transition metals. As examples, Ni is highly concentrated in uranium (U) ores such as those found in northern Saskatchewan in Canada; Co is enriched in the ores of the Zambian and Katangan Copper Belt; and PGE are concentrated in certain parts of the Kupferschiefer in Poland and in the Saskatchewan uranium ores. These mechanisms have usually involved several intermediate steps in the movement of the metals from the mantle to their present sites in the crust. With the exception of Co in the Copper Belt, however, they have not contributed significantly to the production of these metals.

The relative contributions that different types of deposits have made to the production of Ni in the past and contribute to our present resources of Ni are illustrated in Figure 1.2. Sulfides have made the greatest contribution, but laterites constitute the major part of the resources and Mn nodules are also a significant resource. The energy required to extract a kilogram of Ni from a typical ore adds a considerable amount to the cost and varies in the ratio of about 1:2:3 for sulfide, Mn nodules, and laterite ores, respectively. On the other hand, sulfide ores are the most costly to mine, and Mn nodules suffer disputes over the ownership and their commercial-scale mining has yet to be tested. The contribution of the three ore types to future production will depend on how mining costs, particularly wages, increase relative to energy costs, and on technologic progress in the extraction of Ni from its ores. The slackening in demand and prices and the increase in the cost of energy that has affected the metal mining industry in the 1970s and early 1980s, has led to a re-emphasis of the importance of magmatic sulfide ores with respect to the other types of deposits during this period.

1.3 ARGUMENT IN SUPPORT OF A MAGMATIC ORIGIN FOR NI-CU-PGE ORES

We are concerned here with deposits believed to have formed as a result of the magmatic process outlined earlier. The process is widely, but not universally, accepted. Wandke and Hoffman (1924) and Fleet (1977) have suggested that the nearly massive deposits of Ni–Cu sulfides at Sudbury, Ontario, are hydrothermal concentrations; Lusk (1976) has suggested that ores related to ultramafic lavas (komatiites) in Western Australia are exhalative; Barrett et al. (1977) proposes that the same ores owe much of their economic potential to metamorphic processes that concentrate pre-existing, rather dispersed sulfides of magmatic origin. Stumpfl (1986), Johan and Watkinson (1985), and Boudreau (1987) believe that aqueous fluids have played a major role in the concentration of PGE in the Merensky and J–M Reefs of the Bushveld and Stillwater Complexes, respectively.

Naldrett (1979) presented one of the more recent reviews of magmatic *versus* nonmagmatic arguments. He argued in favor of a magmatic origin, citing the observations that:

1. "Where they (ores mined primarily for their Ni and Cu as opposed to their PGE) occur in relatively undeformed rocks, they are almost invariably concentrated toward the stratigraphic base of the mafic or ultramafic unit within which they are associated";
2. "The elements concentrated in the ores are precisely those which would partition into the sulfide phase under magmatic conditions and" conversely "some elements (such as Zn) which would undoubtedly be concentrated hydrothermally are present in conspicuously less concentrations (in magmatic sulfide ores)";
3. "The common occurrence of sulfides interstitial to the silicates of magmatic rocks—particularly those that are relatively undeformed" supports a magmatic origin; and
4. "With regard to both Wandke and Hoffman's (1924) and Fleet's (1977) emphasis on hydrothermal alteration at Sudbury 'with the exception of (small amounts of) blue amphibole, hydrothermal alteration minerals are no more prevalent in ore-bearing than in ore-free norite at Sudbury.' This is to say that the hydrous alteration resulting from the interaction of hydrothermal fluids and the wall-rocks of a deposit is lacking at Sudbury, which makes it unlikely that large quantities of aqueous fluid have been involved in ore formation."

Naldrett (1979) also cites the "billiard-ball model" (discussed in Chapter 4) in support of a magmatic origin.

Arguments that a given Ni–Cu ore is not magmatic can be countered for all except a few examples, whereas alternative models of ore genesis, which would rule out the gravitational settling of sulfide as an important process of ore concentration, do not provide a satisfactory explanation for many of the fundamental characteristics of most of the deposits. Among the exceptions, Sherlock Bay (Miller and Smith, 1975; Keays et al., 1982) and the Ni–U ores of northern Saskatchewan are convincing evidence that Ni can be mobilized within a hydrothermal medium and precipitated as a sulfide. The PGE concentrations in the Polish Kupferschiefer (Kucha, 1982) and the Saskatchewan uranium deposits are evidence that PGE can be similarly mobilized. Undoubtedly, some remobilization has occurred at many Ni sulfide deposits as a result of deuteric processes during cooling, or subsequently during metamorphism. In most of the undeformed deposits, the original magmatic origin is clear. Readers are invited to form their own opinion after reading this book.

1.4 OUTLINE OF THIS BOOK

A number of attempts have been made to classify magmatic sulfide ores, including those of Naldrett (1973), Besson et al. (1979), Marston et al. (1981), and Hammerbeck (1984). Given the magmatic model discussed earlier, the nature of the associated mafic or ultramafic rocks and the magma that produced them has been regarded as important to their genesis and thus to any genetically based classification scheme. Chapter 2 provides a classification of these rocks, together with a brief discussion of their origin and consideration of the types of rocks or magmas that are most likely to host these ore deposits. Because it is difficult to think conceptually about the formation of these deposits without an understanding of (1) the physical chemistry governing their melting and crystallization, (2) the dissolution of sulfur in magmas, and (3) the scavenging ability of liquid sulfides for different metals, these three topics are the focus of Chapter 3. The heart of the book is found in Chapters 4 and 5. In Chapter 4, descriptions of a wide range of Ni–Cu deposit types are given and their modes of origin discussed. The distinctly different problems presented by the PGE-rich deposits of the Bushveld and Stillwater Complexes are described in Chapter 5. The compositions of magmatic sulfide deposits are described in Chapter 6. Also included in Chapter 6 is a discussion of the interaction between sulfide and silicate liquids and the implications of this interaction with respect to the composition of the sulfide ores. Finally, in Chapter 7, geologic and geochemical aspects that have particular implications for exploration are emphasized.

2. CLASSIFICATION OF MAFIC AND ULTRAMAFIC ROCKS AND RELATED MAGMATIC SULFIDE DEPOSITS

A number of attempts have been made to classify bodies of mafic and ultramafic rocks, including those of Wyllie (1967), Naldrett (1973), Naldrett and Cabri (1976), and Besson et al. (1979). As with most other igneous rocks, this is best done in terms of the petrology and chemistry of the rocks, and the tectonic setting in which they were emplaced. A major problem with Archean rocks in this respect is that the tectonic setting of their emplacement is usually not very clear. The scheme in Table 2.1 is adapted from those of Naldrett and Cabri (1976) and Besson et al. (1979). In it, bodies are divided initially into whether they were emplaced in a rifted continental margin–ocean basin environment (category II), a continental environment (category III), or an active orogenic belt (category IV). Archean greenstone belts represent, for the most part, an enigma in terms of present day plate tectonics. For example, were komatiites erupted through continental crust (Arndt, 1986a, Compston et al., 1986) or do they represent the floor of a primitive ocean (de Wit et al., 1987)? Although geochemical evidence can prove continental contamination in certain instances, the absence of such contamination is not proof of an oceanic environment, as Huppert and Sparks (1985b) have shown that the extent of contamination is related to the rate of eruption. Because of these uncertainties, a separate category (category I) has been created for synvolcanic activity in this environment.

2.1 SYNVOLCANIC MAFIC AND ULTRAMAFIC BODIES (LARGELY IN ARCHEAN GREENSTONE BELTS)

Experience in Archean greenstone belts has shown that synvolcanic ultramafic rocks form parts of bodies that fall into two main classes, komatiites and tholeiites, and that the tholeiites have two distinctive subclasses, one picritic and the other largely gabbroic.

2.1.1 Komatiites

Some confusion arose over what constitutes a komatiite. The name was proposed by Viljoen and Viljoen (1969) for rocks well represented in the Komati River area of the Barberton Mountain Land. Certain aspects of their initial definition proved to be somewhat too restrictive. Arndt and Nisbet (1982), reporting a consensus reached after the 1979 Penrose conference on komatiites, proposed that the term *komatiite* be redefined to include any ultramafic volcanic rock, including fragmentals but excluding cumulates from basaltic magma. The lower limit for inclusion as ultramafic would be set arbitrarily at 18 wt percent MgO. The term *komatiitic basalt* is to be applied to those rocks containing less than 18 wt percent MgO and showing a definite spatial and chemical link with komatiites.

Both extrusive and intrusive variants of komatiitic rocks are generally present in a given area, in some cases with the extrusive variant predominant (e.g., Munro Township, Arndt et al., 1977) and in others with the intrusive variant predominant (e.g., the Dumont Sill, Duke, 1985). There is clear evidence for the existence of peridotitic liquids containing up to 33, but more commonly 20 to 30, wt percent MgO. Some of these liquids may have carried phenocrysts in suspension; accumulation of these phenocrysts together with crystals forming subsequent to extrusion in the lower part of the flow (Arndt, 1986b), coupled with upward movement of less dense fractionated liquid (Turner et al., 1986), and possibly some adcumulus growth, is partially responsible for the dunites and more magnesian peridotites. It has also been appreciated that some of the more magnesian cumulates represent the crystallization and accumulation of olivine from magma flowing along channels or conduits, representing lava rivers or tubes (Barnes et al., 1983; Lesher et al., 1984).

Platy and skeletal growth of olivine and clinopyroxene crystals, known as spinifex texture, characterize the upper parts of some flows and near-surface intrusions (Nesbitt, 1971; Pyke et al., 1973; Arndt et al., 1977). In other flows a zone of spinifex texture is lacking, although skeletal overgrowths to phenocrysts and interstitital skeletal pyroxene are present. Spinifex texture was initially attributed to inhibited nucleation and undercooling (Pyke et al., 1973; Donaldson, 1982), but recently it has been proposed (Turner et al., 1986) that the more platy variants of spinifex texture are due to the growing crystals having had to extend from nucleation sites on the roof, down below accumulations of low density fractionated

TABLE 2.1. *Classification of Mafic and Ultramafic Bodies*

Class of Body	Examples
I. Synvolcanic Environment (largely Archean greenstone belts)	
1. Komatiites	
a. Lava flows	
	Barberton, Southern Africa Viljoen & Viljoen (1969)
	Kambalda area, W. Australia Nesbitt (1971) Groves et al. (1986)
	Munro, Ontario, Canada Pyke et al. (1973) Arndt et al. (1977) Arndt & Naldrett (1986)
	General Arndt & Nisbet (1982) Huppert & Sparks (1985a) Turner et al. (1986)
b. Dunite–Peridotite lenses	
	W. Australia Binns & Groves (1976) Naldrett & Turner (1977) Groves & Keays (1979) Donaldson et al. (1984) Gole et al. (1986)
	Canada Dumont Sill, Quebec Duke (1985,1986)
2. Tholeiites	
a. Picritic bodies	
	Kakagi Sill, Ontario, Canada Ridler (1966) Dundonald Sill, Ontario, Canada Naldrett & Mason (1968) Centre Hill Complex, Ontario, Canada McRae (1969)
	Kaapmuiden, Barberton Viljoen & Viljoen (1970)
	Kalgoorlie–Norseman, W. Australia Hallberg & Williams (1972)
b. Gabbroic bodies	
	Dore Lake Complex, Quebec, Canada Allard (1970) Bell River Complex, Quebec, Canada Sharp (1965) Scott (1980)
	Big Trout Lake Complex, Ontario, Canada Sage and Troop (1974) Borthwick & Naldrett (1984) Mulcahy Complex, Ontario Morrison et al. (1986)
	Bird River Sill, Manitoba, Canada Trueman (1971) Scoates (1983)
	Windimurra Intrusion, W. Australia Parks & Hill (1986)
II. Associated with Rifted Plate Margins and Ocean Basins	
1. Floored by and Closely Associated with Continental Crust	
a. Komatiitic	
	Cape Smith, Quebec, Canada Wilson et al. (1969) Hynes & Francis (1982 a, b) Barnes et al. (1982) Dillon-Leitch et al. (1986)

Class of Body	Examples
	Thompsom Belt, Manitoba, Canada Quirke et al. (1970) Coats et al. (1972) Peredery (1979)
b. Largely gabbroic	
	Fox River Sill, Manitoba, Canada Scoates and Eckstrand (1986)
	Labrador Trough, Quebec-Labrador, Canada Baragar and Scoates (1981)
	Kemi–Koillismaa Belt, Finland Soderholm and Inkinen (1982) Alapieti et al. (1979) Alapieti (1982) Alapieti and Lahtinen (1986)
	Skaergaard, Thulean Province
2. Not floored by Continental Crust a. Ophiolite complexes	
	New Caledonia Prinzhofer et al. (1980) Dupuy et al. (1981)
	Newfoundland Church & Riccio (1977) Casey & Karson (1981) Elthon et al. (1982)
	Cyprus Gass (1967) Gass and Smewing (1973) Robertson & Woodcock (1980)
	Turkey Juteau et al. (1977)
	U.S.A. Coleman and Irwin (1977)
	Greece Moores (1969) Bebien et al. (1980) Ross et al. (1980) Jackson et al. (1975)
III. Intrusions in Cratonic Areas 1. Intrusions Related to Flood Basalts	
	Duluth Complex, Minnesota, U.S. Weiblen & Morey (1980)
	Noril'sk Talnakh, U.S.S.R. Godlevskii (1959) Zolutuchin & Vasilev (1967) Genkin et al. (1981)
	Dufek Intrusion, Antarctica Ford & Boyd (1968) Ford (1970) Himmelberg & Ford (1976)
	Insizwa–Ingeli Complex, South Africa Maske (1966) Lightfoot & Naldrett (1983) Lightfoot et al. (1984)
	Ontario Nipissing Diabase, Canada Hriskevitch (1968) Card & Pattison (1973) Lightfoot et al. (1986)
2. Large-layered Complexes With no Documented Relation to Flood Basalt a. Sheetlike i. With repetitive layering	
	Bushveld Complex, South Africa Hall (1932)

TABLE 2.1. *Classification of Mafic and Ultramafic Bodies*

Class of Body	Examples
	Wager & Brown (1968)
	Wisser & von Gruenewaldt (1970)
	Hunter (1976)
	von Gruenewaldt (1979)
	von Gruenewaldt et al. (1985)
	Stillwater Complex, Montana, U.S.
	Jones et al. (1960)
	Hess (1960)
	Jackson (1961)
	Page (1977)
	Todd et al. (1982)
	Segerstrom & Carlson (1982)
	Irvine et al. (1983)
	Czamanske & Zientek (1985)
	Muskox Complex, Canada
	Irvine & Smith (1967)
	Irvine (1970)
	Irvine (1980)
ii. Without repetitive layering	
	Sudbury, Canada
	Collins (1934, 1936)
	Naldrett et al. (1970)
	Pye et al. (1984)
	Naldrett et al. (1986)
b. Dykelike	
	Great Dyke, Zimbabwe
	Worst (1960)
	Wilson (1982)
	Wilson & Prendergast (1987)
	Prendergast (1988)
	Jimberlana, W. Australia
	Campbell (1977)
	Keays & Campbell (1981)
IV. Bodies Intruded in Orogenic Belts	
1. Synorogenic Intrusions	
	Aberdeenshire Gabbros, Scotland
	Wadsworth (1970)
	Boyd & Munro (1978)
	Ashcroft & Munro (1978)
	Rona, Norway
	Boyd & Mathieson (1979)
	Seiland Province, Norway
	Robins & Gardner (1974)
	Sturt et al. (1978)
	Robins (1982)
	Fongen-Hyllingen Complex, Norway
	Wilson (1987)
	Moxie Intrusion, U.S.A.
	Boone (1974)
	Thompson & Naldrett (1984)
2. Alaskan-type Complexes	
	Duke Island, Alaska, U.S.A.
	Irvine (1974)
	Union Bay, Alaska, U.S.A.
	Ruckmick & Noble (1959)
	Tulameen, British Columbia, Canada
	Findlay (1969)
	Northern California, U.S.A.
	Gray et al. (1986)
	Urals
	Columbia
	Venezuela

Class of Body	Examples
3. Alkalic Bodies	
	Gardar Province, Greenland Ferguson (1964) Emeleus & Upton (1976) Parsons (1979, 1981) Bailey et al. (1981)
	Kola Peninsula, U.S.S.R. Vlasov et al. (1959) Atamanov et al. (1962) Gerassimovsky et al. (1966)
	a. Kimberlites Numerous
	b. Carbonatites Numerous

liquid, which ponded below the chilled crust forming the roof. The chilled upper margin is thought to represent the initial composition of a flow (Arndt, 1986b). Margins containing 23 wt percent MgO have been shown to melt at about 1,425°C (Shima and Naldrett, 1975), those with 33 wt percent MgO around 1,650°C (Green et al., 1975).

When a flow keeps moving while it is cooling, turbulent convection is thought to prevent the formation of anything but the thinnest crust and to keep most of the crystallizing olivine in suspension until it amounts up to 50 volume percent. At this stage the increase in viscosity due to the suspended load of crystals results in the flow congealing without the further settling of olivine, thus giving rise to a nearly uniform flow with no spinifex texture (Arndt, 1986b).

2.1.2 Tholeiites

2.1.2.1 PICRITIC SUBCLASS

Ultramafic rocks belonging to the picritic subclass occur as basal accumulations in differentiated sills or flows. They also occur as zones of hyaloclastite or aphanitic to fine-grained pyroxenite capping the flows, or as vent breccias probably related to the sills. A typical sill of this subclass, such as the Dundonald Sill, consists of a basal, 320-m-thick layer of peridotite overlain by 130 m of augite pyroxenite, capped by as much as 260 m of gabbro, within which zones of granophyre up to 60 m thick occur toward the top (Naldrett and Mason, 1968). A similar stratigraphy is observed in the lower portions of differentiated flows belonging to this subclass, although in these flows the gabbro grades upward into a fine-grained pyroxenite capped by a hyaloclastite of uncommonly MgO-rich (13–15 wt %) composition (Arndt et al., 1977). The compositions of hyaloclastites of this type (50 wt % SiO_2, 14 wt % MgO, 0.9–1.52 wt % TiO_2) are thought to represent the initial composition of the magma responsible for the flows and sills. They contain 10 to 20 percent normative olivine and can be described as olivine tholeiite and picrite.

2.1.2.2 GABBROIC SUBCLASS

The Dore Lake Complex is one of the most completely studied of this subclass. (Allard, 1970, 1976; Allard et al., 1972). Others include the Bell River Complex (Sharp, 1965; Scott, 1980), the Big Trout Lake Complex (Sage and Troop, 1974; Borthwick and Naldrett, 1984), the Bird River Sill (Scoates, 1984; Scoates and Eckstrand, 1986), the Mulcahy Complex (Morrison et al., 1986), and the Kamiskotia Complex (Wolfe 1970; Barrie, 1987). The bodies consist predominantly of gabbroic anorthosite, anorthositic gabbro and gabbro, and many are characterized by layers rich in titaniferous magnetite. Ultramafic rocks are relatively uncommon, in some cases (e.g., Dore Lake Complex) being restricted to layers rich in cumulus pyroxenite. The general conformity of the layering of most of these bodies to folding in the enclosing volcanic rocks suggests that they were intruded before folding. Trace element (Kamiskotia Gabbro Complex, Barrie, 1987) and isotopic evidence (Mulcahy Gabbro Complex, Morrison et al., 1986; Davis et al., 1982) suggest that these intrusions are closely related genetically to the volcanic rocks that they intrude.

2.2 BODIES ASSOCIATED WITH RIFTED PLATE MARGINS AND OCEAN BASINS

2.2.1 Floored by or Closely Associated with Continental Crust

A number of intrusive and, in some cases, extrusive bodies occur at the borders of Archean cratonic blocks, emplaced within Proterozoic sedimentary sequences. These belts include the Circum–Superior belt, which comprises the

Thomson belt, the Cape Smith belt, the Fox River belt, and the Labrador trough and the Kemi and Koillismaa schist belts (interpreted by Gaal, 1985, as originally being a single continuous belt) in Finland.

Baragar and Scoates (1981) argue that the Circum–Superior belt is a rifted continental margin. In support of their arguments, they point out that the earliest rocks in the Mistassini–Otiah, Labrador Trough, and Belcher Island portions of the belt were continental redbeds with and without K-rich basalts. These rocks were followed by continental sedimentation, which gave way to mafic volcanism, some of it komatiitic and, except for the southern margin, all of it low-K oceanic in type. This last observation has been confirmed in the Labrador Trough by Besson et al. (1986), who also interpret it as having been emplaced in a rift zone.

The Kemi and Koillismaa belts separate Svecokarelian granitoids (the central Lapland granitoid province) and Lapponian supracrustal rocks to the north from Archean granitoids to the south (Papunen and Vorma, 1985). The Archean rocks were cratonized 2.5 Ga ago. Layered intrusions with zircon dating from 2.44 Ga (Simonen, 1980) were emplaced along the Archean–Proterozoic contact or within the Archean gneisses adjacent to the contact. Gaal (1985) infers that four individual intrusions were emplaced along a deep-seated tensional fracture, and were subsequently disrupted by transcurrent faulting. The rift zone was filled with epicontinental sediments followed, in some parts, by turbidites. Both Gaal (1985) and Alapieti and Lahtinen (1986) suggest that the structure may have some connection with an aulacogene.

In all of the preceding examples the intrusions and any associated volcanics occur close to the Archean block that they border, and are associated with sediments of continental derivation.

The Thulean igneous province is characterized by numerous well-studied igneous bodies, including the Skaergaard and Rhum intrusions, which, although either floored by (Skaergaard) or intruding (Rhum; Emeleus, 1987) the continental crust, are part of the igneous activity accompanying the opening of the North Atlantic.

Thus, all mafic and ultramafic igneous bodies of this association appear to be part of a rifting event that coincides with a continental margin or former continental margin.

2.2.2 Not Floored by Continental Crust—Ophiolite Complexes

Ophiolite complexes are interpreted as portions of the oceanic crust and the upper mantle and, in the best preserved examples, are characterized by a basal zone of metamorphic-textured peridotite and dunite overlain by (1) a sequence of cumulates ranging in composition from dunite through peridotite to gabbro; (2) a sequence of pillow lavas and associated feeder dikes, and (3) a capping of marine sedimentary rocks. In some places (for example, Oman), emplacement has been accomplished with a minimum of disruption and the full ophiolitic sequence is well preserved; in others the ultramafic rocks occur as isolated blocks within a melange.

Initially, most ophiolites were interpreted as portions of lithosphere generated at a midoceanic ridge. This remains the favored interpretation for the Oman (Hopson et al., 1981; Pallister and Hopson, 1981; Pallister and Knight, 1981) and the New Caledonia (Dupuy et al., 1981) ophiolites. The presence of lavas with compositions consistent with those for forming oceanic crust near island arcs (e.g., Troodos, Schminke et al., 1983; Vuorinos, Beccalura et al., 1984), the presence of sediments indicating shallower water than the deep oceans and the proximity of calc-alkaline centers (e.g., Del Puerto, Evarts, 1977), and the high temperature (interpreted as indicative of young age) of some ophiolites during emplacement (e.g., Bay of Islands, Malpas, 1979) has led, however, to the suggestion that many ophiolites represent lithosphere newly formed in basins occurring between island arcs and continental margins.

2.3 INTRUSIONS IN CRATONIC AREAS

2.3.1 Intrusions Related to Flood Basalts

Intrusions related to flood basalts include the very large (6–7 km thick and more than 8,000 km^2 in area) Dufek Intrusion in Antartica, the 240 × 80 km Duluth Complex in Minnesota, the Insizwa–Ingeli Complex of South Africa, and the smaller (up to 1 × 10 km) intrusions of the Noril'sk–Talnakh area in Siberia. They are associated with tholeiitic basalts of the flood type. Flood basalts are characterized by a higher proportion of incompatible elements than many other tholeiitic types. De Paolo and Wasserburg (1979) suggested that their strontium (Sr) and neodymium (Nd) isotope ratios indicate derivation from previously unmelted mantle, although most examples of this class examined thus far are regarded as having been contaminated by crustal rocks.

2.3.2 Large Stratiform Complexes

As shown in Table 2.1, large stratiform complexes are subdivided into sheetlike and dikelike bodies. Sheetlike bodies include the Bushveld, Stillwater, Muskox, and Sudbury complexes (Sudbury differs from the other examples in lacking the repetitive layering of rock types, and is grouped separately). Dikelike bodies include the Great Dyke of Zimbabwe. The overall compositions of these complexes tend to be mafic rather than ultramafic, although ultramafic zones are present in each of them except in Sudbury. Even at Sudbury, however, numerous ultramafic inclusions ranging in composition from dunite to olivine metagabbro occur in the ore (Rae, 1975; Scribbins et al., 1984) and are thought to have been derived from a series of hidden, but related, layered intrusions (Naldrett, 1986).

The magma types of this class of intrusion are somewhat variable. Sharpe (1981) and Harmer and Sharpe (1985) have shown that the Bushveld magma was of two main types, an incompatible element-enriched, high SiO_2 (56 wt %), high MgO (13–14 wt %) magma, and a less incompatible element-enriched magma with a higher $^{87}Sr/^{86}Sr$ initial ratio than the more magnesian type and plagioclase rather than a mafic mineral as its first liquidus phase (see discussion in Chapter 5). Two magmas are also thought to have been involved in the formation of the Stillwater Complex (Todd et al., 1982; Irvine et al., 1983). Naldrett et al. (1986) proposed a flood basalt precursor for the Sudbury Igneous Complex, which became heavily contaminated by country rocks (see Chapter 4).

2.3.3 Other Classes

Other intrusions in cratonic areas include those of subalkaline to peralkaline affinity, such as those of the Gardar province of southern Greenland, kimberlites, and carbonatites.

2.4 BODIES EMPLACED DURING OROGENESIS

2.4.1 Synorogenic Intrusions

Mafic intrusions of the synorogenic type are common in the Appalachian–Caledonian orogenic belt. They have been best described in Aberdeenshire, Scotland (Boyd and Munro, 1978), in Maine (Boone, 1974; Thompson and Naldrett, 1984), and in Norway [Seiland (Robins and Gardner, 1974) and the Fongen–Hyllingen Complex (Wilson, 1987)]. They are characterized by: (1) evidence of syndeformational intrusion (specifically, irregular igneous layering if such layering is present; tectonite fabrics; and an irregular distribution of a wide variety of igneous rock types) and (2) partial metamorphism and deformation in the late stages of tectonism. The widespread occurrence of such intrusions, from Maine to northern Norway in the Caledonides, may be related to a deep level of erosion, particularly in the Norwegian part of the belt. Similar intrusions may occur at deep unexposed levels of other Phanerozoic orogenic belts. In Maine, these intrusions have been related to continental collision (Boone, 1974), whereas a genesis related to a steeply dipping Benioff zone is postulated for the Seiland petrographic province (Robins and Gardner, 1974).

2.4.2 Alaskan-type Ultramafic Complexes

Complexes of the Alaskan-type draw their name from a series of distinctive bodies exposed along the mainland of southeast Alaska. More than 30 complexes occur along the 550-km length of the Alaskan panhandle, most of them in a zone about 40 km wide. Some of the larger complexes show a rough concentric zoning that, where best developed, consists of a dunitic core surrounded by successive shells of olivine clinopyroxenite, magnetite-rich clinopyroxenite, and hornblendite. As a group, Alaskan-type complexes are distinguished from alpine ultramafics or stratiform intrusions in having highly calcic clinopyroxene, no orthopyroxene, abundant hornblende, iron-rich chromite, and magnetite; the last mineral occurs in concentrations that are of local economic interest.

Similar bodies (Table 2.1) are known in the Urals, south–central British Columbia, Columbia, Venezuela (Irvine, 1974), and northern California (Gray et al., 1986). The most extensively studied of all is the Duke Island Complex of Alaska, described by Irvine (1974). Irvine provides an up-to-date general account of bodies of this type and emphasizes their alkalic nature.

2.5 RELATIVE IMPORTANCE OF DIFFERENT CLASSES OF MAFIC AND ULTRAMAFIC BODIES AS HOST FOR ORES

Magmatic sulfide deposits are associated with many of the different classes of mafic and ultramafic bodies distinguished in Table 2.1. Some of the classes have proved to be much more important hosts than others. Naldrett (1973) pointed out that ophiolitic bodies, Alaskan-type bodies, alkalic bodies, kimberlites, and carbonatite-bearing ring complexes contain few, if any, significant magmatic sulfide ores, and this generalization is still valid, despite the discovery of small deposits of Ni–Cu sulfide associated with an ophiolitic sheet at Acoje, Philippines.

Considering first Ni–Cu sulfide resources, the most important classes are:

I. Those associated with the magmatic activity accompanying the early stages of formation of Precambrian greenstone belts and comprising two main types:
   a. Komatiitic lavas and intrusions, especially the more ultramafic variants. Ultramafic komatiites are hosts to important sulfide deposits, in particular, those of the Kambalda camp of Western Australia (Ross and Hopkins, 1975; Gresham and Loftus-Hills, 1981; Gresham, 1986).

   b. Tholeiitic intrusions such as those hosting the ores of the Pechenga Ni camp of the Kola Peninsula, U.S.S.R. (Gorbunov, 1968; Gorbunov et al., 1985a,b) and Lynn Lake, Manitoba, Canada (Ruttan, 1955).

II. Komatiites, occurring in the rifted continental margin environment, are hosts to deposits of the Thompson, Manitoba district (Peredery, 1979, 1982) and to high-grade deposits, as yet unmined, in the northern tip of the Ungava Peninsula of Quebec (Wilson et al., 1969; Barnes et al., 1982; Dillon-Leitch et al., 1986).

III. In the intracontinental environment, three associations have proved important:
   a. The intrusive equivalents of flood basalts associated with intracontinental rifting. Important examples include the Noril'sk camp of Siberia (Glaskovsky et al., 1977; Distler and Genkin, 1980; Genkin et al., 1981) and the as yet unexploited sulfides at the western edge of the Duluth Complex, Minnesota (Wager et al., 1969; Bonnichsen, 1974; Mainwaring and Naldrett, 1977; Matlock and Watowich, 1980; Foose and Weiblen, 1986).
   b. Noritic rocks associated with an astrobleme (scars resulting from meteorite impacts). The sole example of this type is the Sudbury mining camp in Canada.
   c. Disseminated sulfide deposits in large layered intrusions that are mined primarily for their PGE.

IV. Tholeiitic intrusions, generally synchronous with orogenesis in Phanerozoic orogenic belts, are hosts to deposits of lesser importance [e.g., the Rona deposit, Norway (Boyd and Mathiesen, 1979), the Moxie and Katahdin gabbros, northern Maine (Thompson and Naldrett, 1984)].

The relative importance of the different petrotectonic settings as hosts for Ni–Cu sulfide ores is illustrated in Figure 2.1, in terms of their contribution both to past production and to future resources. Four settings dominate in so far as Ni production is concerned: (I) those associated with volcanic rocks in Precambrian greenstone belts, especially those associated with komatiites; (II) those associated with komatiites in the rifted continental margin environment; (IIIa) those associated with flood basalt magma and intracontinental rifting and (IIIb) Sudbury. Another important deposit, at present not assigned to a category, is that of Jinchuan, China. Examples of these deposits are described briefly in Chapter 4.

With regard to deposits mined primarily for their PGE, another class of body, that of the large layered complexes in cratonic regions, has proved to be by far the most important (as seen in Fig. 2.1, this class is also important as a Ni resource). The data in Table 2.2 indicate that the Bushveld Complex accounts for 80 percent, the Stillwater 1.4 percent, and the Great Dyke 10.4 percent of the total world resources of PGE, amounting to a total of 91.8 percent in complexes of this type. Also indicated in Table 2.2, the balance of world production comes as a by-product of Ni–Cu production, particularly that of the Noril'sk-Talnakh

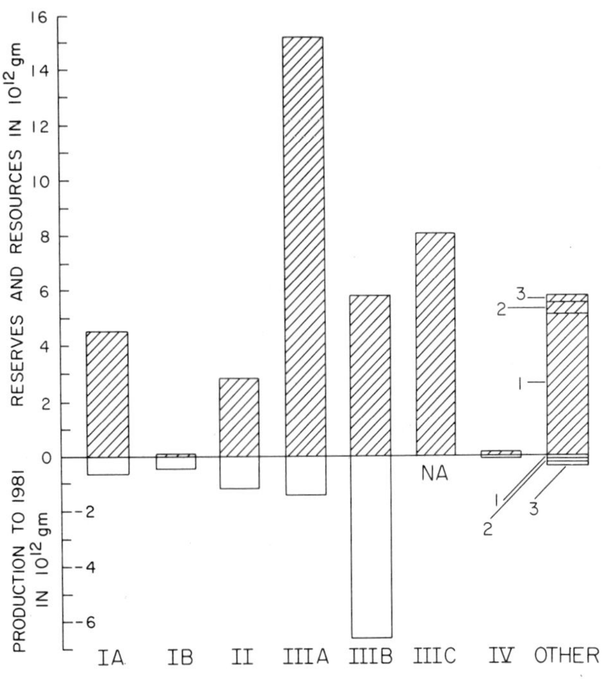

Fig. 2.1. Resources plus reserves of Ni and 1981 production as a function of the petrotectonic association. I to IV as outlined in text. Other: 1 = Jinchuan, China; 2 = Selibe-Phikwe, Botswana; 3 = various deposits, Finland. (Data from DeYoung et al., 1986.)

TABLE 2.2. *Relative Proportions, Grade (in gm/tonne) and Identified Resources (in 10^6 gm) of PGE in Major Deposits*

	Bushveld Complex						Noril'sk–Talnakh USSR		Stillwater J–M Reef		Great Dyke		Sudbury, Canada		Columbia		
	Merensky Reef		UG–2		Plat. Reef												
	Prop	Res	Prop	Res	Prop	Res	Prop	Res	Prop	Res	Prop	Res	Prop	Res	Prop	Res	Total Resources
Pd	25	4,371	35	11,315	46	5,425	71	4,402	66.5	713	37.1	2936	40	112	1		29,274
Pt	59	10,323	42	13,547	42	4,960	25	1,550	19	217	52.8	4130	38	105	93		34,925
Rh	3	527	8	2,573	3	372	3	186	7.6	84	4.7	367	3.3	<31	2		4,109
Ru	8	1,395	12	3,875	4	465	1	62	4.0	43	4.7	367	2.9	<31	—		6,207
Ir	1	186	2.3	744	0.8	93	—	—	2.4	<31	0.6	46	1.2	<31	3		1,069
Os	0.8	155	—	—	0.6	62	—	—	—	—	0.6	46	1.2	<31	1		263
Grade in gm/tonne total PGE	8.1		8.71		7.27		3.8		22.3		4.7		0.9				
Total resources		16,957		32,054		11,377		6,200		1,057		7,892		217			

Prop, proportion; Res, reserves.

Data from Buchanan and Nolan (1979) except for that on the Great Dyke which is from I. G. Anderson (personal communication, 1981).

district. A second class, that of layered complexes in the rifted continental environment, as exemplified by the Penikaat intrusion of the Kemi-Koillisimaa belt (Alapieti and Lahtininen, 1986), appears to have considerable potential.

Concentrations of PGE also occur in conjunction with chromite concentrations in ophiolite complexes, but these contain high Ru, Ir, and Os concentrations and very low Pt and Pd concentrations, and are not viable economically. Exceptions are the deposits of the Unst (Shetland) ophiolite (Prichard et al., 1986), particularly the Cliff deposit, which are extremely rich in Pt and Pd.

3. EXPERIMENTAL STUDIES ON SULFIDE, SULFIDE OXIDE, AND SULFIDE SILICATE SYSTEMS

Students of magmatic sulfide ores are particularly fortunate in comparison with those students of most other types of ores in that five elements (Fe, S, Ni, Cu, and O) account for well over 99 percent of the total composition of the ore. Furthermore, in most cases the ores are thought to have been introduced as melts, in essentially their present compositions, so that the question of solubility in a transporting fluid does not arise.

Despite the simple composition, five components still require a considerable stretching of one's powers of visualization if their mutual phase relations are to be considered simultaneously. It is possible, however, to simplify the discussion by first considering melting relations in the two-component system Fe–S, and then in the three-component system Fe–S–O, and then by considering the effect on the melting relations by introducing small amounts of the two other components, Ni and Cu. Once the ore has crystallized, essentially all of the oxygen is present as magnetite, which partakes in no major reactions with the rest of the ore during cooling. Thus, at subsolidus temperatures the ore can be treated as the four-component system, Fe–Ni–Cu–S. Furthermore, once the bulk of the Cu has exsolved as chalcopyrite above 500–600°C, it is believed not to react with the rest of the ore to any major extent, and a great deal can be learned simply by analogy with the system Fe-Ni–S.

For these reasons, high-temperature phase relations in the systems Fe–S and Fe–O are outlined briefly here as an introduction to a more detailed discussion of the system Fe–S–O. Phase relations in the system Fe–Ni–S are discussed after, followed by pertinent portions of the system Cu–Fe–S and of the quaternary system Cu–Fe–Ni–S. Discussions of these synthetic systems are followed by consideration of factors governing the solubility of sulfur in silicate melts, and finally a discussion of the partitioning of group VIII transition metals plus Cu between sulfide and silicate melts and olivine.

These experimental studies form the basis for discussions in subsequent chapters on: (1) the crystallization of sulfide ores; (2) the subsequent mineralogic changes as they cool; (3) the conditions in a silicate magma that cause them to separate in the first place; and finally (4) the controls that silicate magma compositions exert on the compositions of ores segregating from the magmas.

3.1 THE SYSTEM Fe–S–O

High-temperature phase relations in the "condensed" systems Fe–S and Fe–O are illustrated in Figures 3.1 and 3.2. For reasons given later in this section, natural magmas only fall within a limited range of Fe/S and Fe/O ratios. Within this range, the important features of the system Fe–S are the congruent melting of pyrrhotite at 1,190°C, and the two bounding phase regions, pyrrhotite + liquid + vapor on either side. On the metal-rich side, the three-phase region terminates at the eutectic at 988°C and on the sulfur-rich side at 1,083°C. Magnetite is the dominant oxide mineral crystallizing from high-temperature magmas. More rarely wustite forms, but, because of its lower stability limit of 560°C, it is rarely preserved in the natural environment.

3.1.1 Phase Relations in the System Fe–S–O

Condensed phase relations at 900°C (15°C below the temperature of the first appearance of a liquid in the Fe-rich portion of the system Fe–S–O) are shown in Figure 3.3, which outlines the compositional field occupied by almost all natural ore magmas. Phase relations on the (O + S)-rich side of the magnetite–pyrrhotite tie-lines drawn in the Figure have not been determined experimentally because of the high vapor pressures associated with these compositions. It is very likely, however, that these compositions involve the four-phase field of magnetite + hematite + pyrrhotite + vapor (in which the pyrrhotite is somewhat richer in sulfur than that shown in equilibrium with magnetite, Fig. 3.3) followed by the three-phase field of hematite + pyrrhotite + vapor.

Liquidus relations in the system Fe–S–O (in the presence of vapor) are shown in Figures 3.4A and 3.4B. The dashed lines in Figure 3.4A are based on the data from Hilty and Crafts (1952), the dashed–dotted lines are interpretative, and the solid lines are from the data of Naldrett (1969). Figure 3.4A is characterized by fields of iron, wustite, pyrrhotite, and magnetite. The iron, wustite, and pyrrhotite fields meet at a ternary eutectic where these three phases plus liquid and vapor are

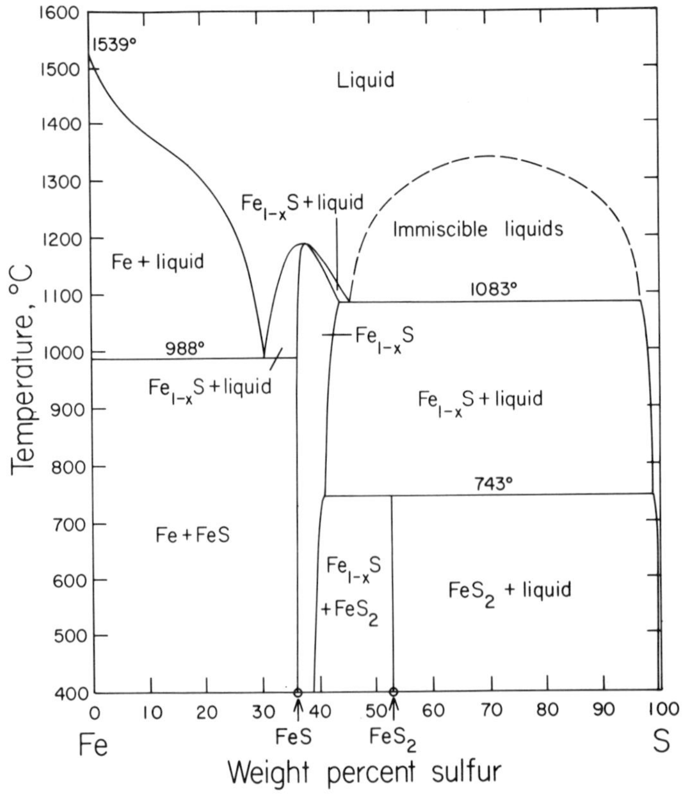

Fig. 3.1. The Fe-S system at its own vapor pressure. (After Kullerud, 1963.)

all stable. The wustite, pyrrhotite, and magnetite fields meet at a ternary reaction point at which magnetite plus liquid react in the presence of vapor to form pyrrhotite (containing 62.8 ± 0.2 wt % Fe) and wustite. Figure 3.4B is an enlargement of a portion of Figure 3.4A.

The relationship between solidus and liquidus is illustrated in Figure 3.5. To the right of the eutectic composition shown on the Figure, the section is drawn along the pyrrhotite–wustite and pyrrhotite–magnetite cotectic lines. Consequently, regions in which one solid phase coexists with a liquid and a vapor phase do not appear in this section. In addition, many of the data points do not represent experiments with bulk compositions in the plane of the section, but have been projected onto this plane along tie-lines.

The ternary eutectic, Fe–wustite–troilite, was reported by Naldrett (1969) to occur at 915 ± 2°C and to have the composition 68.2 wt percent Fe, 24.3 wt percent S, and 7.5 wt percent O. The ternary reaction point occurs at 934 ± 2°C. As shown in Figure 3.5, the solidus rises very steeply away from the reaction point in the direction of decreasing Fe content (a mixture of pyrrhotite containing 62.5 wt % Fe and magnetite would start to melt at 1,010°C) and then flattens out as the pyrrhotite in equilibrium with the magnetite becomes less Fe-rich.

Wendlandt and Huebner (1979) investigated the effect of pressure on the position and temperature of the ternary eutectic in the Fe–S–O system. At 30 kb pressure, it has a temperature of 1,000 ± 10°C and a composition containing 72 to 73 wt percent Fe and less than 0.75 wt percent 0. At this pressure, it is very close to the binary Fe–FeS eutectic in terms of both temperature and composition.

Brett and Bell (1969), Ryzhenko and Kennedy (1973), and Usselman (1975) studied the effect of pressure on the Fe–FeS eutectic in the Fe–S system up to 60 kb. All workers found that the pressure had either no or just a slight effect on the temperature of the eutectic, although the composition became more Fe-rich with an increase in pressure [Ryzhenko and Kennedy (1973) report it to contain 22 wt % S at 60 kb]. Sharp (1969) and Ryzhenko and Kennedy (1973) investigated the effect of pressure on the congruent melting of pyrrhotite and have found the melting point to increase at the rate of 13°C/kb up to 65kb. Naldrett (1973) used this figure to approximate the effect of pressure on the melting of Fe–S–O liquids.

3.1.2 Effect of Other Components on Solidus Temperatures

Naldrett (1969) pointed out that at temperatures immediately below the solidus, a pyrrhotite–pentlandite–chalcopyrite–magnetite ore in which the sulfides contained less than 15 wt percent Ni and 4 wt percent Cu would consist of two phases, a nickeliferous–cupriferous–pyrrhotite solid–solution and magnetite. Craig and Naldrett (1967) investigated the effect of the substitution of Ni and Cu for Fe in the pyrrhotite on the solidus temperatures of pyrrhotite–magnetite mixtures. They found that the substitution of up to 20 wt percent Ni for Fe on an atom-for-atom basis did not lower melting temperatures measurably below those established for the pure system Fe-S-O. The substitution of 2 wt percent Cu on a similar basis lowered solidus temperatures 15 to 20°C.

Naldrett and Richardson (1967) investigated the extent to which water lowered the melting temperature of pyrrhotite–magnetite mixtures. Within the accuracy of their experiments (±10°C), they concluded that water did not act as a flux for sulfide ores or dissolve appreciably in a sulfide–oxide melt.

The above data indicate that, ignoring the pressure effect, melting temperatures in the Fe-S-O system are probably within 20°C of the melting temperatures of natural low-Cu (<2 wt %) sulfide–oxide magmas.

3.1.3 Variation in fO_2, fS_2, and aFeO

The composition of an Fe-sulfide–oxide liquid at any given temperature is a function of its O and S fugacities (Naldrett, 1969). This point is illustrated in Figures 3.6A and 3.6B in which isobars of S and O fugacity are shown at 1,200 and 1,450°C over the same portion of the F-S-O system as shown in Figure 3.4B. The isobars are drawn on the basis of the data of Rosenqvist (1954), Bog and Rosenqvist (1958), and Nagamori and Kameda (1965). In all of these studies, the investigators equilibrated an Fe-S-O liquid with three O_2- and S_2-bearing gases (for example, CO, CO_2, and SO_2) in proportions defining a series of different values of fO_2 and fS_2 at the temperatures in question. The composition of the liq-

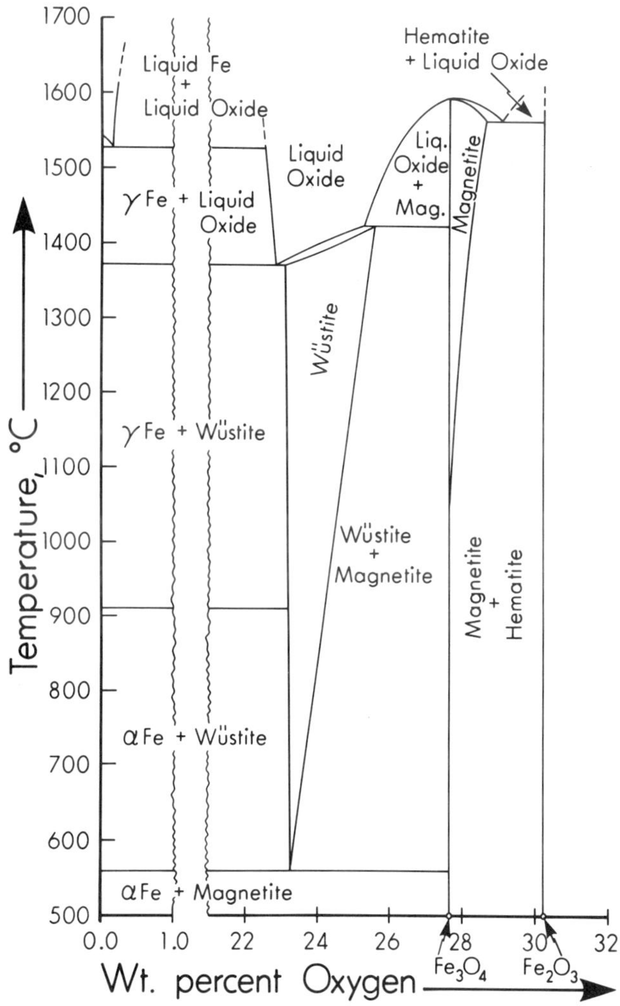

Fig. 3.2. The Fe–O system at 1 atm total pressure. (After Darken and Gurry, 1953, and Phillips and Muan, 1960.)

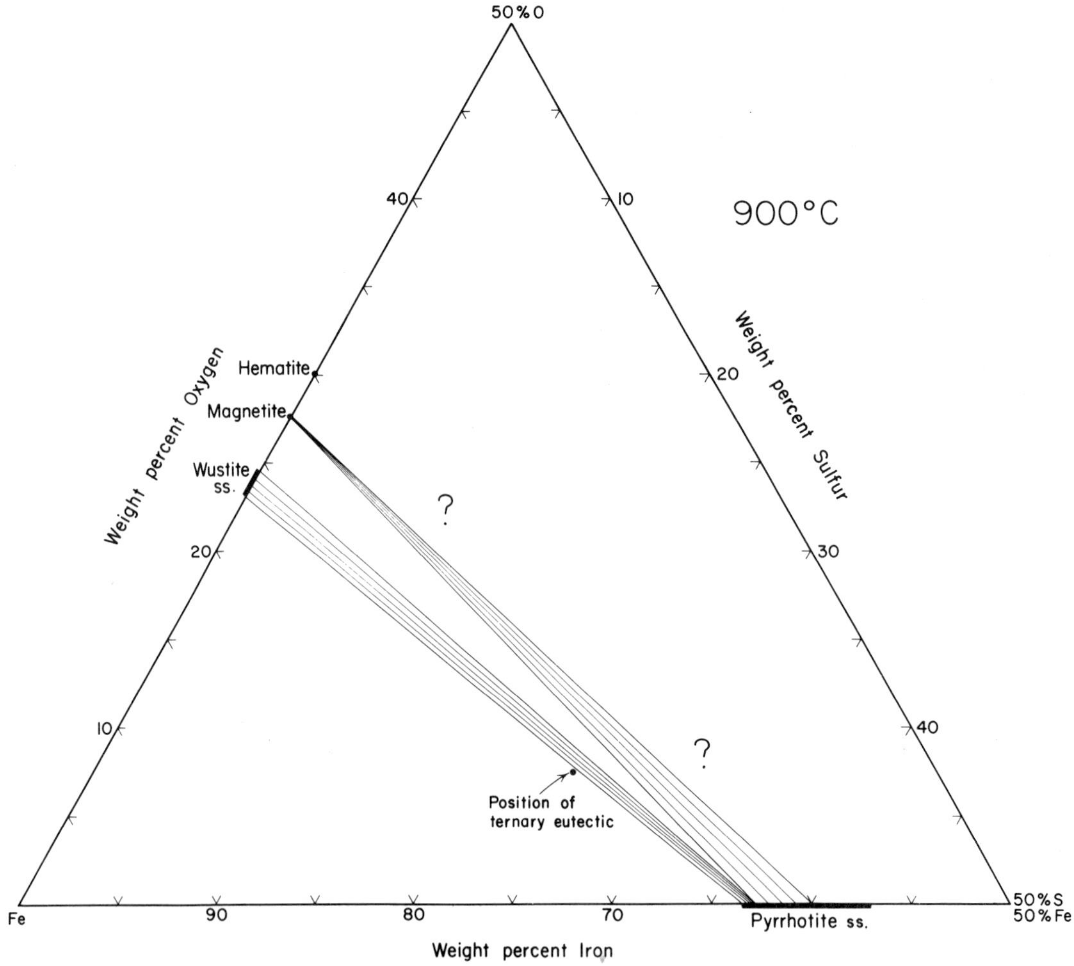

Fig. 3.3. The Fe-rich portion of the Fe–S–O system at 900°C. (From Naldrett, 1969.)

uids corresponding to the specific values of fO_2 and fS_2 were then determined at equilibration.

Knowing the chemical potentials (or activities) as a function of composition for two of the components in a three-component system, the Gibbs–Duhem relation allows calculation of the chemical potential of the third component. This calculation has been done for the data given earlier on the system Fe–S–O (Shima and Naldrett, 1975) to obtain the activity of Fe. In turn, as the standard free energy for the reaction

$$Fe + \tfrac{1}{2} O_2 = FeO_{liquid} \qquad (3.1)$$

is known, it is possible to combine activity data for Fe and fugacity data for O_2 to calculate the activity of FeO. These data then serve as the basis for the *a*FeO contours in Figures 3.6A and 3.6B, for which the standard state of FeO is taken to be a supercooled liquid of stoichiometric FeO composition at 1,200°C and a liquid of the same composition at 1,450°C.

3.1.4 Application of the System Fe–S–O to Natural Ore Magmas

3.1.4.1 CONTROLS ON THE S AND O CONTENT OF ORE MAGMAS

Consider a basaltic magma at 1,200°C, devoid of Ni, Cu, and Co, only saturated in sulfide, so that a small number of sulfide droplets are segregating from it and are in equilibrium with it. Because the droplets are few in number and widely dispersed throughout a large body of magma, their compositions are controlled by the composition of the host magma. This magma has a certain FeO content and a certain Fe^{3+}/Fe^{2+} ratio, which, together with other major elements, define the values of

Fig. 3.4. ***A.*** Liquidus diagram of the Fe-rich portion of the Fe–S–O system. ***B.*** Enlargement of the area outlined in ***A.*** (Both diagrams from Naldrett, 1969.)

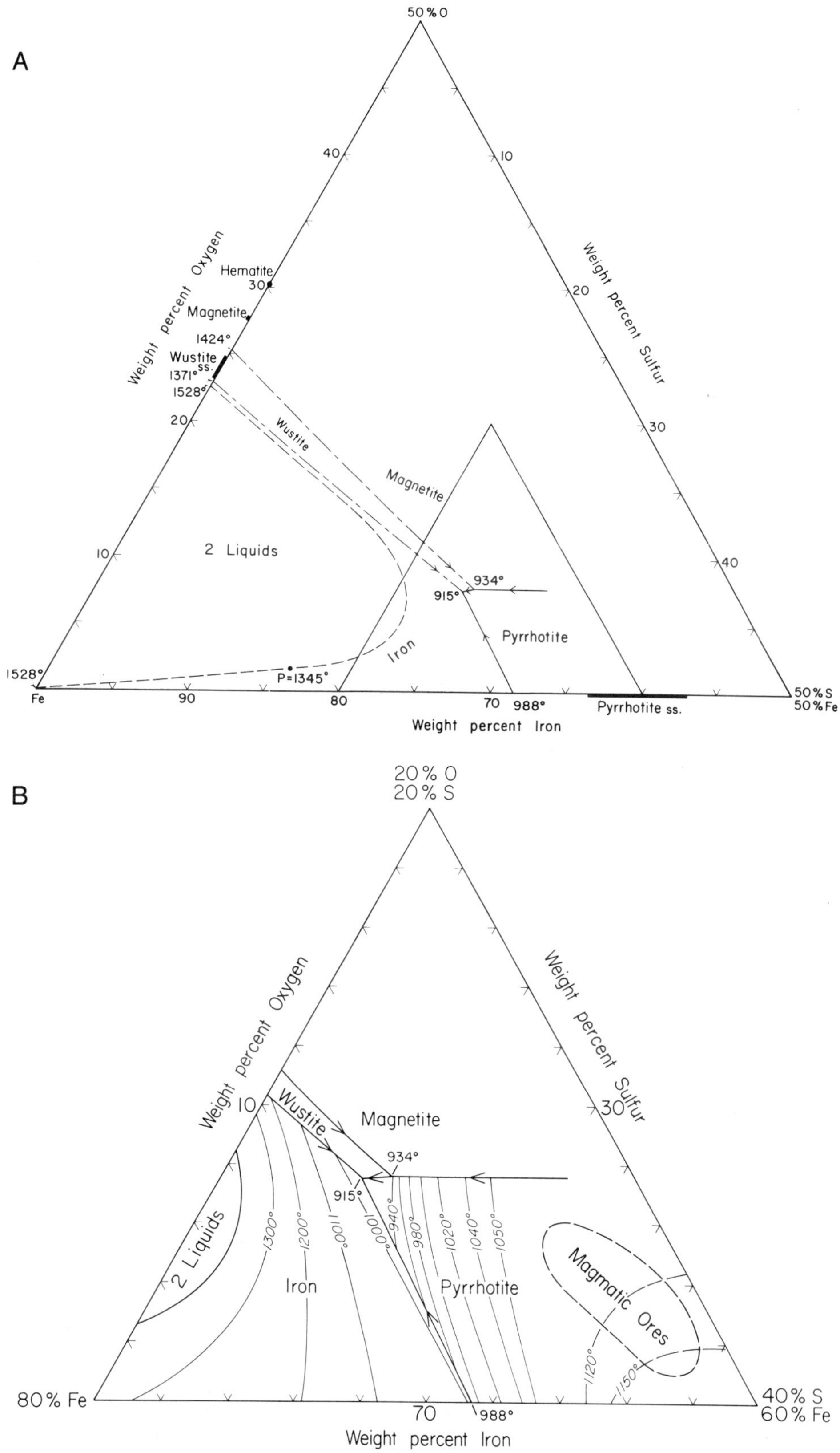
A
50%O
40
10
Weight percent Oxygen
Weight percent Sulfur
Hematite
30
Magnetite
1424°
Wustite ss.
1371°
1528°
20
Wustite
Magnetite
30
10
2 Liquids
934°
915°
40
Pyrrhotite
Iron
1528°
P=1345°
Fe
90
80
70
988°
Pyrrhotite ss.
50%S
50%Fe
Weight percent Iron
B
20% O
20% S
Weight percent Oxygen
Weight percent Sulfur
10
Wustite
Magnetite
30
934°
915°
2 Liquids
1300°
1200°
1100°
1000°
940°
980°
1020°
1040°
1050°
Magmatic Ores
Iron
Pyrrhotite
1120°
1150°
80% Fe
70
988°
40% S
60% Fe
Weight percent Iron

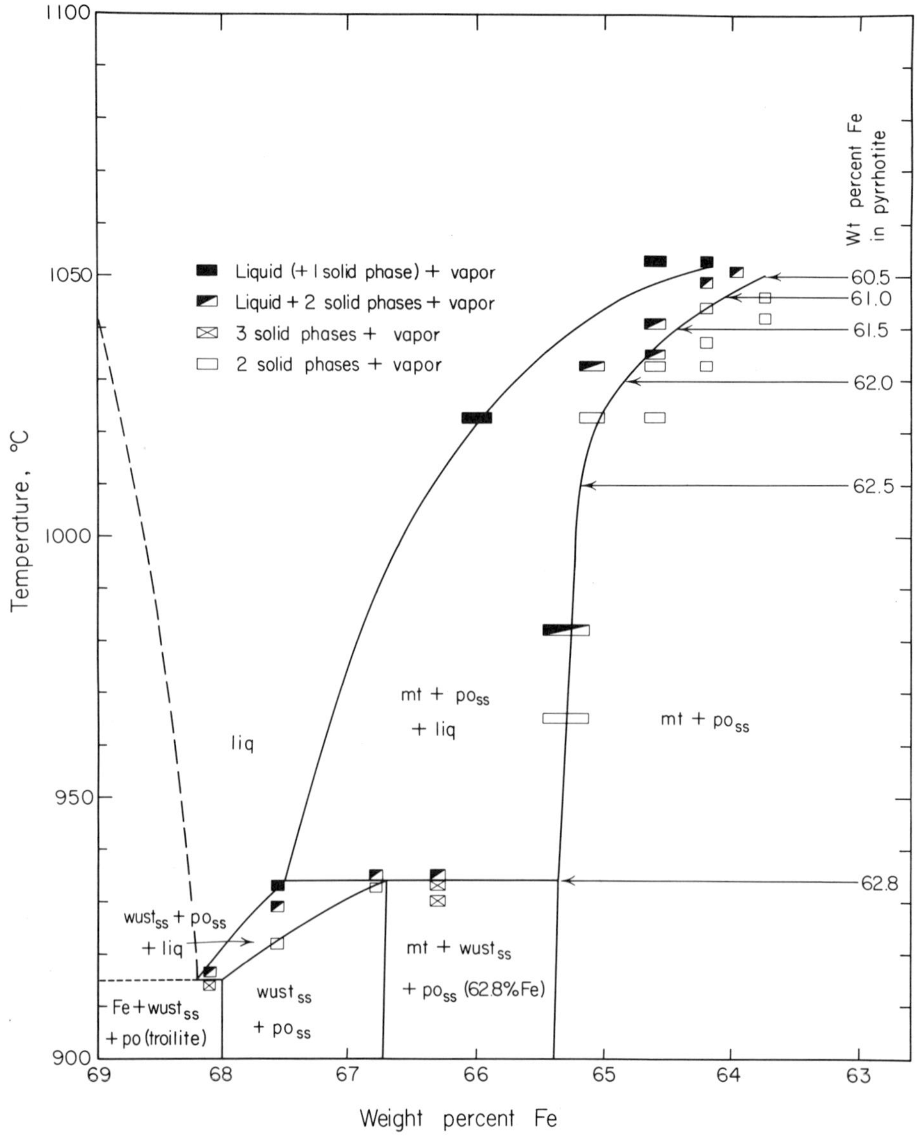

Fig. 3.5. A T–X section showing the relationship between solidus and liquidus in the Fe–S–O system. The section is drawn along the pyrrhotite–wustite and pyrrhotite–magnetite cotectic lines. Many of the data points do not represent bulk compositions in the plane of the section, but have been projected to the plane along tie-lines. (From Naldrett, 1969.)

*a*FeO (obtainable from Roeder, 1974, or according to the relationship of Doyle and Naldrett, 1986) and fO_2 in the magma. Because at equilibrium, the same values of *a*FeO and fO_2 must apply to the sulfide droplets, the composition of the droplets must correspond to the values in Figure 3.6A; i.e., in the hypothetical case under consideration, the composition of the basalt magma, in particular its FeO content and Fe^{3+}/Fe^{2+} ratio, defines the composition of the segregating sulfide ore and the fS_2 prevailing at the time of segregation. Because most natural magmas have relatively restricted ranges of FeO content and oxidation state, the compositions of magmatic ores are similarly restricted in their S and O contents.

In nature, Ni, Cu, and Co are also present in the basaltic magma and partition into the sulfide droplets, as discussed in Chapter 6. This tendency complicates the simple relationship presented in this section, but only does so to a limited extent. Control of the composition of the sulfide liquid by the *a*FeO and fO_2 prevailing in the host magma will still hold.

Doyle and Naldrett (1986) pointed out that if, as Fudali (1965) has argued on the basis of experiments, the fO_2 of the basaltic sources to magmatic sulfide ore magmas are in the region of

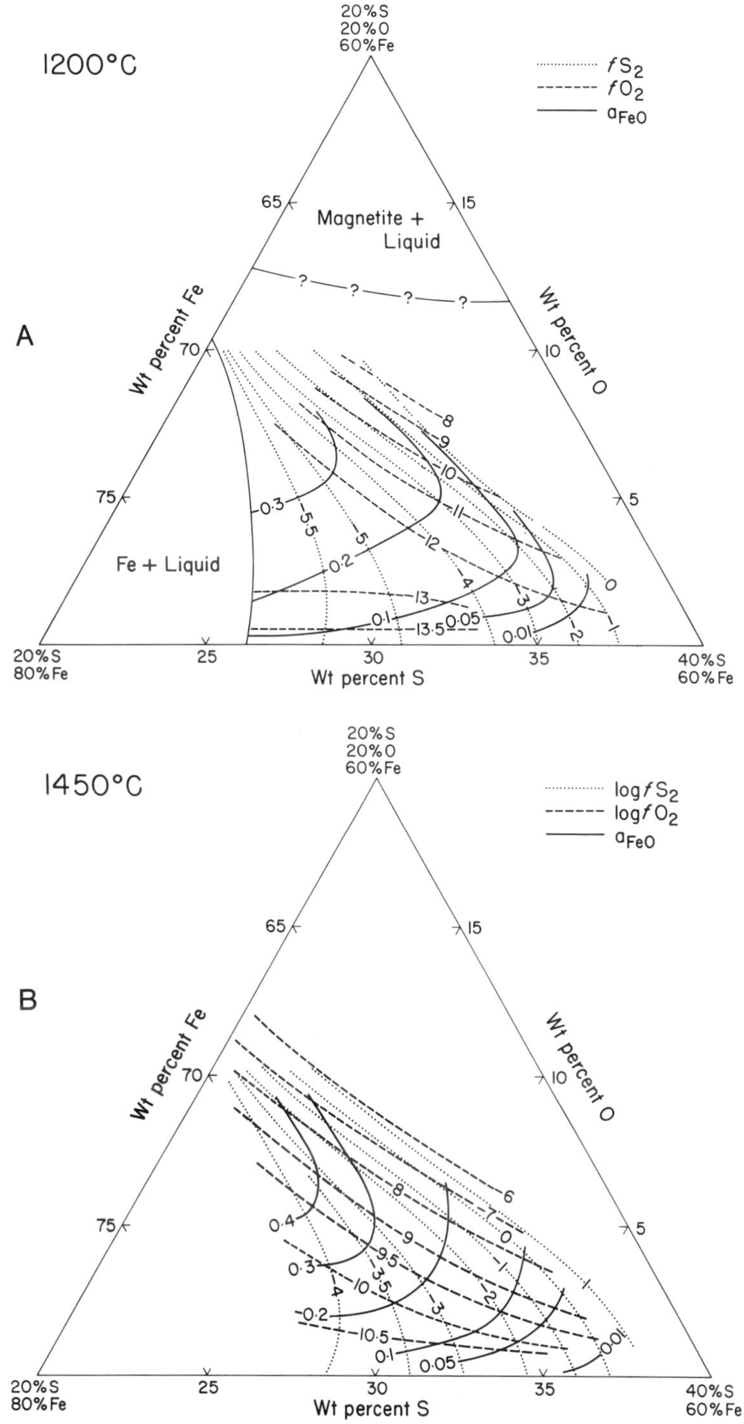

Fig. 3.6. Contours of fO_2, fS_2, and aFeO on the same portion of the Fe–S–O system as in Fig. 3.4B. ***A.*** at 1200°C and ***B.*** at 1450°C. (From Shima and Naldrett, 1975.)

10^{-8}, the compositions of the ore magmas should fall on the oxygen-rich side of the magnetite–pyrrhotite cotectic in the Fe–S–O system, implying that they should crystallize magnetite before pyrrhotite. It is clear from polished section study that the majority crystallize pyrrhotite before magnetite.

A possible answer to this problem lies in the work of Celmer (1987), who noted that, at a given oxygen fugacity, the oxygen content of an Fe–Ni–

Cu–S matte depends strongly on the Fe/(Ni + Cu) ratio of the matte. When the fO_2 is 10^{-4} at 1,250°C, the oxygen content decreases from 6 wt percent in Fe-rich mattes to 0.1 wt percent in Fe-free mattes. Celmer's data show that the substitution of 9 wt percent Ni, 1 wt percent Cu for Fe in a typical matte cuts the dissolved oxygen by nearly one-third. The presence of Cu and Ni in natural sulfide magmas may lower the oxygen content, with the result that their bulk compositions fall in the pyrrhotite rather than the magnetite field.

3.1.4.2 CRYSTALLIZATION OF SULFIDE ORES

The temperatures of the beginning of crystallization and final solidification of a sulfide ore are important when considering how far the sulfides can move away from their host intrusion as an ore magma, and whether they are likely to be mobilized as a liquid during high-grade metamorphism.

The region representing the compositions of most magmatic ores is superimposed on the Fe-S-O system in Figure 3.4B. Although detailed information is not available on liquidus temperatures, most ores start to crystallize between 1,160 and 1,120°C. Because ores, like silicate magmas, can probably move as a mixture of crystals and liquid, the solidus temperature of an ore provides a minimum temperature for its mobilization in a partly liquid state. Figure 3.5 shows that the solidus temperature varies with the Fe content of the pyrrhotite forming the deposit. Almost all ores are made up of pyrrhotite containing between 62.5 and 60.5 wt percent Fe (due to the controls on their composition discussed earlier) so that, on the basis of the system Fe-S-O, one would predict that their solidus temperatures would be between 1,010 and 1,050°C. As previously reported, the substitution of up to 20 wt percent Ni for Fe has no measurable effect on the solidus temperature. The substitution of 2 wt percent Cu lowers the solidus temperature by 20°C. The substitution of more than 2 wt percent Cu has a greater effect, therefore, temperatures are distinctly lower than those estimated for Cu-poor ores; exactly how much lower is not known. At one stage it was thought (Smith, 1963, pp. 304–307) that water could have a marked fluxing effect on an ore, but this suggestion now seems unlikely based on the experimental studies described in Section 3.1.2.

3.2 RELEVANT SULFIDE SYSTEMS

3.2.1 The System Fe–Ni–S

The bounding binary systems, Fe-S and Ni–S, to the ternary Fe-Ni-S system have been described by Kullerud and Yoder (1969), Kullerud (1967), and Chuang et al. (1985); and Kullerud and Yund (1962) and Sharma and Chang (1980), respectively. The most comprehensive study of the ternary is that of Kullerud et al. (1969). More detailed studies include those of Kullerud (1963) and Bell et al. (1964) on pentlandite stability, those of Naldrett et al. (1967), Misra and Fleet (1973), and Barker (1983) on the limits of the $Fe_{(1-x)}S$–$Ni_{(1-x)}S$ or monosulfide solid solution (Mss), those of Craig (1971) and Misra and Fleet (1973) on violarite stability, those of Craig (1973) and Misra and Fleet (1974) on pentlandite–pyrite equilibria, and that of Lenz et al. (1978) on liquidus relations. Craig and Scott (1974) provide a summary of much of the earlier relevant data. The isothermal sections of the condensed Fe–Ni–S system shown in Figure 3.7 have been selected to illustrate many of the more important aspects of the system that have a bearing on the crystallization of Ni sulfide ores.

Apart from alloys along the Fe–Ni join, the system is composed entirely of liquids at 1,200°C, a metal-rich liquid spanning the center of the system and a S-rich liquid containing probably less than 2 wt percent metal at the sulfur apex. $Fe_{(1-x)}S$ appears on the Fe–S join at 1,190°C and spreads farther into the Fe–Ni–S system at lower temperatures (as shown at 1,100°C in Figure 3.7A), forming the Mss. The early crystallizing phase in the system is Fe-rich and Ni-poor with respect to the liquid with which it is in equilibrium; thus, continued fractional crystallization of Mss results in Ni-enrichment in the later fractionated sulfide liquid. The metal-rich liquid draws completely away from the Fe–S join at 1,083°C on the S-rich side and at 988°C on the S-poor side of $Fe_{(1-x)}S$. The Mss reaches the Ni–S join at 999°C, dividing the system in two. This barrier remains in effect, isolating S-rich from metal-rich phases to below 250°C.

Vaesite (NiS_2) appears on the Ni–S join at 1,022°C and pyrite on the Fe–S join at 743°C; both phases take a few percent of the other into solid-solution at these temperatures. An important development in the system is the formation of a pyrite–vaesite tie-line at 728°C (Fig. 3.7B). Above this temperature, Mss is in equilibrium with liquid S. This observation implies that a wide range of bulk compositions consist of Mss plus liquid S. Liquid S at this temperature is extremely reactive and would extract Fe from most Fe-bearing silicates to form Fe sulfide (Naldrett and Gasparrini, 1971). Because magmatic ores are intimately related to their host magmas and subsequently to their host rocks, this reaction would essentially preclude the existence of liquid S, and hence these bulk compositions, at these high temperatures (i.e., magmas that subsequently crystallize to form large proportions of pyrite or vaesite, or both, are highly unlikely in nature). Deposits with this mineralogy are most unlikely to have

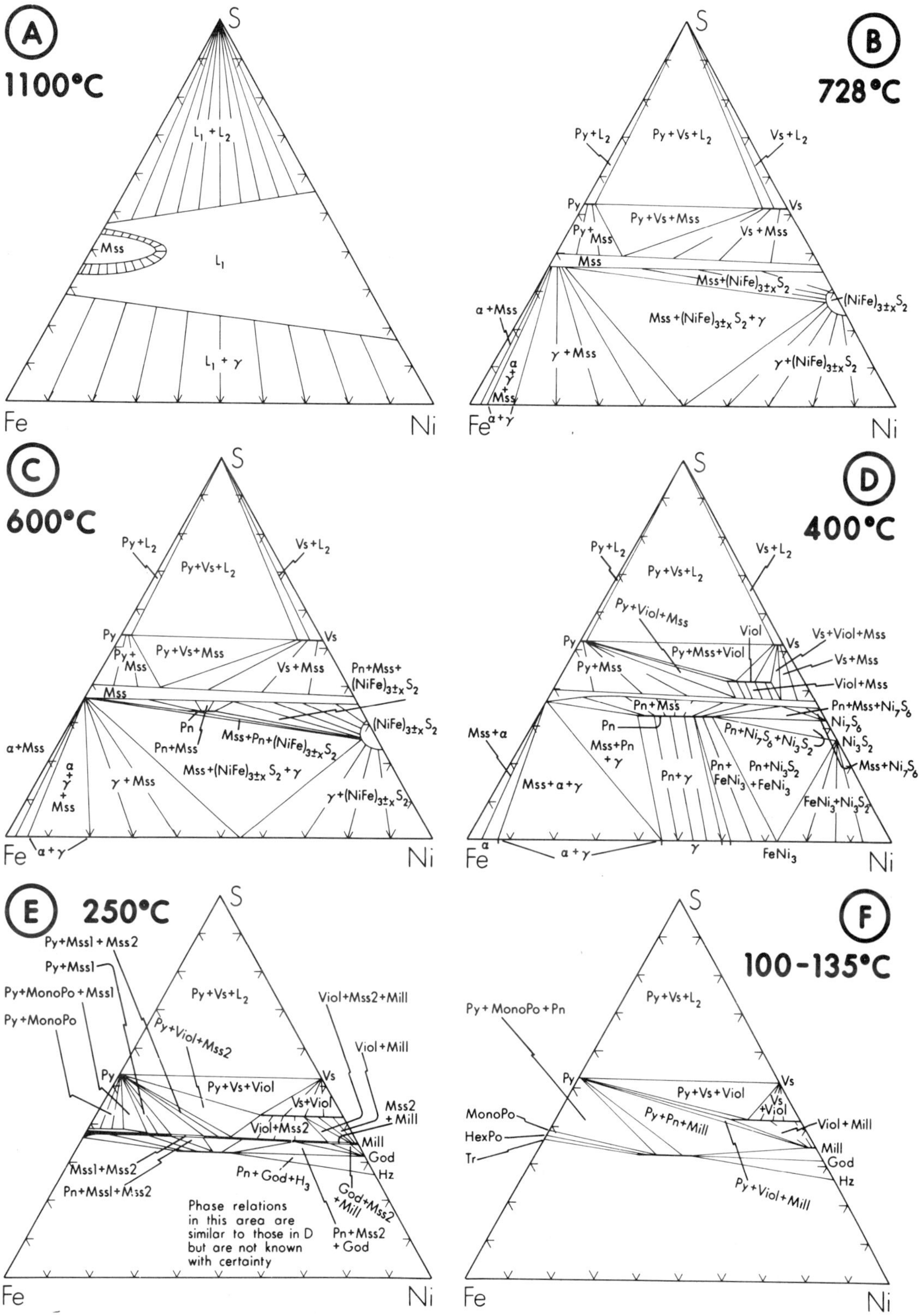

Fig. 3.7. Isothermal sections through the "condensed" Fe–Ni–S system. (Data from Kullerud et al. [1969], Kullerud [1963], Bell et al. [1964], Naldrett et al. [1967], Misra and Fleet [1973], Craig [1971], Misra and Fleet [1974], and Craig and Scott [1974].)

formed with their present bulk compositions from a sulfide magma, and are much more likely to have formed through the alteration of an original, normal high-temperature assemblage of minerals, or to have been deposited hydrothermally.

The 600°C isotherm (Fig. 3.7C) illustrates the appearance of pentlandite, which Kullerud (1963) determined to appear in the system at 610°C from a reaction between $(NiFe)_{(3\pm x)}S_2$ and Mss. Genkin (personal communication, 1983) disagrees with Kullerud, believing pentlandite to be stable to significantly higher temperatures. Whatever its upper stability limit, pentlandite rapidly increases in Ni content below 600°C, varying from 25 to 49 wt percent Ni at 500°C (Shewman and Clark, 1970). The study by Misra and Fleet (1973) differed from that of Kullerud in that they did not observe the three-phase field pentlandite–high-temperature heazlewoodite–Mss or tie-lines between high-temperature heazlewoodite and Mss, but instead concluded that pentlandite is in equilibrium with taenite at 600°C. They also found a much higher Fe content (up to 15 wt %) in the high temperature form of heazlewoodite than did Kullerud.

Most Ni sulfide ores have bulk compositions that, when projected into the Fe-Ni-S system, lie within but toward the S-rich side of the Mss at 600°C. Thus, ignoring the effect of Cu, one would not expect pentlandite to occur as a phase in these ores at these temperature; however, Cu has a considerable effect as discussed later. With falling temperature, the Mss narrows, drawing back from both its S-rich and metal-rich sides but, as most ores lie toward the S-rich side, it is the S-rich phases pyrite, vaesite, or violarite that exsolve first from the Mss, not pentlandite. Violarite first appears as a phase within the Fe–Ni–S system at 461°C (Fig. 3.7D), whence its Ni/Fe ratio extends to reach the Ni–S join at 356°C where it has the mineral name polydymite.

Craig (1973) concluded that the Mss breaks away from the Ni–S join between 300 and 250°C, but that tie-lines still connect it to NiS (millerite). He found that by 250° (Fig. 3.7E), Mss has separated into two phases plus millerite; however, all phases are connected by tie-lines, therefore, pentlandite is still isolated from bulk compositions on the S-rich side of the Mss. Craig concluded that it is only between 250 and 200°C that these tie-lines break down and new tie-lines are established between pyrite and pentlandite. By this stage, an S-rich monoclinic and an S-poor hexagonal variant of the Fe-rich Mss have separated from one another. Presumably these phases both contain Ni, but draw back rapidly with decreasing temperature to the Fe–S join, exsolving pentlandite and pyrite.

In contrast to Craig, Misra and Fleet (1973) found that the Mss starts to break down by 300°C. Barker (1983) also interpreted his results to imply that Mss was not continuous at this temperature.

The slow reaction rates in the system at low temperature have precluded experimental investigations so far. Figure 3.7F is based on the studies of Graterol and Naldrett (1971), Misra and Fleet (1973), and Craig (1973), and is a reconstruction of possible phase relations in the system at about 140°C. The low temperature reconstruction by Misra and Fleet differs from Figure 3.7F in that they concluded that godlevskite is not a stable phase at low temperature and that millerite and heazlewoodite coexist stably.

3.2.2 The Systems Cu–Fe–S and Fe–Cu–Ni–S

Phase relations in the Cu–Fe–S system have been studied by Merwin and Lombard (1937), Schlegel and Schiller (1952), Hiller and Probsthain (1956), Yund and Kullerud (1966), Kullerud et al. (1969), Mukaiyama and Isawa (1970), Cabri (1973), Barton (1973), and Ueno (1981), among others. The nature of the phase relations at 600°C are well established and are illustrated in Figure 3.8A. The system is dominated by three solid–solutions: the bornite, the intermediate, and the pyrrhotite solid–solutions. The pyrrhotite solid–solution is capable of dissolving several percent Cu, and Craig and Kullerud (1969) have shown that Cu is similarly soluble in nickeliferous Mss. Thus, the compositions of many low-Cu–Ni sulfide ores fall within the Cu-bearing Mss at 600°C.

It can be deduced from comparison of the Cu–Fe–S and Fe–Ni–S systems that, above 1,000°C, the sulfide liquid in equilibrium with Mss is enriched in both Cu and Ni with respect to Mss. Thus, fractional crystallization of Mss from this liquid leads to a Ni- and Cu-rich residual liquid. Craig and Kullerud (1969) have shown that the liquid in equilibrium with Mss at 850°C is enriched in Cu but depleted in Ni with respect to Mss. Thus, fractional crystallization of Cu-rich liquids at these low temperatures is likely to reverse the trend of increasing Ni and leads to liquids richer in Cu, but poorer in Ni than their predecessors in the fractionation process.

Based on their observations of natural assemblages at Noril'sk–Talnakh, Distler and Genkin (personal communication, 1984) believe that liquid immiscibility occurs within Fe–Cu–Ni–S liquids in nature, leading to co-existing Cu-rich and Mss-rich liquids. This suggestion remains to be demonstrated experimentally.

Below 600°C, phase relations in the Cu–Fe–S system are less well established, except for the recent unpublished study of Ueno (1981). Figure 3.8B is a reconstruction of possible phase relations at 300°C by Craig and Scott (1974) based "on data and discussions of several workers." The data of Ueno (1981) are in general agreement with this, although the bounds of the intermediate

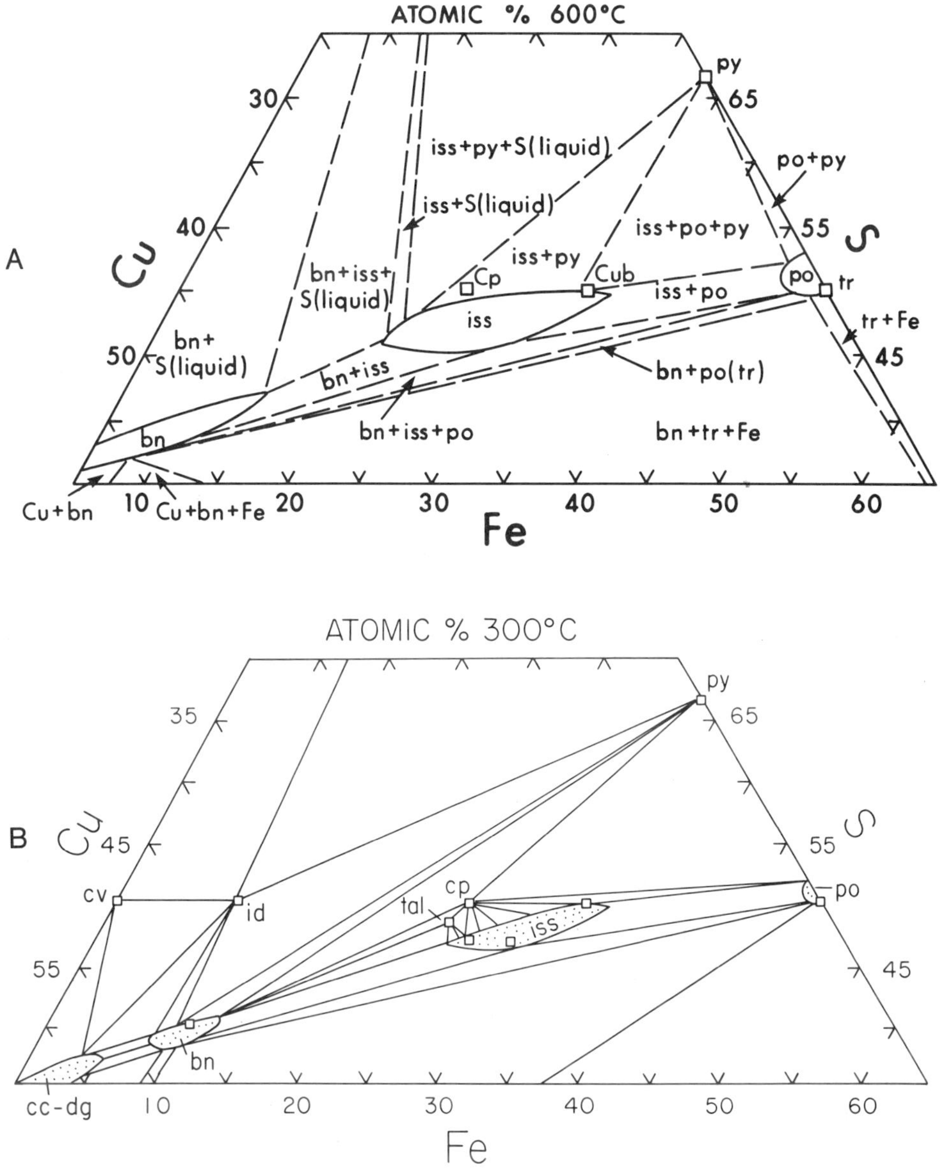

Fig. 3.8. *A.* Isothermal section at 600°C through the central part of the Cu–Fe–S system. (From Cabri, 1973.) *B.* Isothermal section at 300°C through the central part of the Cu–Fe–S system. (From several authorities as compiled by Craig and Scott, 1974.)

solid-solution (Iss) and the compositions of the solid-solution participating in adjacent three-phase assemblages are somewhat different. The crystallization of most of the Cu-bearing mineral phases in Ni–Cu ores is, therefore, not well understood. Judging from the Cu–Fe–S system, it would seem that Iss exsolves from a cupriferous Mss and that in the more sulfur-rich (pyrite-bearing) assemblages, the major part of this exsolution occurs between 700 and 400°C, depending on Cu content. Iss presumably breaks down with falling temperature to chalcopyrite and pyrrhotite; to chalcopyrite, pyrrhotite, and cubanite; or to pyrrhotite, cubanite, and one of the S-deficient intermediate phases: talnakhite, mooihoekite, or haycockite. The effect of Ni on the phase relations of the Cu–Fe–S system is not known.

Hill (1984) has studied the effect of Cu on phase relations in the Fe–Ni–S system at 600°C. Mss in equilibrium with Iss can contain up to 1 wt percent Cu at this temperature. Both pyrite (on the S-rich side) and pentlandite (on the metal-rich side) can coexist with Cu-bearing Mss in equilibrium with chalcopyrite at temperatures higher

(150°C in the case of pyrite, 300°C in the case of pentlandite) than would be predicted from the pure Fe–Ni–S system. The pentlandite coexisting with Mss and chalcopyrite contains up to 2 wt percent Cu, twice that of the coexisting Mss.

3.3 THE SOLUBILITY OF S IN SILICATE MELTS

Knowledge of the solubility of S in mafic and ultramafic silicate melts is important in understanding how magmatic sulfide deposits form and also in evaluating the potential of igneous bodies as hosts for ore deposits of this type. Early work on S solubility was undertaken by metallurgists, notably Richardson and his co-workers (Fincham and Richardson, 1954; Abraham et al., 1960). Although the melt compositions studied by these workers were those of smelter slags rather than natural magmas, Fincham and Richardson's (1954) study provided valuable insight into the mechanism of S solution and its important implications in nature. They showed that at low O fugacities (less than 10^{-6} atm in melts studied at 1,400 and 1,500°C) S dissolves primarily as sulfide and that a function termed the *sulfur capacity** of the melt, C_s, is constant for melts of the same composition and obeys the relationship:

$$C_s = \frac{(\text{wt \% S}) \times (pO_2)^{1/2}}{(pS_2)^{1/2}} \quad (3.2)$$

C_s increases with rise in temperature and changes with composition, generally increasing with increasing FeO, MgO, and CaO contents, and decreasing with increasing SiO_2 and Al_2O_3 contents.

MacLean (1969), in his study of the system Fe–S–O–SiO_2, found that the S content of a silicate melt in equilibrium with a separate sulfide-rich liquid decreases with increasing O content of the system (this S content is henceforth referred to as the sulfur content at sulfide saturation, or SCSS). He attributed this observation to the fact that sulfur dissolves by displacing oxygen bonded to Fe^{2+} and that increasing oxygen results in an increase in Fe^{3+} at the expense of Fe^{2+} in the melt. Connolly and Haughton (1972) confirmed this point.

Shima and Naldrett (1975) studied an ultramafic (komatiitic) melt and obtained strong evidence supporting MacLean's contention as to the mechanism of S solution.

Considering the reaction:

$$FeO_{(\text{in melt})} + \tfrac{1}{2}S_2 = FeS_{(\text{in melt})} + \tfrac{1}{2}O_2 \quad (3.3)$$

*It should be noted that the function sulfur capacity is somewhat misleading. It does not represent the overall capacity of a melt to dissolve S, but is somewhat akin to the equilibrium constant for reaction (3.1). It merely relates the amount of sulfur that will dissolve in a given melt in response to imposed S and O fugacities.

the activities of the reactants are related by the expression:

$$aFeS = \gamma_{FeS}N_{FeS} = K \times aFeO \times \frac{fS_2^{1/2}}{fO_2^{1/2}} \quad (3.4)$$

thus,

$$\log N_{FeS} = \tfrac{1}{2}\log fS_2 + [\log K + \log aFeO - \tfrac{1}{2}\log fO_2 - \log \gamma_{FeS}], \quad (3.5)$$

where γ_{FeS} is the rational activity coefficient for FeS in the melt, N_{FeS} is the mole fraction of FeS, K is the thermodynamic equilibrium constant, and fS_2 and fO_2 are the fugacities of S and O, respectively. If, for small changes in N_{FeS}, γ_{FeS} is assumed to remain constant and also if the amount of FeO formed in a reaction such as (3.1) is sufficiently small that its formation has no appreciable effect on aFeS*, it follows from expression (3.4) that a plot of $\log N_{FeS}$ against $\log fS_2$ (fO_2 and T kept constant) should produce a straight line relationship with a slope of ½, provided that sulfur is dissolving according to reaction (3.1). Figure 3.9, a plot of Shima and Naldrett's (1975) data, indicates that this relationship is obeyed. Furthermore, Figure 3.9 shows that as $\log fO_2$ increases, $\log N_{FeS}$ decreases. This is also in accordance with expression (3.4).

The reader should appreciate that the preceding discussion and data apply to the solution of S in a silicate melt from a superjacent gas phase in response to changes in the fugacity of S in the gas. With the exception of Shima and Naldrett's experiment at $\log fO_2 = -10.4$ and $\log fS_2 = -2$, saturation in Fe sulfide liquid was not achieved.

Haughton et al. (1974) were the first to provide extensive data on the solubility of S in a silicate melt of naturally occurring composition in equilibrium with a sulfide liquid (i.e., SCSS as it is referred to in this book). They found a strong correlation of SCSS with FeO and, to a lesser extent with TiO_2 content, but no systematic change with the varying fO_2 of their experiments. Buchanan and Nolan (1979) performed a similar series of experiments and were able to demonstrate the effect of varying fO_2 (Fig. 3.10). The data at high FeO contents in this figure are not directly applicable to natural basalts, as FeO was increased in the experiments at the expense of all other elements, and as a result, at high FeO contents, SiO_2 fell well below the level to be expected in even the most Fe-rich natural basalts.

Mathez (1976) studied the S content of the glassy margins of submarine basalt pillows from the Atlantic and the Pacific that had been quenched at depths of more than 1000 m. He concluded that these were saturated or nearly satu-

*This is a reasonable assumption, as the amount of Fe interacting with sulfur amounts to 0.2 and 0.6 in most melts containing 8 to 12 wt percent FeO.

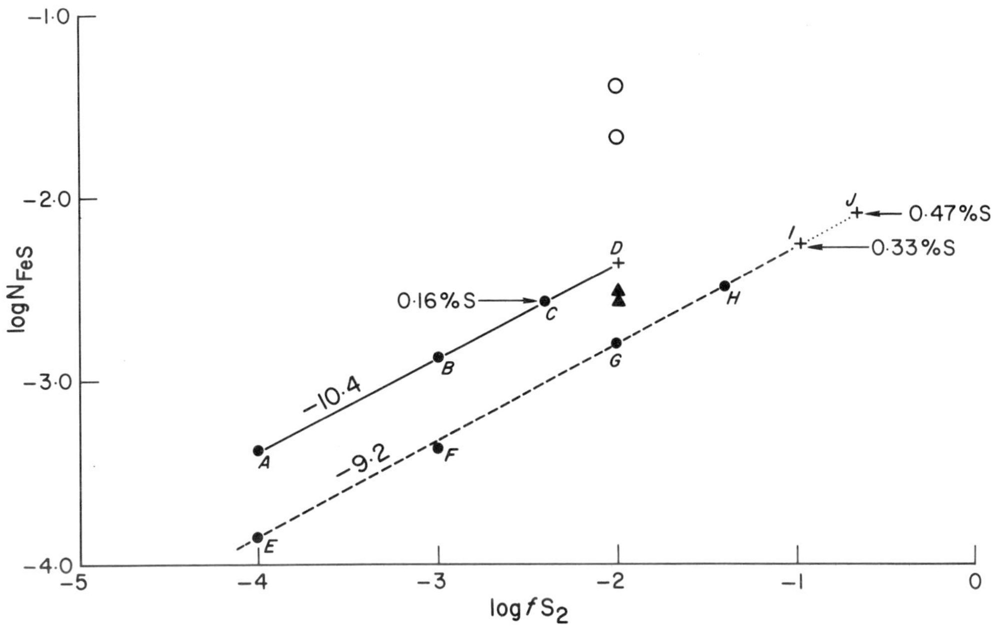

Fig. 3.9. Relationship between N_{FeS} and fS_2 in a given silicate melt at two values of fO_2. Solid dots are experimental values for unsaturated melts. Open circles are values for melt with immiscible sulfide droplets. Solid triangles are values for the S content of melts containing immiscible droplets from which the droplets themselves have been removed. (From Shima and Naldrett, 1975.)

rated when they were quenched. Their S content increases markedly and linearly with increasing FeO content from about 0.105 wt percent S at 9.0 wt percent FeO to 0.18 wt percent S at 12.9 wt percent FeO. These S concentrations are substantially higher than the experimental results illustrated in Figure 3.10, an observation that Mathez attributed to the high pressure at which the submarine basalts were quenched, but which, in the light of Buchanan and Nolan's (1979) data, could be due to fO_2.

3.3.1 Effect of Temperature

Buchanan et al. (1983) determined the solubility of S as a function of fS_2 in a basaltic melt containing 17 wt percent FeO at 1,200, 1,300, and 1,400°C. Data were obtained in the field of undersaturation, indicating that at constant fO_2 and fS_2, the dissolved S content increases by a factor of eight and a half times per 100°C at 1,000°C, but at 1,400°C this factor is only three times per 100°C. Buchanan et al. also attempted to predict the saturation values of the S content (i.e., SCSS), but their inability to achieve saturation in most of their samples indicates that their predicted values should be treated with caution.

Relatively few data exist on the variation of SCSS with temperature. There seems to be little doubt that SCSS increases with temperature. Haughton et al. (1974) found that tholeiitic melts containing 10 wt percent FeO dissolve about 0.09 wt percent S when saturated with Fe sulfide, whereas Shima and Naldrett (1975) found that an ultramafic melt containing the same FeO content dissolves between 0.16 and 0.27 wt percent S at an fO_2 of $10^{-10.4}$. This comparison is rendered inaccurate by differences in melt composition and fO_2. FeO, however, is consistently the compositional variable that has the greatest effect on saturation with FeS. Wendlandt (1982) found that the increase in temperature of a basaltic melt containing close to 8 wt percent FeO from 1,300 to 1,460°C at 20 kb total pressure and an fO_2 close to the C–CO_2–CO–O_2 buffer (about 1 log unit above the quartz–fayalite–magnetite [QFM] buffer according to Wendlandt) caused an increase in SCSS from 0.09 to 0.16 wt percent, corresponding to an increase of 0.05 wt percent per 100°C. Wendlandt's experiments on silicate melts of differing FeO contents produced similar results, given the errors inherent in his data. Possibly an increase in SCSS of two and a half times from 1,200 to 1,450°C is of the order of that to be expected in nature.

3.3.2 Effect of Pressure

Huang and Williams (1980) investigated portions of the system Fe–Si–S–O at 32 kb and found that the miscibility gap between sulfide and silicate liquids expands with increasing pressure. Wendlandt (1982) studied the variation in SCSS in two basalts and an andesite at pressures be-

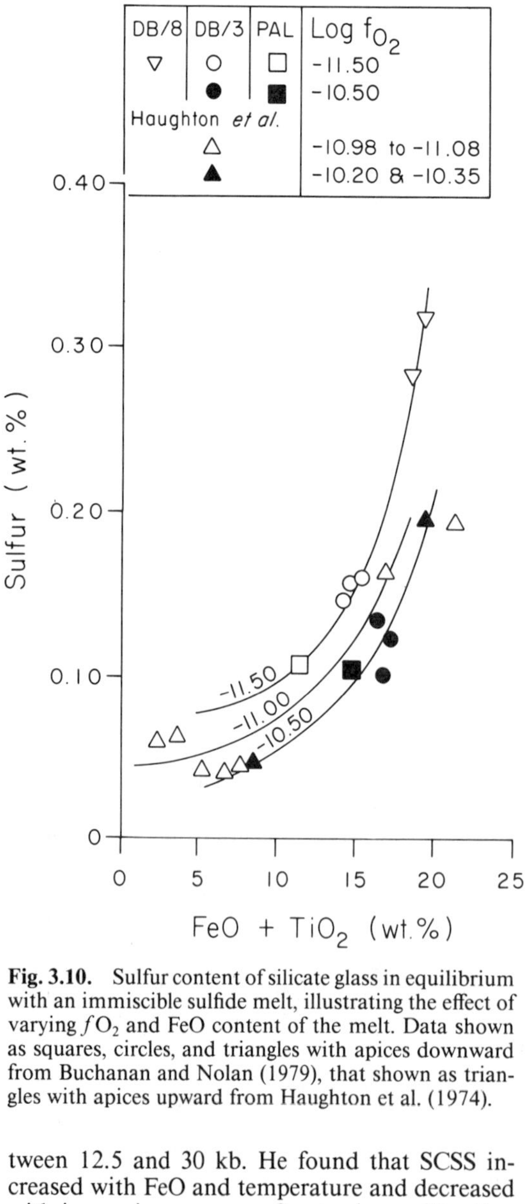

Fig. 3.10. Sulfur content of silicate glass in equilibrium with an immiscible sulfide melt, illustrating the effect of varying fO_2 and FeO content of the melt. Data shown as squares, circles, and triangles with apices downward from Buchanan and Nolan (1979), that shown as triangles with apices upward from Haughton et al. (1974).

tween 12.5 and 30 kb. He found that SCSS increased with FeO and temperature and decreased with increasing pressure. Comparison with other studies was difficult because he relied on the C–CO_2–CO–O_2 buffer, which varied with changing, uncalibrated conditions between his experiments. His studies, however, would seem to indicate that under natural conditions increasing pressure has a depressing effect on SCSS. This implies that as a magma rises to the surface, its ability to dissolve S increases and, therefore, it is not likely to approach saturation with sulfide. In Figure 3.11, Wendlandt's data on the effects of pressure and FeO content have been combined (solid lines) to show the effect of pressure on SCSS in a magma containing constant FeO (Wendlandt found that due to the design of his experiments, the FeO content of the silicate portion of a particular starting composition did not remain constant); these relationships have been modified in Figure 3.11 (dashed lines) to show the variation to be expected in a basalt rising from a depth equivalent to 30 kb, but that loses heat and stays at its liquidus temperature (assumed to vary from 1,500°C at 30 kb to 1,300°C at 12.5 kb). The reduction in SCSS due to cooling would largely counterbalance the effect of pressure. The uncertainties in the data are considerable, however, and any genetic hypothesis based on this discussion will bear these uncertainties.

3.3.3 Experiments with Melts in Equilibrium with H–O–S and H–C–O–S Fluids

Gorbachev and Kashirceva (1986) have conducted a series of experiments on basaltic to picritic melts in equilibrium with (H–O–S)- and (H–C–O–S)-bearing fluid phases. Their melts were in equilibrium with and enclosed in peridotite sheaths, that were themselves encased within doubled-walled Pt capsules. The O fugacity was controlled by the assemblage olivine + orthopyroxene + chromite. Calculations indicate that the fO_2 was between 10^{-8} and 10^{-9} atm. Their data indicate that at 15 kbars, and in equilibrium with both an H–O–S fluid and a Fe sulfide–oxide liquid, the dissolved S in a melt containing 12.5 wt percent of both FeO and MgO increases from 0.45 wt percent at 1,250°C to 0.65 wt percent at 1,350°C. They found that in the presence of both an H–O–S and an H–O–C–S fluid, S solubility in the melt increased slightly with increasing pressure up to 15 kb total pressure, but decreased progressively with increasing pressure above this.

The results of Gorbachev and Kashirceva (1986) are compared with those of Wendlandt (1982) in Figure 3.12. The latter's experiments were not conducted at pressures much below 15 kb, therefore, given the errors involved, the results neither confirm nor refute Gorbachev and Kashirceva's finding of a change in the dependence of S solubility on total pressure at 15 kb. The wide range of variables in Gorbachev's experiments preclude any exact analysis of the reasons for the variable dependence on pressure, but they open up an interesting area for future experimentation.

3.4 PARTITIONING OF CHALCOPHILE ELEMENTS BETWEEN SULFIDES AND SILICATE MELTS AND SILICATE MINERALS

It is common to discuss the partitioning of a trace or minor element between two phases, A and B, in terms of a partition coefficient. Referring to equilibria between sulfide melts and silicate mag-

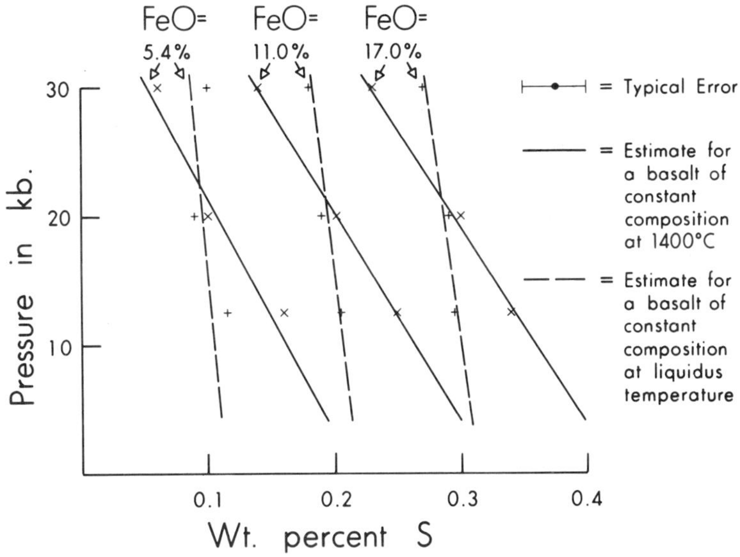

Fig. 3.11. Solubility of S in basaltic melts of different FeO content estimated as a function of pressure at 1,400°C (solid line) and at the liquidus temperature of the basalt (dashed line). (Based on the data of Wendlandt, 1982.)

mas, the Nernst partition coefficient D_i for a metal i would be defined as:

$$D_i^{(\text{Sul. melt/Sil. magma})} = \frac{\text{wt \% of metal } i \text{ in sulfide melt}}{\text{wt \% of metal } i \text{ in silicate melt}} \quad (3.6)$$

3.4.1 Partitioning of Ni Between Sulfide and Silicate Liquids

In the case of metals, such as Fe, Ni, Cu, and Co, it is believed (see Shimazaki and MacLean, 1976; Rajamani and Naldrett, 1978) that they are bonded to O in the silicate magma and to S in the sulfide melt. Thus, the reaction considered is (for example, in the case of Ni):

$$NiO^{(\text{Sil. magma})} + \tfrac{1}{2}S_2 = NiS^{(\text{Sul. melt})} + \tfrac{1}{2}O_2 \quad (3.7)$$

The equilibria involved in this reaction are related to the thermodynamic equilibrium constant K_3 by the expression:

$$K_3 = \frac{a\text{NiS}}{a\text{NiO}} \times \frac{fO_2^{1/2}}{fS_2^{1/2}}, \quad (3.8)$$

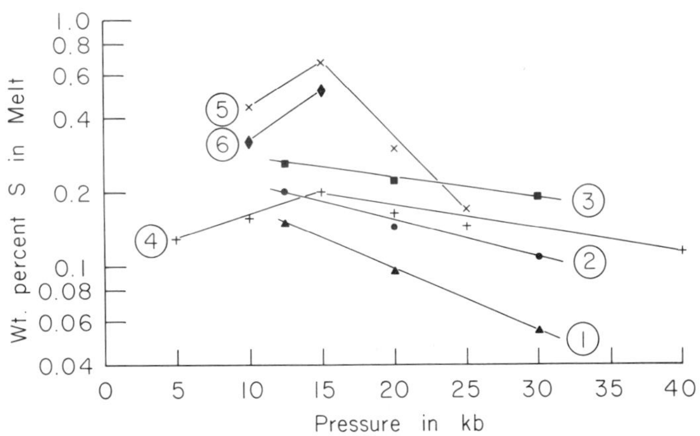

Fig. 3.12. Comparison of the data of Gorbachev and Kashirceva and Wendlandt (1982) from Gorbachev and Kashirceva (1986). (1), (2) and (3) = experiments of Wendlandt (1982) (see legend to Fig. 3.12), (4) = basalt + H_2O + CO_2 at 1,250°C, (5) = basalt + H_2O at 1,350°C. The composition of the melt changes with increasing pressure from about eighteen wt percent MgO, ten wt percent FeO at low pressure to ten wt percent MgO, fourteen wt percent FeO at high pressure, due to the lower degree of melting at the higher pressures. (6) = similar experiments to (5) at 1,300°C.

$$= \frac{\gamma_{NiS} \times N_{NiS}}{\gamma_{NiO} \times N_{NiO}} \times \frac{fO_2^{1/2}}{fS_2^{1/2}} \quad (3.9)$$

where a refers to the respective activities, γ to the activity coefficients, N to the mole fractions, and fO_2 and fS_2 are the fugacities of the gases. From comparison of expression (3.6) and (3.9), one would expect $D_i^{\text{Sul. melt/Sil. magma}}$ for all of the metals Ni, Fe, Cu, and Co to be a function of fO_2 and fS_2 in addition to being a function of the temperature and pressure and the compositions of the two phases in question. Reactions, such as (3.3) and (3.7), however, may be combined to give exchange reactions of the type:

$$NiO^{(\text{Sil. magma})} + FeS^{(\text{Sul. melt})} = NiS^{(\text{Sul. melt})} + FeO^{(\text{Sil. magma})} \quad (3.10)$$

The equilibrium constant for this reaction can be expressed by:

$$K_4 = \frac{aNiS \times aFeO}{aNiO \times aFeS}, = \frac{\gamma_{NiS} \times \gamma_{FeO}}{\gamma_{NiO} \times \gamma_{FeS}} \times \frac{N_{NiS} \times N_{FeO}}{N_{NiO} \times N_{FeS}}. \quad (3.11)$$

The equilibrium constant of this reaction is independent of the fugacities of S and O although these may affect the activities of the components in the reaction. For example, fS_2 will affect the activity coefficients of NiS and FeS. Citing the work of Scott et al. (1974) on the $Fe_{(1-x)}S$–$Ni_{(1-x)}S$ solid-solution, Rajamani and Naldrett (1978) suggested that in liquids of the same composition as monosulfide solid solution, γ_{NiS} and γ_{FeS} would have very similar values despite the fact that both would decrease with increasing fS_2, and vice versa. Thus, the ratio of these two functions would be close to 1 and show little change with variations in fS_2. Variations in fO_2 can affect N_{FeO} by changing the Fe^{+3}/Fe^{+2} ratio of the magma, but as long as the value of fO_2 remains below about 10^{-8} atm (see Fudali, 1965), the effect on the Fe^{+3}/Fe^{+2} ratio of basaltic magmas is small. Doyle and Naldrett (1986) emphasize that variations in fO_2 also affect the oxygen content of sulfide-oxide liquids, and hence activity composition relationships within them. Provided that the fO_2/fS_2 ratio (and hence the O/(O + S) ratio in the sulfide–oxide liquid) remains approximately constant, this effect will not be noticeable with regard to element partitioning. An assumption of constant fO_2/fS_2 ratio is made in the following discussion. The effect of variable fO_2 will be considered later.

If a function, K_D (the exchange partition coefficient), is defined (note the use of mole fractions in the definition) as:

$$K_D = \frac{N_{NiS} \times N_{FeO}}{N_{NiO} \times N_{FeS}}, \quad (3.12)$$

then from equations (3.11) and (3.12) and noting that, as discussed earlier, $\gamma_{NiS}/\gamma_{FeS} \simeq 1$,

$$K_D = K_4 \times \frac{\gamma_{NiO}}{\gamma_{FeO}} \quad (3.13)$$

As K_4 is constant at given temperature and pressure, K_D can be calculated from thermochemical data for any silicate magma and iron-rich sulfide melt, as long as the values of γ_{NiO} and γ_{FeO} are known for the silicate melt in question. Furthermore, if K_D is determined for one particular silicate melt, knowledge of the temperature and compositional dependence of γ_{NiO} and γ_{FeO} and of the temperature dependence of K_4 permits new values of K_D to be computed for melts of different composition and temperature.

Rajamani and Naldrett (1978) obtained values for K_D for Ni, Cu, and Co using liquids with compositions of natural basalts, picrites, and andesites (Table 3.1). Subsequently, Boctor and Yoder (1983) obtained similar values. The latter also demonstrated a dependence of K_D on fO_2.

Rajamani and Naldrett's (1978) data on sulfide–oxide liquid–silicate magma partitioning (with K_D values for Ni of about 40) are compatible with the magmatic hypothesis, leading, as they showed (and as shown in Chapter 6), to reasonable predictions of the actual compositions of specific Ni–sulfide ores. The O_2 fugacity in their experiments was not buffered, but they predicted from the starting compositions of their materials that it was close to the quartz–fayalite–magnetite buffer, i.e., within the range of natural magmas. Boctor and Yoder's (1983) data confirm K_D values of around 40 to 50 at temperatures of 1,250 to 1,300°C and fO_2 values of 10^{-8} to 10^{-9} atm. Boctor and Yoder's data also demonstrate that K_D decreases with increasing temperature of equilibration and increasing maficity of the silicate magma, as predicted by the calculations of Rajamani and Naldrett (1978) and Duke and Naldrett (1978), and the experimental studies of Campbell et al. (1979).

Celmer (1987) studied the distribution of Ni between Ni–Cu–Fe matte and fayalite-rich slags. He found that the Nernst distribution coefficient for Ni was about 160 in SiO_2 saturated slags at 1,250°C, an fO_2 of about 4×10^{-9}, and an fS_2 of about 5×10^{-3}. The coefficient decreased with decreasing Fe/Ni ratio in the matte and with increasing fS_2 and fO_2, and increased by a factor of 1.5 to 2 with the addition of up to 15 wt percent Al_2O_3 and 10 wt percent CaO to the slag. The slag compositions are far from those of basaltic melts, therefore the comparison of absolute values is not relevant; however, it is probable that Nernst coefficients for natural ores will vary in a similar direction to changes in the fugacities of O_2 and S_2 as in slag/matte combinations.

TABLE 3.1. *Experimental K_D Values for Transition Metal Partitioning Between Sulfide Melts and Silicate Magmas*

	K_D	Temperature °C	Pressure	$\log fO_2$	Melt. Comp.
Ni					
Rajamani & Naldrett (1978)	42	1255	very low	unknown[a]	basalt
	38	1305	very low	unknown[a]	basalt
	34	1325	very low	unknown[a]	olivine basalt
	59	1255	very low	unknown[a]	andesite
Boctor & Yoder (1983)	49.4 ± 3	1300	1 atm	−9.0	basalt
	50.9 ± 2.6	1300	1 atm	−9.0	basalt
	43.1 ± 3.2	1400	1 atm	−8.0	basalt
	35.9 ± 2.5	1400	1 atm	−8.0	basalt
	22.4 ± 2.3	1460	1 atm	−8.55	mafic 21% MgO
Gorbachev (personal communication, 1988)	35[b]	1250	13 kb	Ni–NiO	Mafic 10.8 at % MgO 7.1 at % FeO
	25[b]	1250	15 kb	Ni–NiO	Mafic 21.9 at % MgO 5.4 at % FeO
	27[b]	1250	25 kb	Ni–NiO	Mafic 11.8 at % MgO 7.7 at % FeO
Gorbachev (personal communication, 1988)	32[c]	1250	15 kb	Ni–NiO	Mafic 12.7 at % MgO 8.0 at % FeO
	35[c]	1300	10 kb	Ni–NiO	Mafic 14.6 at % MgO 9.2 at % FeO
	42[c]	1300	15 kb	Ni–NiO	Mafic 14.3 at % MgO 10.0 at % FeO
	36[c]	1350	25 kb	Ni–NiO	Mafic 16.4 at % MgO 9.4 at % FeO
Cu					
Rajamani & Naldrett (1978)	35	1255	very low	unknown[a]	basalt
	24	1305	very low	unknown[a]	basalt
	48	1325	very low	unknown[a]	olivine basalt
	34	1255	very low	unknown[a]	andesite
Co					
Rajamani & Naldrett (1978)	15	1255	very low	unknown[a]	basalt
	9	1305	very low	unknown[a]	basalt

[a]Rajamani and Naldrett (1978) state that from the initial compositions and proportions of their starting materials, they believe the O_2 fugacities in their experiments to have been close to the quartz–fayalite–magnetite buffer.

[b]These data were obtained with excess H_2O and CO_2.

[c]These data were obtained with excess H_2O.

3.4.2 Partitioning of Ni and Fe Between Sulfide Liquid and Olivine

Nickel in a silicate magma also partitions preferentially into olivine according to the reaction:

$$\mathrm{NiO}^{(\mathrm{Sil.\ magma})} + \mathrm{FeSi_{0.5}O_2^{(Olivine)}} = \mathrm{FeO}^{(\mathrm{Sil.\ magma})} + \mathrm{NiSi_{0.5}O_2^{(Olivine)}}, \quad (3.14)$$

where the equilibrium constant K_5 is defined as:

$$K_5 = \frac{a\mathrm{FeO} \times a\mathrm{NiSi_{0.5}O_2}}{a\mathrm{NiO} \times a\mathrm{FeSi_{0.5}O_2}}, \quad (3.15)$$

and because $NiSi_{0.5}O_2$ and $FeSi_{0.5}O_2$ behave almost ideally in olivine (Campbell and Roeder, 1968),

$$K_5 = \frac{\gamma_{\mathrm{FeO}}}{\gamma_{\mathrm{NiO}}} \times \frac{N_{\mathrm{FeO}}}{N_{\mathrm{NiO}}} \times \frac{N_{\mathrm{NiFe0.5O2}}}{N_{\mathrm{NiFe0.5O2}}}. \quad (3.16)$$

Reactions (3.10) and (3.14) can be combined to express the partitioning between a sulfide melt and olivine:

$$\mathrm{FeS}^{(\mathrm{Sul.\ melt})} + \mathrm{NiSi_{0.5}O_2^{(Olivine)}} = \mathrm{NiS}^{(\mathrm{Sul.\ melt})} + \mathrm{FeSi_{0.5}O_2^{(Olivine)}}, \quad (3.17)$$

where the equilibrium constant K_6 is defined as:

$$K_6 = \frac{K_4}{K_5} = \frac{\gamma_{\mathrm{NiS}}}{\gamma_{\mathrm{FeS}}} \times \frac{N_{\mathrm{NiS}}}{N_{\mathrm{FeS}}} \times \frac{N_{\mathrm{FeSi0.5O2}}}{N_{\mathrm{NiSi0.5O2}}}, \quad (3.18)$$

The exchange partition coefficient for the reaction is, therefore, equal to the equilibrium constant multiplied by the ratio of the activity coefficients of the two sulfide components in the sulfide melt. Values for the partition coefficient obtained in a variety of experiments involving Fe–Ni sulfides and olivine by Clark and Naldrett (1972), Fleet et al. (1977), Boctor (1981, 1982),

TABLE 3.2. *K_D Values for Ni Partitioning Between Sulfide Melts and Olivine*

	K_D	Temperature °C	Log fO_2
Experimental data			
Clark & Naldrett (1972)	33.2 ± 3.4	930	QFM
Fleet et al. (1977)	33	1,160	unknown
Boctor (1981)	13.4 ± 1.4	1,400	−9.0
Boctor (1982)	8.2 ± 1.4	1,400	−8.0
	18.8 ± 1.4	1,300	−9.0
	18.7 ± 0.8	1,300	−9.0
Fleet & MacRae (1983)	27.2 ± 3.5	1,200	unknown
Fleet & MacRae (1987)	32–34	1,395	−8.77
	30–38	1,385	−8.87
	25–30	1,300	unknown
	25–27	1,200	unknown
Gorbachev & Kashirceva (1986)[a]	25 ± 3	1,250	Ni–NiO
Natural data from Thompson et al. (1984) and Fleet et al. (1983)			
Katahdin	3–9		
Burnt Nubble	7–8		
Big Indian Pond	8–16		
Black Narrows	5–8		
Stillwater PGE Zone	7–14		
Dumont Intrusion	7–24 (av. 12.0)		
Water Hen	6		
Great Lakes Nickel	8		
Perseverence W. Australia[b]	23 ± 4		
Renzy Lake, P.Q.[b]	21.6 ± 3.4		

[a]Gorbachev's experiments were at pressures of 5 and 10 kbar in the presence of an H–O–S fluid phase.
[b]Metamorphosed to upper amphibolite or higher grade and probably represents metamorphic equilibration.

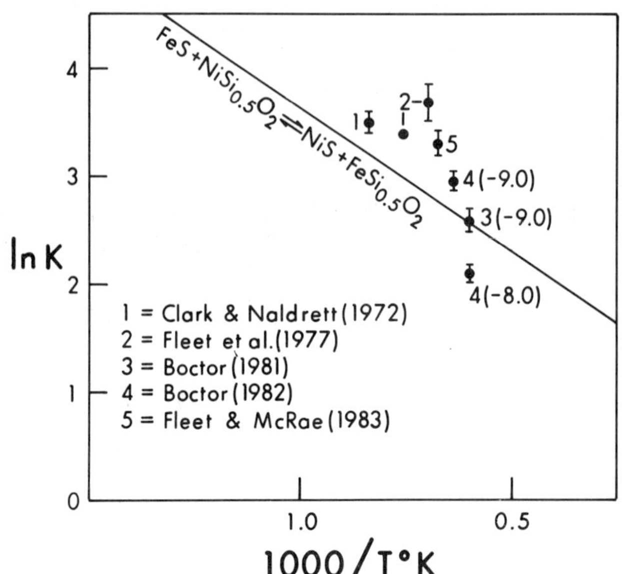

Fig. 3.13. Experimental determinations of the distribution of Ni and Fe between olivine and sulfide. The sulfide phase was a solid-solution in the experiment of Clark and Naldrett (1972); in all other cases it was molten. The error bars represent plus or minus 2 standard deviations of the mean of several determinations of the distribution coefficient at a given T and fO_2. Where O_2 fugacities are known precisely log fO_2 is shown in parenthesis. The dashed line is the variation of the equilibrium constant for the exchange reaction with T, calculated with respect to the pure solid reactants as standard states. (After Thompson et al., 1984.)

and Fleet and MacRae (1983, 1987) are given in Table 3.2, along with some data from actual deposits.

There is a fundamental divergence between the experimental data. Fleet et al. (1977) and Fleet and MacRae (1983, 1987) obtained values between olivine and liquid sulfide ranging from 27 to 40, which are similar to those obtained by Clark and Naldrett (1972) between olivine and solid sulfide at a much lower temperature (915°C). (Fleet's most recent data, M. E. Fleet, personal communication 1987, is consistent with his earlier data.) Their data have led Fleet and MacRae (1987) to suggest that K_D does not vary appreciably with temperature.

On the other hand, Thompson et al. (1984) have derived a relationship for K_6 (Fig. 3.13) from thermochemical data for solid sulfides and silicates that suggests that there should be a temperature dependence. If one assumes that K_D approximates to K_6 (i.e., that the ratio of the activity terms in equation (3.16) is close to unity) and that the exchange reaction is not changed greatly by the involvement of a liquid rather than a solid monosulfide solid-solution, analogy with Figure 3.13 indicates that K_D should also decrease with increasing temperature.

As with Boctor and Yoder's (1983) later experiments on coexisting sulfide and silicate liquids, Boctor's (1982) experiments on coexisting sulfide liquid and olivine indicate a dependence of K_D on fO_2, with K_D increasing from 8.2 to 13.4 at 1,400°C with a decrease in fO_2 from 10^{-8} to 10^{-9} atm. Because the sulfide liquid is the common factor in both the sulfide–silicate liquid and sulfide–olivine experiments, it is logical to look to this for an explanation of the fO_2 dependence. As was stated at the outset of this discussion, the fact that O is present and the possible effect that it might have on the activities of the components, such as NiS and CuS, in the sulfide (–oxide) liquid have been ignored up to this point.

Doyle and Naldrett (1987) summarized the available experimental data on the Fe–S–O system. They found that at 1,200°C the function, $-\log(fO_2/fS_2)$, increases approximately linearly with an increase in the molar ratio O/(O + S) and that at O_2 fugacities of the order of 10^{-8} atm, the atomic proportions of O in many Fe–S–O liquids equal or exceed the proportions of S. The experimental data indicate that as O increases at the expense of S, at a constant atomic proportion of Fe, the activity of Fe is not constant; i.e., the activity coefficient of Fe is a function of the O/(O + S) ratio, and thus of the fO_2/fS_2 ratio. This can be seen from Figure 3 of Shima and Naldrett (1975). Because of the near parallelism of the fO_2 and fS_2 isobars, a decrease in fO_2 from 10^{-8} to 10^{-9} atm at a constant fS_2 of 10^{-2} atm involves a change in composition corresponding to a decrease in aFeO in the coexisting sulfide and silicate melts from 0.4 to 0.15. This decrease in aFeO can cause an increase in aFeS by a factor of 1.2. Analogous changes can be expected with respect to aNiS and the exchange partitioning will be reflected in the net sum of these changes.

Doyle and Naldrett observed that, when taken together, Fleet and MacRae's (1983) and Boctor's (1981, 1982) values of K_D for Ni–Fe exchange between olivine and sulfide–oxide liquids increased progressively with decreasing O/(O + S) ratio of the liquids. This led them to suggest that, as noted previously, the activity coefficient of Ni is also affected by the composition of the sulfide–oxide liquid, the coefficient for Ni decreasing more than that of Fe for a given decrease in the O/(O + S) ratio, therefore, causing K_D for the exchange reaction between Ni and Fe to decrease.

Fleet and MacRae accept that the addition of oxide to a sulfide–oxide liquid will lower K_D, but challenge the extent of the lowering. Their recent data indicate that K_D decreases to 22 in a liquid containing 10 wt percent O_2 instead of the value of about 10 predicted by Doyle and Naldrett (1987) from Boctor's (1982) data. Fleet and MacRae argue, contrary to the conclusions of Fudali (1965) but in accordance with the recent suggestion of Christie et al (1986), that the O fugacities in the igneous magmas giving rise to layered intrusions are two orders of magnitude lower than those of the Fayalite-Magnetite-Quartz (FMQ) buffer assumed by Doyle and Naldrett. If correct, this would result in the O content of natural sulfide liquids being much less than assumed by Doyle and Naldrett (1987), reducing even more the lowering of K_D.

Observations on sulfides and olivines coexisting in nature (Table 3.2) reveal K_D values from 3 to 24 with most falling in the 5 to 15 range. Notable exceptions are the two deposits in Table 3.2 that have been subjected to the highest grades of metamorphism; these have values in excess of 20. Natural values in the range of 5 to 15 are consistent with the experimental results of Boctor (1981, 1982), but are different from those of Fleet and his co-workers. If the latter researchers are correct, and if K_Ds are of the order of 30, problems arise for the magmatic model, which is what these authors have long maintained.

On the other hand, if Boctor's values are correct for natural conditions, and if the thermodynamic arguments of Thompson et al. (1984) as to the temperature dependence of K_6 and, therefore, K_D are correct, a higher value is to be expected for sulfide–olivine equilibria that have been achieved at metamorphic than at magmatic temperatures. This is consistent with the hypothesis that the Perserverance and Renzy Lake deposits have reequilibrated during metamorphism, as Binns and Groves (1976) claimed for Perserverance, but that reequilibration has not occurred in the other deposits.

3.4.3 General Remarks on the Partitioning of Ni Between Sulfide Liquids and Silicate-rich Phases

At the time of writing conflicting data exist for the exchange partition coefficients that apply to Fe–Ni exchange between sulfide liquids and olivine. There is no direct conflict with respect to partitioning between sulfide and silicate liquids, although, as discussed later, the data for sulfide–olivine equilibria also carry implications with respect to sulfide–silicate liquid partitioning. In addition, variations in fO_2 are likely to affect partitioning between sulfide and silicate melts. As discussed earlier, Celmer (1987) found that an increasing fO_2 led to a decrease in the Nernst coefficient for matte–slag combinations, but this aspect has yet to be investigated for basaltic compositions. The uncertainty as to the magnitude of any possible effect should, therefore, be borne in mind when the values tabulated in Tables 3.1 and 3.3 are put to use.

Naldrett (1979) pointed out that there should be an interrelationship between the K_Ds for Ni–Fe exchange between olivine and silicate melts, olivine and sulfide melts, and sulfide and silicate melts. Experimental studies (Arndt, 1977) on the distribution of Ni between olivine and a range of sulfide-free silicate melts indicate that the K_D varies from about 2 for a komatiitic to 8 to 14 for a basaltic melt. Given a K_D between sulfide melt and basalt of about 50, the K_D between sulfide and olivine should be about 5 to 7, which is far from the value of about 30 determined by Fleet and his co-workers. It is possible that (as Fleet and MacRae, 1983, point out) the presence of sulfur in a silicate melt favors the retention of Ni in the melt more than Fe, thus reducing the olivine–silicate melt K_D below the values determined for sulfide-free systems, but Celmer looked for this in his matte–slag experiments and was unable to document any such effect. The divergence between the K_Ds of Fleet and his co-workers and natural observations thus poses a problem that extends beyond the area of coexisting olivine and sulfide melts and raises questions about the equilibria between sulfide and silicate melts.

A further complexity has been introduced by Cowden and Woolrich (1987) with respect to the effect of oxidation on sulfide melt–silicate magma partitioning. They point out that, as the oxidation of a silicate magma is likely to change the number of octahedral sites (by reducing the Fe^{2+} content) in the magma that are available for divalent cations, the tendency for these cations to enter octahedral sites in sulfide melts will be increased, particularly those with high octahedral site preference energies. They predict, therefore, that oxidation should increase $K_D^{(\text{Sul. melt/Sil. magma})}$ between Ni and Fe. This is, of course, opposite to the effect of oxidation highlighted by Doyle and Naldrett (1987) for sulfide liquid–olivine partitioning. Cowden and Woolrich (1987), however, assume a constant S/metal ratio for the sulfide melt, which Doyle and Naldrett showed not to be the case, and consider only the implications of oxidation on the silicate magma, and not on the sulfide melt.

Despite the fact that all of the experimental studies have been on the exchange reactions discussed previously, because of simplicity in application it has been the custom in the literature involved with the modeling of sulfide liquid–silicate liquid partitioning to use the results of the experiments converted to simple Nernst-type coefficients. Naldrett (1981a) warned of the errors involved in using Nernst coefficients at high concentrations of the elements involved, and where the partitioning also involves appreciable amounts of some third component such as Cu. These warnings should also be borne in mind when using the numbers tabulated in Table 3.3; these numbers represent the preferred values of the Nernst coefficients, derived from the exchange partition coefficients of Table 3.1.

3.4.4 Partitioning and the Magma/Sulfide Ratio

When a very small amount of sulfide melt segregates from a silicate magma, the concentration of any metal i in the sulfide melt (Y_i) is related to the initial concentration in the silicate magma (X_i) by the partition coefficient $D_i^{\text{Sul. melt/Sil. magma}}$ according to the expression:

$$Y_i = D_i^{\text{Sul. melt/Sil. magma}} \times X_i, \quad (3.19)$$

which is merely a rearrangement of expression (3.6) where Y_i is the wt percent of metal i in the sulfide melt and X_i the wt percent of i in the silicate melt. Provided that equilibrium is always attained between sulfide melt and silicate magma, an expression of this kind always holds, provided that X_i is taken to be the concentration in the silicate magma after the attainment of equilibrium.

It is often more useful to be able to model the composition of the sulfide melt in terms of the ini-

TABLE 3.3. *Preferred Nernst Partition Coefficients*

	Ni	Cu	Co	Pt	Pd
Komatiitic magma					
27 wt % MgO	100	250	40	10^5	10^5
19 wt % MgO	175	250	58	10^5	10^5
Basaltic magma	275	250	80	10^5	10^5

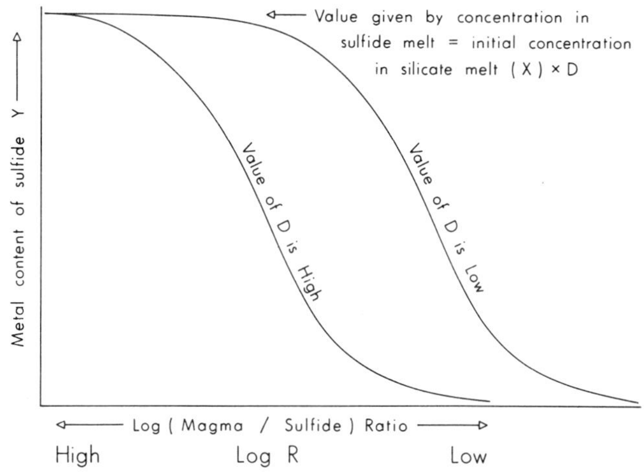

Fig. 3.14. Schematic representation of the variation of the metal content of a sulfide melt as a function of R (magma/sulfide ratio) and D (Nernst distribution coefficient).

tial composition of the silicate magma (i.e., before the segregation of, or reaction with, sulfide occurs). Where the ratio of silicate magma to sulfide melt is very large, expression (3.19) provides a satisfactory answer. As this ratio decreases, however, a stage is reached where the sulfide is concentrating so much of the metal present in the whole system that it causes a significant drop in the concentration of this metal in the silicate magma with which it is equilibrating. Campbell and Naldrett (1979) have shown that under these circumstances, it is necessary to use the more general expression:

$$Y_i = \frac{D_i \times X_{oi} \times (R + 1)}{(R + D_i)}, \qquad (3.20)$$

where R is the ratio of the mass of silicate magma to the mass of sulfide melt, and X_{oi} refers to the initial concentration of metal in the silicate magma. Figure 3.14 illustrates schematically the effect that variations in R and D have on Y_i.

If the sulfides in a particular magma have been added to the magma from an external source, the added sulfide produces an additional dilution so that expression (3.20) then becomes:

$$Y_i = \frac{D_i \times X_{oi} \times R}{(R + D_i)}. \qquad (3.21)$$

It is only when R becomes very low that the values of Y_i obtained from expressions (3.20) and (3.21) differ appreciably.

4. SELECTED NICKEL–SULFIDE DEPOSITS: GENETIC IMPLICATIONS

In this chapter, selected Ni–sulfide deposits are described according to the petrotectonic settings outlined in Chapter 2, namely (I) those associated with Precambrian (largely Archean) synvolcanic volcanism, (A) associated with komatiitic volcanism and (B) associated with intrusions related to tholeiitic volcanism; (II) those associated with rifted continental margins; and (IIIA) those associated with flood basalts resulting from intracontinental rifting and (IIIB) the Sudbury deposits, associated with magmatism resulting from meteorite impact.

4.1 SETTING IA: DEPOSITS IN PRECAMBRIAN GREENSTONE BELTS ASSOCIATED WITH KOMATIITES

Ore deposits of this setting fall naturally into two groups:

Group 1: Ore bodies occurring at the base of ultramafic flows, generally small (1 to 5 $\times$ 10^6 tonnes) and of high grade (1.5–3.5 wt % Ni). Examples include the deposits of the Kambalda district (Ross and Hopkins, 1975; Gresham and Loftus-Hills, 1981; Gresham, 1986) and the Widgiemooltha dome (Fisher, 1979; McQueen, 1981), Western Australia; and the Langmuir mine of the Abitibi belt, Canada (Green and Naldrett, 1981). The Agnew deposit (in excess of 40 $\times$ 10^6 tonnes and a grade of 2.0 wt % Ni) is now thought (Hill et al., 1985; Gole et al., 1988) to belong to this group and is much larger and of somewhat lower grade than the remainder.

Group 2: Very large (100–250 $\times$ 10^6 tonnes), low-grade (0.6 wt % Ni) deposits of finely disseminated sulfide in dunitic lenses that are interpreted as intrusive, sill-like bodies feeding komatiitic volcanism. Examples include the Six-Mile and Mt. Keith deposits near Yakabindi, Western Australia (Burt and Sheppy, 1975; Naldrett and Turner, 1977); and the Dumont nickel deposits of Quebec (Duke, 1985; Eckstrand, 1975).

4.1.1 Group 1 Extrusion-related Deposits

4.1.1.1 KAMBALDA

The extrusive association of the komatiite-related deposits was first recognized and described (Woodall and Travis, 1969) in the Kambalda–St. Ives mining camp of Western Australia (referred to hereafter as the Kambalda mining camp). Since mining started in 1967, 16.6 $\times$ 10^6 tonnes of ore had been mined until June 1983 with a grade of 3.13 wt percent Ni, and the reserves in June 1983 stood at 33.9 $\times$ 10^6 tonnes at 3.78 wt percent Ni (Gresham, 1986).

The most comprehensive data about this deposit type have since been accumulated from study of the Kambalda camp. Twenty-four separate ore shoots have been recognized (Gresham and Loftus-Hills, 1981), with a total premining reserve of 31 $\times$ 10^6 tonnes grading 3.6 wt percent Ni and 0.28 wt percent Cu. Individual ore shoots range in size from very small to over 5 $\times$ 10^6 tonnes. Of the 18 shoots described by Gresham and Loftus-Hills, five appear to be over 2.8 $\times$ 10^6 tonnes, four between 1.4 and 2.8 $\times$ 10^6, and seven less than 500 $\times$ 10^3 tonnes.

4.1.1.1.1 Geology of the Kambalda District

The Kambalda district lies within the Eastern Goldfields, which comprise the eastern half of the Yilgarn Archean block of Western Australia. In particular, it forms part of the 800-km long Norseman–Wiluna greenstone belt, which Groves et al. (1984) interpret as a fault-bounded rift zone, 200 km in width. Gemuts and Theron (1975) described the stratigraphy of the Norseman–Kalgoorlie section of this belt (Fig. 4.1) as consisting of two volcanic cycles, each composed of a lower basaltic and ultramafic sequence, an intermediate felsic volcanic sequence, and an upper sedimentary sequence, derived in part from and stratigraphically equivalent to the felsic volcanics. All of the important Ni deposits are restricted to the mafic–ultramafic sequence (known as the Kambalda sequence) at the base of the lower cycle.

The earliest folding in the Norseman–Wiluna belt is thought to have been a stage of recumbent folding and nappe development (Archibald et al., 1978; Platt et al., 1978), but the most pronounced folding is on the north–northwest axes, which have been refolded on the east–west axes so that the stratigraphy is delineated in a series of north–northwest trending basins and domes. The Kambalda deposits are located around one of these domes, the Kambalda dome (Fig. 4.2). The volcanic and sedimentary rocks have been intruded by Archean granites, felsic and intermediate por-

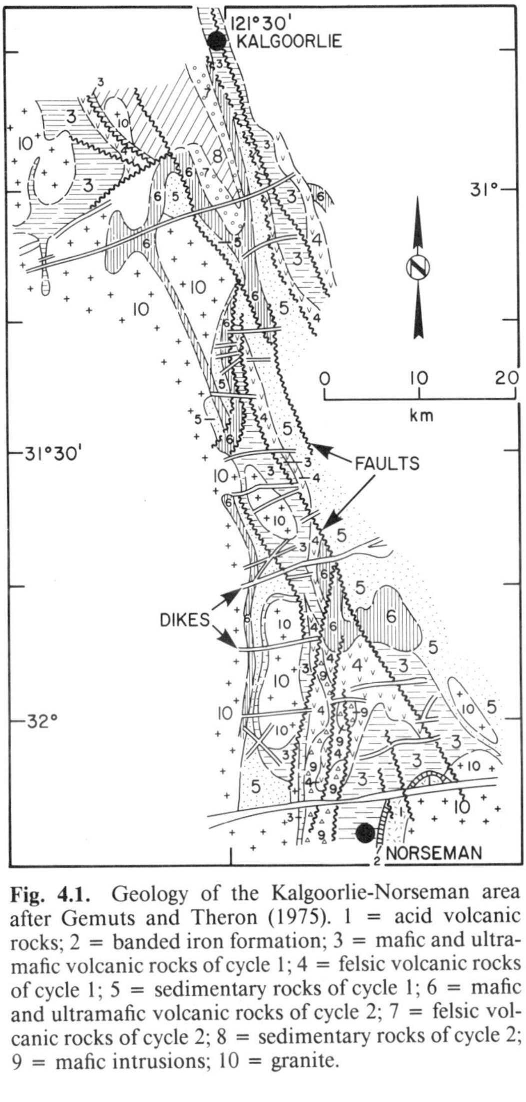

Fig. 4.1. Geology of the Kalgoorlie-Norseman area after Gemuts and Theron (1975). 1 = acid volcanic rocks; 2 = banded iron formation; 3 = mafic and ultramafic volcanic rocks of cycle 1; 4 = felsic volcanic rocks of cycle 1; 5 = sedimentary rocks of cycle 1; 6 = mafic and ultramafic volcanic rocks of cycle 2; 7 = felsic volcanic rocks of cycle 2; 8 = sedimentary rocks of cycle 2; 9 = mafic intrusions; 10 = granite.

phyries, mafic dikes and sills, and Proterozoic dikes. Metamorphic studies indicate peak temperatures slightly in excess of 500°C and pressures of around 2.5 kb (Barrett et al., 1977; Bavington, 1979).

At Kambalda, the ore-bearing mafic–ultramafic sequence consists of a lower (+200-m) unit of pillowed basalt, the "footwall basalts," overlain by an ultramafic unit, which is up to 1,000 m in thickness. Chauvel et al. (1985) have dated the sequence as 2.7 Ga old. Gresham and Loftus-Hills (1981) emphasize the difference between the general succession at Kambalda and the succession developed immediately above the ore deposits. They subdivide the latter into a lower member making up one-fifth to one-third of the total thickness (Fig. 4.3) and consisting of one or more thick, very MgO-rich units, overlain by an upper member. The upper member consists of an unordered succession of komatiitic flows, many with spinifex texture, interspersed with more magnesian, olivine peridotite units that either lack, or have a much lower proportion of spinifex-textured material within them. Away from the ore environment, one or more thin, olivine-rich flow units (in most cases separated by thin-bedded, "exhalite-type" sediments, consisting of a mixture of clastic, tuffaceous, and chemically precipitated components; Bavington, 1981) are overlain by an upper member consisting of a regular succession of komatiitic flows. These flows are between 1- and 10-m thick and show polyhedrally brecciated flow tops, spinifex-textured upper portions, and cumulus-textured lower zones as first described by Pyke et al. (1973). Sediments are commonly lacking in the near-ore environment, which overall shows a much less regular succession of rock types.

The ultramafic unit is overlain by the hanging-wall basalts, which are themselves divided into a lower member consisting of pillowed basalts (some with felsic ocelli) and an upper massive, more magnesian member. The two members are separated by a zone of exhalite-type sediment.

Broad distinctions in the major element composition of the mafic and ultramafic rocks are brought out in Table 4.1 and Figure 4.4. Here the highly magnesian nature (41.1 wt % MgO) of the olivine peridotite associated with the ore at Lunnon shoot and at other Kambalda ore shoots (average of 41.7 wt % MgO) stands out in contrast with the lower MgO content (average 31.8 wt % MgO) of the peridotite in the cumulus portion of the thin komatiitic flows that flank the ore environments. Even higher MgO contents characterize the ore-bearing units at Scotia (45.4 wt %) and Wannaway (44.6 wt %).

4.1.1.1.2 Ore Deposits

Ore occurs in three environments at Kambalda:

1. Contact ore at the base of the lowermost unit, in contact with the underlying football basalt;
2. Hanging-wall ore, either at the base of the overlying units (in most cases the second or third unit up from the base of the ultramafic sequence) or, much more rarely, as zones of blebby sulfide within the basal unit;
3. As offset ore that has been structurally displaced from either an original contact or hanging-wall position.

4.1.1.1.2.1 Ore Control. Contact ores account for 77 percent, hanging-wall ores for 18 percent, and offset ores for 5 percent of the total reserves plus past production at Kambalda (calculated from data of Gresham, 1986). The contact ores are localized by basinal depressions in the ultramafic-underlying basalt contact or, more commonly, by north–northwest trending troughs. These are narrow elongate depressions with

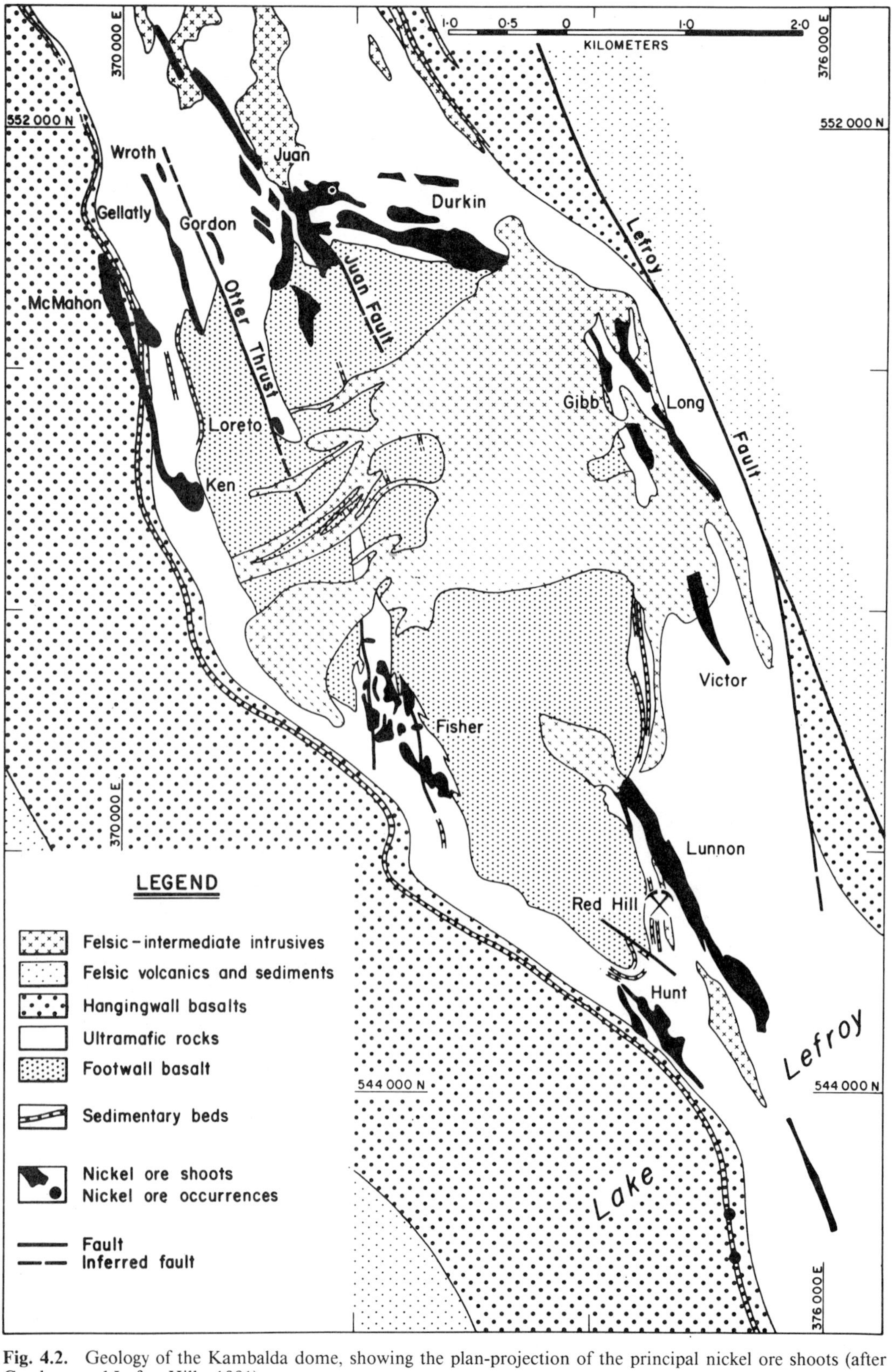

Fig. 4.2. Geology of the Kambalda dome, showing the plan-projection of the principal nickel ore shoots (after Gresham and Loftus-Hills, 1981).

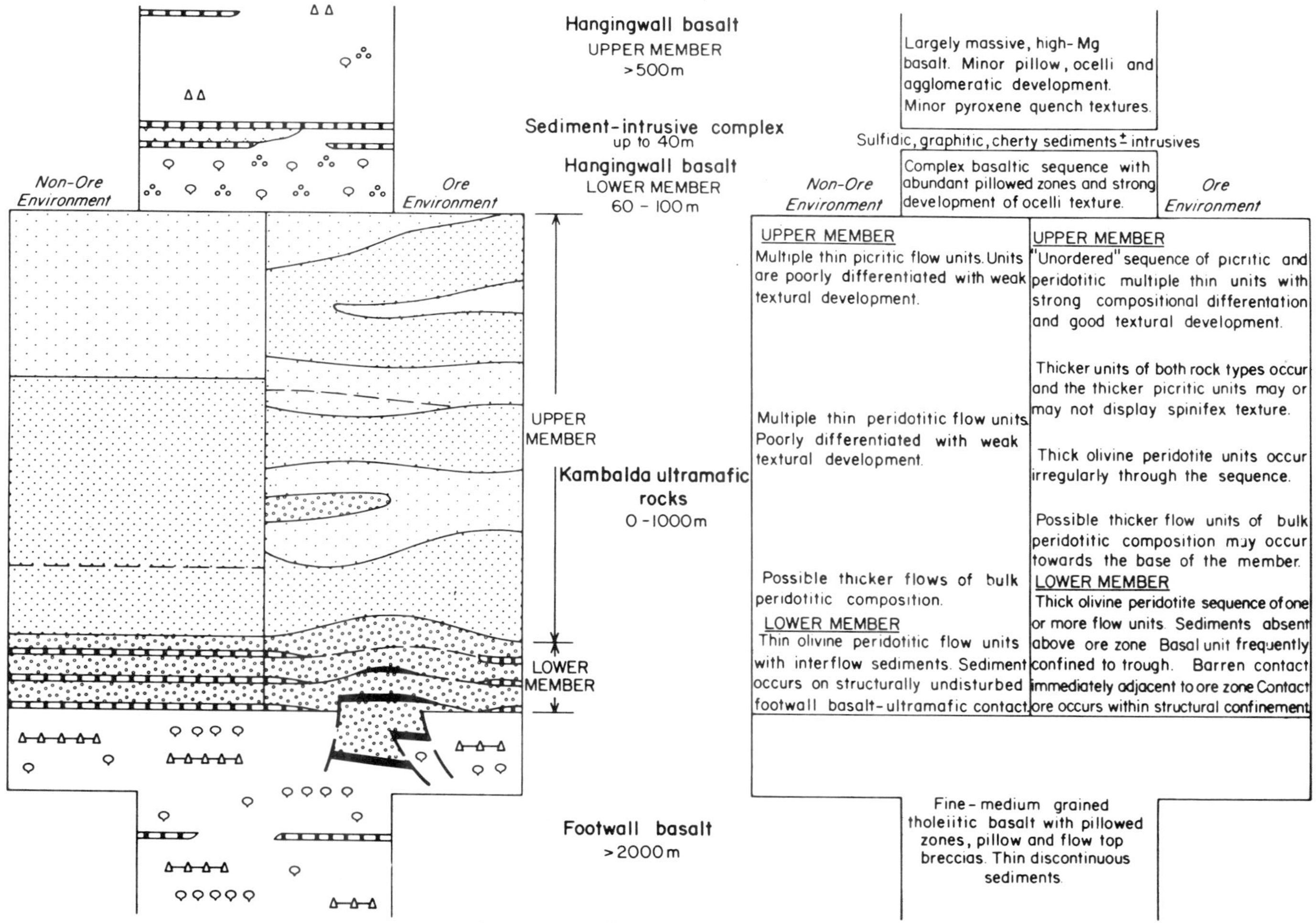

Fig. 4.3. Schematic sections through the ultramafic sequence of the Kambalda dome emphasizing the contrast between the prism of rocks above ore-bearing and barren basalt contact (from Grehsam and Loftus-Hills, 1981).

TABLE 4.1. *Major Element Analyses of Units from Kambalda and the Adjacent Eastern Goldfields (in wt%)*

Elements	1	2	3	4	5	6	7	8	9	10	11	12	13	14	15
SiO_2	47.7	45.1	44.3	45.7	42.8	49.2	49.3	42.2	44.1	43.6	46.9	52.1	53.12	49.7	49.1
TiO_2	0.34	0.14	0.14	0.33	0.17	0.40	0.20	0.09	0.43	0.11	0.27	0.68	0.62	0.65	1.07
Al_2O_3	7.87	3.23	2.81	8.01	4.02	9.40	4.92	1.63	8.34	2.53	6.04	14.70	12.23	12.48	14.99
Fe_2O_3				2.64	3.76	2.15	2.06	9.32	12.81	4.61	1.79		1.43	12.05	13.37
FeO	11.5	8.80	9.20	8.59	5.93	10.11	6.65			4.14	9.86	10.70	9.23		
MnO	0.20	0.12		0.14		0.15	0.15	0.14	0.16	0.11	0.18	0.07	0.20	0.20	0.23
MgO	23.9	41.1	41.7	26.9	37.9	26.5	31.8	45.4	28.2	44.6	26.5	8.20	11.37	12.92	7.03
CaO	7.79	1.41	1.69	5.95	2.68	6.81	3.89	0.51	5.92	0.51	6.33	11.00	9.53	9.39	11.68
Na_2O	0.15	0.06	0.03	0.12	0.06	0.22	0.11			<0.01	0.11	2.10	1.53	2.22	2.21
K_2O	0.06	0.02	<0.01	0.06			0.07	0.02	0.05	0.02	0.03	0.36	0.45	0.40	0.30
P_2O_5	0.08	0.05	<0.01	0.02	0.01	0.03	0.02	0.09	0.43	0.01	0.03		0.07	0.08	0.10
S	0.39	0.30		0.08	0.55	0.72	0.65						0.02		

1 = Average spinifex-textured or equivalent flow top material from basal unit at Lunnon shoot. (Gresham & Loftus-Hills, 1981, Table 3, No. 1.)
2 = Average olivine peridotite from basal unit at Lunnon shoot. (Gresham & Loftus-Hills, 1981, Table 3, No. 3.)
3 = Average olivine peridotite from basal units at Hunt, Fisher, Gibb, Long, Otter, Juan and Durkin shoots (Gresham & Loftus-Hills, 1981, Table 3, No. 4.)
4 = Average spinifex-textured or equivalent flow top material from second unit above base at Lunnon shoot. (Gresham & Loftus-Hills, 1981, Table 3, No. 5.)
5 = Average olivine peridotite from same unit as 4. (Gresham & Loftus-Hills, 1981, Table 3, No. 7.)
6 = Average spinifex textured or equivalent flow top material from the upper flows at Lunnon shoot. (Gresham & Loftus-Hills, 1981, Table 3, No. 8.)
7 = Average peridotite from thin upper flows at Lunnon shoot. (Gresham & Loftus-Hills, Table 3, No. 9.)
8 = Average ore-bearing dunite, Scotia. (Stolz & Nesbitt, 1981, Table 2, No. 7.)
9 = Spinifex-textured peridotite, from Scotia. (Stolz & Nesbitt, 1981, Table 3, No. 5.)
10 = Ore-bearing dunite, Wannaway. (McQueen, 1981, Table 5, No. 84520.)
11 = Spinifex-textured rock from same unit as No. 10. (McQueen, 1981, Table 5, No. 84517.)
12 = Average footwall basalt, Lunnon. (Gresham & Loftus-Hills, 1981, Table 2, No. 1.)
13 = Upper average hanging-wall komatiitic basalt, Lunnon. (Gresham & Loftus-Hills, 1981, Table 2, No. 8.)
14 = Average basalt, Scotia. (Stolz & Nesbitt, 1981, Table 2, No. 3.)
15 = Average tholeiitic basalt, Scotia. (Stolz & Nesbitt, 1981, Table 2, No. 4.)

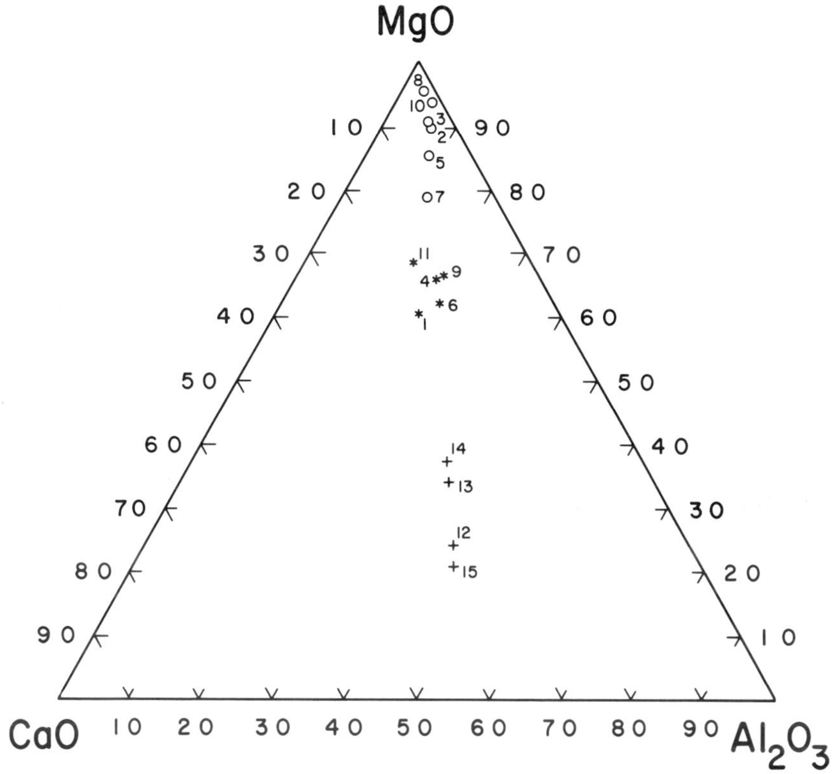

Fig. 4.4. Variation in the relative proportions of MgO, CaO, and Al_2O_3 in volcanic rocks of the Widgiemooltha–Kambalda–Scotia area of the Eastern Goldfields of Western Australia (in wt percent). Numbers refer to Table 4.1.

length-to-width ratios of about 10:1 and depths below the basalt flanks ranging from a few meters to as much as 100 m. A typical example is shown in Figure 4.5. The boundaries of the troughs are usually interpreted as steeply dipping reverse or normal faults, as seen in the vertical section through Lunnon shoot illustrated in Figure 4.6.

Ross and Hopkins (1975) pointed out that the faults may have been active during deposition, because they appear to offset the basal contact of the lowermost flow unit more than its upper contact. On the other hand, Gresham and Loftus-Hills (1981) report from drilling at the northwestern end of Foster shoot that the trough structure may be underlain by an unfaulted sedimentary layer within the basalt, indicating that the deep trough here was a feature of the preultramafic topography, unrelated to faulting. Lesher (1983) has shown that the fabric of ultramafic rock immediately adjacent to the basalt walls of the troughs is different from material within the centers of the troughs and that the marginal ultramafic rocks themselves are less magnesian than those in the center, suggesting that the troughs predated the extrusion of the komatiite. He has also documented stratiform features (including layering of pillowed and unpillowed basalt beneath the troughs) that have not been offset by faulting, but that should have been if such faulting was responsible for the troughs.

The hanging-wall ores occur in the lower three flows, generally where contact ores are also developed, and grade laterally into sediments that themselves carry nickeliferous sulfides in some cases. In general they are characterized by lower grades, but by sulfides with higher Ni tenor than the more massive contact ores (Gresham, 1986).

The offset ores occur along shears and comprise sulfides that have been sheared off from their original basal or interflow contact positions, so that they now lie in atypical intraflow or intrabasalt positions.

4.1.1.1.2.2 Ore Types. Figure 4.7 illustrates a typical section through the ore at the Lunnon shoot. Minor, generally Cu-enriched, stringers penetrate between pillows of the footwall basalts and are overlain by massive ore. This ore has a banded appearance due to the presence of thin (1 cm thick) lenses (10 cm long) of pentlandite that is thought to have exsolved at low temperature from pyrrhotite and to have nucleated in response to the prevailing stress field. Bands of pyrite up to 30 cm thick may occur in this ore. Massive ore commonly shows deformation textures and swells and thins over short distances, giving the appear-

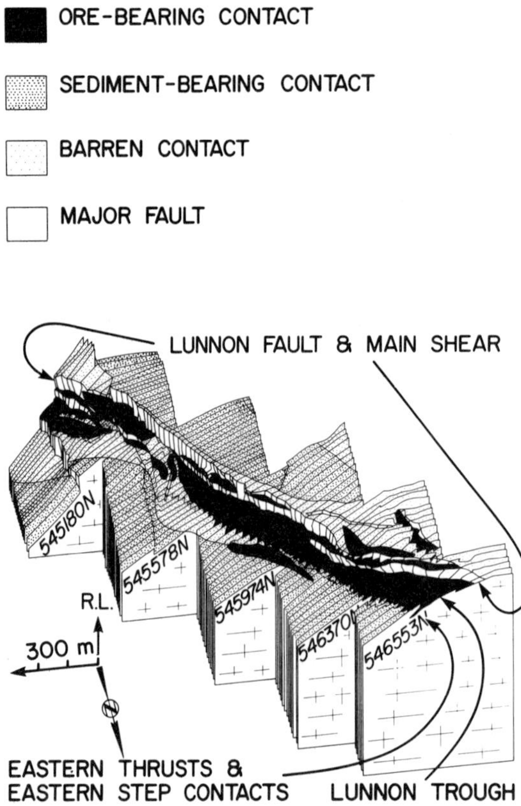

Fig. 4.5. An isometric view of the Lunnon ore structure (after Gresham and Loftus-Hills, 1981).

ance of having been remobilized over short distances in response to stress.

The massive ore is overlain in sharp contact by net-textured ore (referred to locally as matrix ore) in which a network of sulfides encloses olivine grains. This ore, in turn, grades abruptly upward into low-grade ore. Contacts between these ore types tend to be parallel to what is interpreted as the original horizontal. Concentrations of a Cr-rich, Mg- and Al-poor, spinel occur at the base of the massive and net-textured ore, or, in some cases, within the massive ore.

Most of the hanging-wall ores consist of net-textured and disseminated sulfides, although zones of massive sulfide are present in some deposits. An unusual type of ore found in the hanging-wall is spinifex ore; in this ore, sulfides appear to have replaced or displaced the matrix between blades of olivine in the spinifex-textured zone of the underlying flow.

As previously mentioned, a clear antithetic relationship generally exists between sediments and Ni–sulfide ore. This relationship is well illustrated in Figure 4.5, and is also apparent from Marston and Kay's (1980) reconstruction of the Juan area (Fig. 4.8). Ore and sediments, however, are intermixed in some localities. One such area is the Jan shoot (Fig. 4.9), where some of the sediments are mined as ore. Here the sulfides appear to have replaced the sediments, penetrating them along fractures and along their planes of fissility. Field observation suggests that this replacement might have been due to impregnation of the cherty sediments by sulfide magma-wetting (it is known from experiments that sulfide liquids have a great wetting ability when they come into contact with silica glass); however, the extremely high Pd/Ir ratios of the sediment-hosted ore in comparison with normal massive ore indicates that the process is more complex than simple impregnation. Paterson et al. (1984) ascribe the Ni, Pd, and other economic elements in the sulfides of the sediments to diffusion and hydrothermal remobilization out of the magmatic ore during metamorphism.

The Windarra deposits (270 km north–northeast of Kalgoorlie) are another area where sedimentary rocks are mineralized with Ni–sulfide ore. Much massive ore is associated with pods of highly magnesian komatiite and is interpreted as having settled from this (Schmulian, 1984). Schmulian points out, however, that several ore shoots are "unbrecciated equivalents of nickel-free sulfide-facies BIF (banded iron formation) with no evidence of sulfide enrichment within the ore zone." The sulfur isotope data of Seccombe et al. (1977) are consistent with a sedimentary origin for the sulfur in these zones, and the S/selenium (Se) ratios demand a significant contribution of sedimentary sulfur. Schmulian (1984) proposes that the ore is due to the diffusion of Ni and Cu from adjacent ultramafic-hosted ore into nickel-poor sulfide-facies BIF.

The principle ore minerals at Kambalda are, in order of decreasing abundance, pyrrhotite, pentlandite, pyrite, oxides (magnetite and chromite), chalcopyrite, millerite, and violarite. Pyrrhotite and millerite are mutually exclusive, as would be predicted from stability relations in the Fe–Ni–S system. The mineralogic variations reflect chemical variations; chemical analyses are compiled in Table 6.1, from which it is seen that the average Cu content of the various Kambalda shoots (those listed between Lunnon and Jan) varies from 0.8 to 1.4 wt percent and the Cu/(Cu + Ni) ratio from 0.062 to 0.076. Cobalt concentrations range from 0.2 to 0.3 wt percent in most shoots (Woolrich et al., 1981).

Figure 4.10 is based on the data of Woolrich et al. (1981) and shows that with the exception of the Gibb shoot, the deposits plot either within the field of monosulfide solid–solution, or within estimated analytical precision of the Mss field. This behavior is to be expected, in the light of the discussion in Chapter 3 emphasizing the control that the composition of the host silicate magma exerts on sulfides reaching equilibrium with it. The composition of the sulfides at the Gibb shoot has

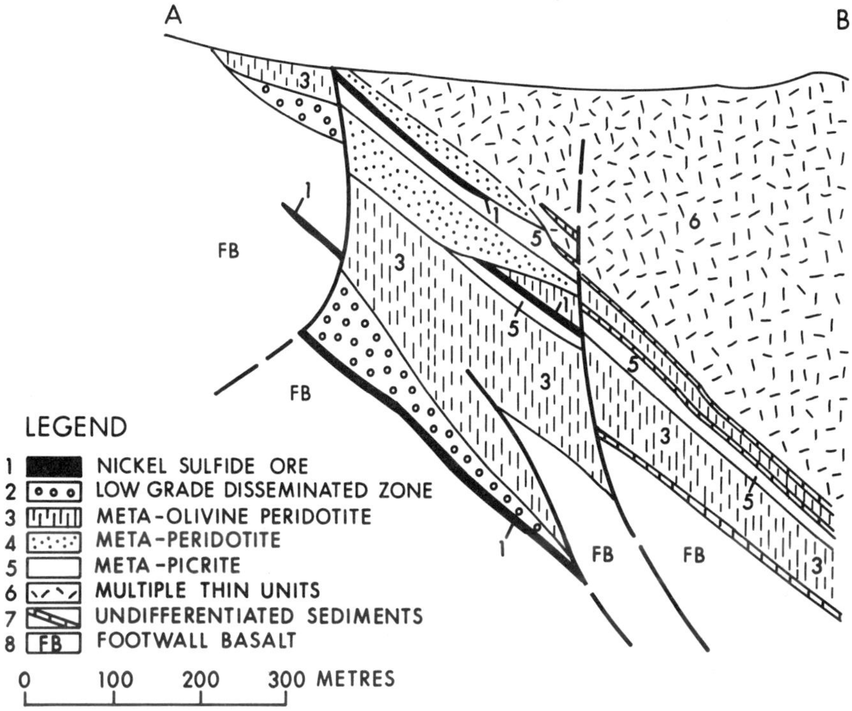

Figure 4.6 Vertical section through the Lunnon Shoot, Kambalda, Western Australia (modified after Ross and Hopkins, 1975).

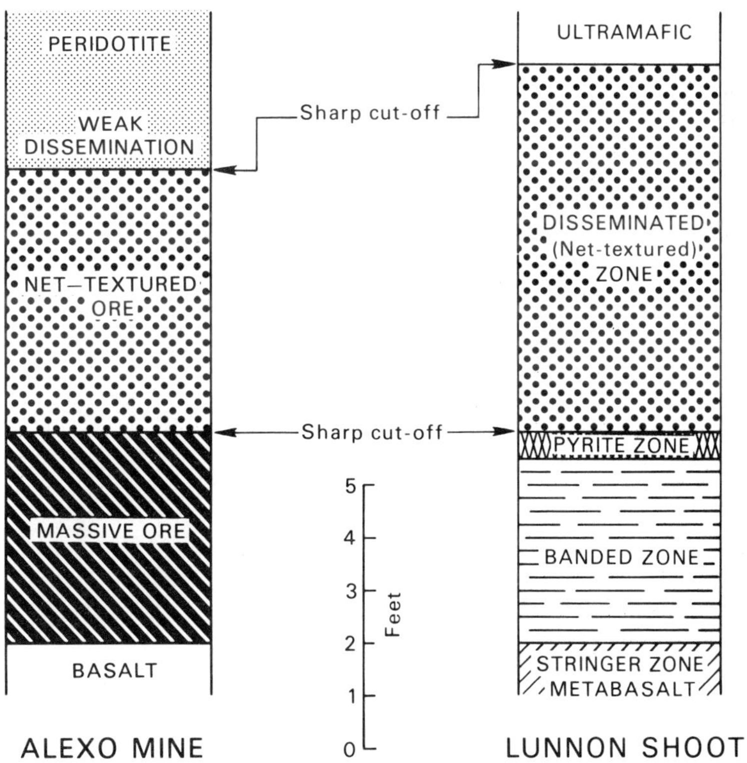

Fig. 4.7. Comparison of typical profiles through ore zones at Alexo and Lunnon shoot (from Naldrett, 1973).

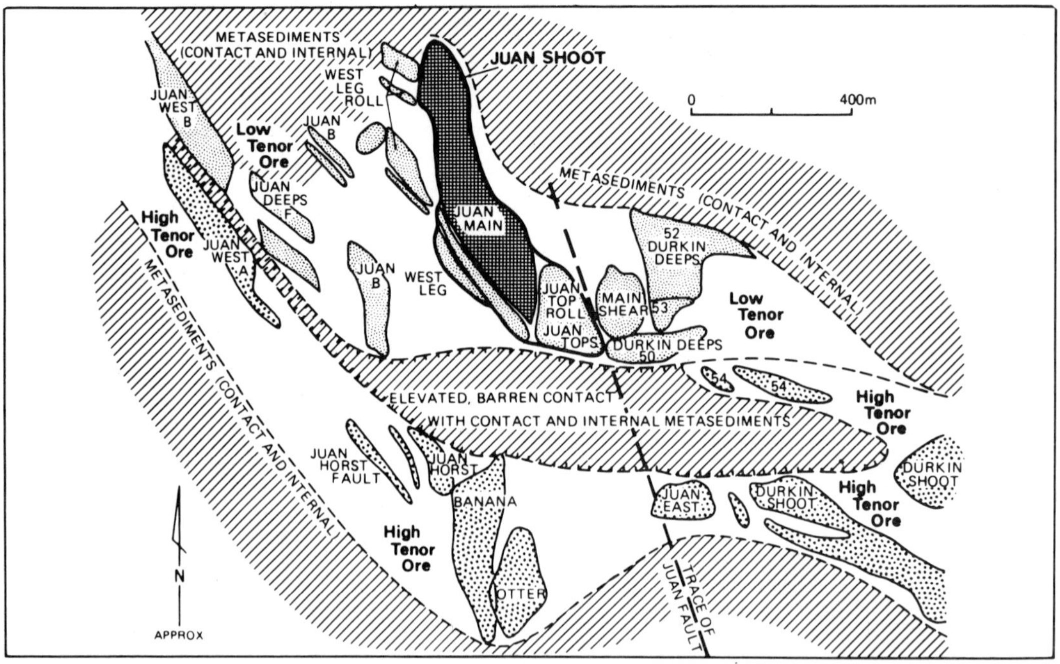

Fig. 4.8. Reconstruction of the contact between the ultramafic and football basalt units in the vicinity of Juan shoot, Kambalda, showing the extent of sedimentary rocks at the contact and their relationship to the ore deposits. Also note the separation between ore shoots of high and low Ni tenor (from Marston and Kay, 1980).

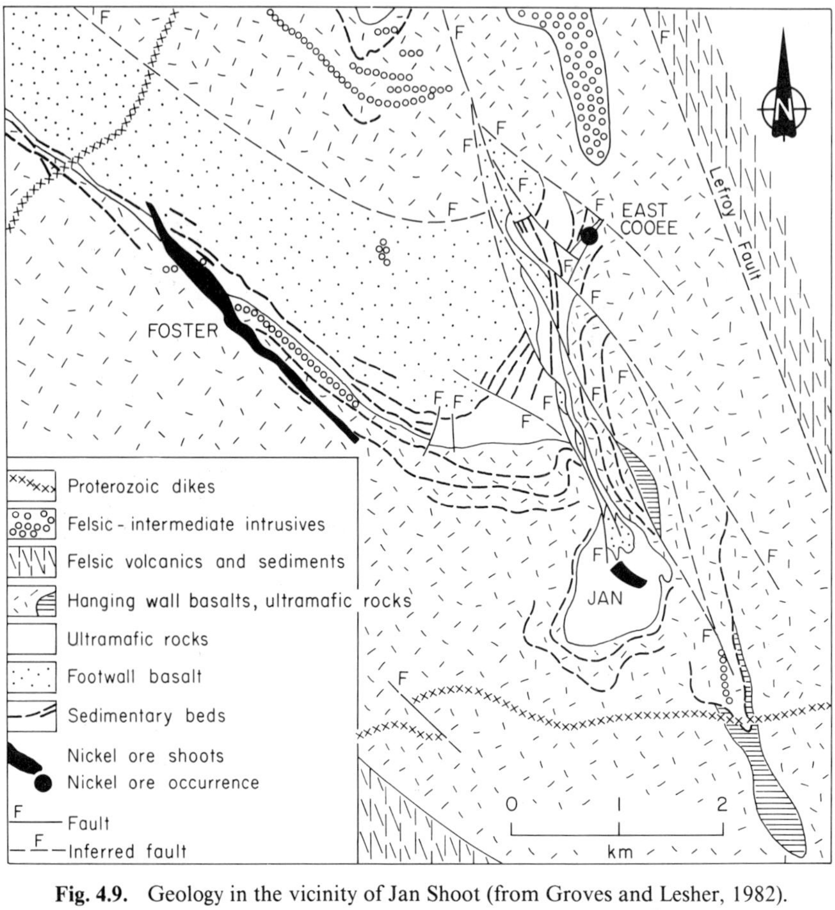

Fig. 4.9. Geology in the vicinity of Jan Shoot (from Groves and Lesher, 1982).

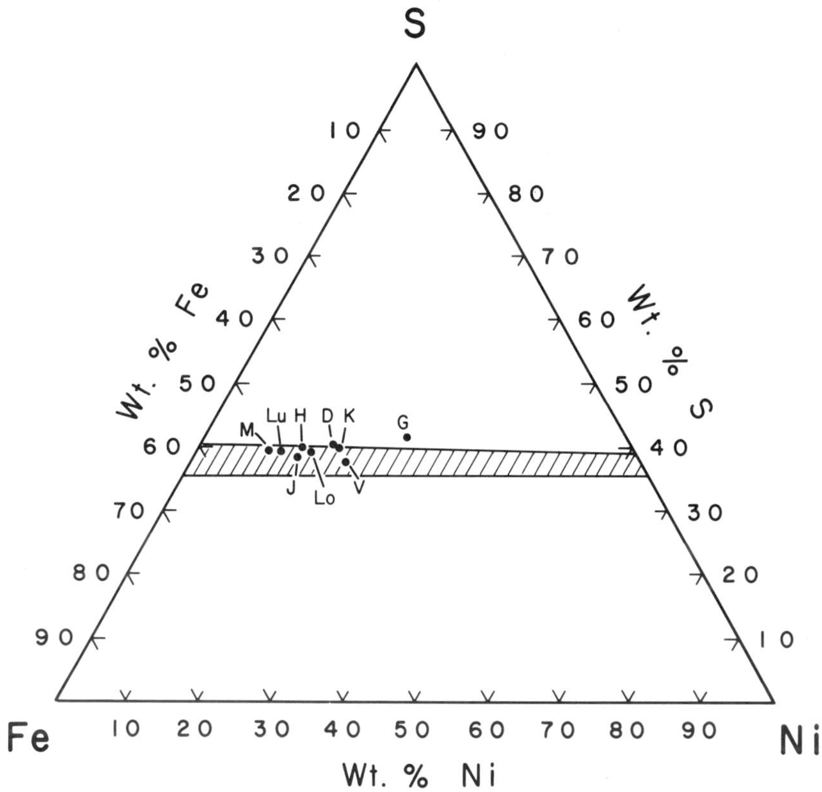

Fig. 4.10. The composition of different Kambalda ore shoots projected into the Fe–Ni–S system and compared with the composition of the Mss (diagonal cross-hatching) at 600°C (Compositional data from Woolrich et al., 1981) M = McMahon, Lu = Lunnon, J = Juan Main Shoot, H = Hunt, Lo = Long, D = Durkin, K = Kelley, V = Victor, G = Gibb.

probably been altered subsequent to their initial crystallization and the high S and Ni contents are probably the consequence of this alteration (see Eckstrand, 1975).

4.1.1.2 SHANGANI, ZIMBABWE

The Shangani deposit is associated with komatiites in the Sebakwian Group of the Zimbabwian shield. It consists of 16×10^6 tonnes of ore grading 0.92 wt percent Ni. The data from Naldrett (1981) indicate a Cu/(Cu + Ni) ratio for massive ore of 0.028 and a Co content of 0.16 wt percent. Massive ore is, however, only a minor constituent of the total ore reserves.

As shown in Figure 4.11, the geology of the area consists of a sequence of felsic tuffs overlain by komatiites and komatiitic basalts and intruded by a layered peridotite–gabbro complex. The felsic volcanic rocks are interdigitated with komatiites toward their base. One local aspect of this interdigitation is a sequence of ultramafic and mafic rocks (mushroom-shaped in cross section, and elongated in the down-plunge direction) that occurs well within the felsic volcanics (Fig. 4.12). The original form of this sequence is uncertain, but it seems to have formed on the sea floor, perhaps as a volcanic pool overlying its feeder. The original interpretation was that the structure was intrusive (Viljoen et al., 1976) but the description by Williams (1979) of the interdigitation of ultramafic rocks with komatiitic basalts and sediments toward the top of the sequence argues strongly for an extrusive origin.

The ore is localized at the undersides of the "cap" of the mushroom on both sides of the "stalk." Only one side appears to be mineralized in Figure 4.12, but both sides have been mined up-plunge, where the structure intersects the surface. For the most part, the ore is disseminated and consists of a typical pyrrhotite–pentlandite assemblage.

4.1.1.3 LANGMUIR NO. 1 AND NO. 2

The Langmuir No. 1 and No. 2 deposits (Green, 1978; Green and Naldrett, 1981) occur in the Abitibi belt of the Superior Province of the Canadian Shield, approximately 40 km southeast of Timmins, Ontario. They occur on the southeastern flank of the Shaw dome (Fig. 4.13). Basalt and intrusive granite occur within the core of the dome. The basalt is overlain by a sequence of felsic pyroclastic rocks and minor andesitic flows, which

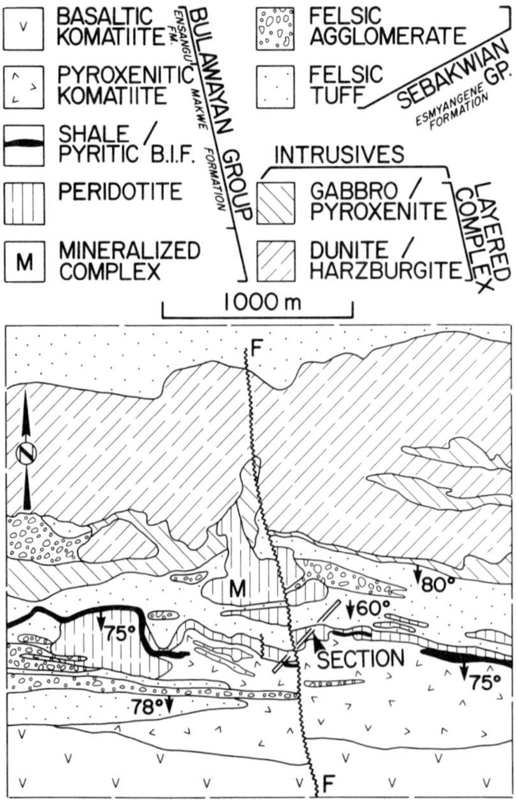

Fig. 4.11. Geology in the vincity of the Shangani (Zimbabwe) komatiite-related Ni deposit (from Williams, 1979).

are, in turn, overlain by a relatively continuous, thick (30–50 m) horizon of banded Fe formation. Komatiitic lava flows occur interdigitated with the felsic rocks and increase in proportion upward to become the dominant rock type. The komatiites are themselves overlain by komatiitic basalts, which are capped by a very thick sequence of Fe-rich tholeiitic basalts. Irregular masses of peridotite cut the felsic volcanics and basalts in the core of the dome and have been interpreted as feeders to the komatiitic volcanism.

The Langmuir No. 2 deposit has been studied in more detail than the No. 1 deposit, and is the only deposit to have been mined. The ore and ore-bearing flows occur adjacent to a discordant mass of peridotite that cuts downward through the underlying andesitic lavas (Fig. 4.14); this peridotite has been interpreted as a feeder by Green and Naldrett (1981).

The mineralization, particularly that at the base of the ultramafic sequence, is controlled by faults and topographic irregularities in very much the same way as seen at Kambalda. Figure 4.15 illustrates the troughlike structure in which the ore is localized, and the postore faults that have emphasized the trough.

4.1.1.4 THE AGNEW DEPOSIT

The Agnew deposit occurs in the Wiluna–Agnew greenstone belt, which differs from the environment of Kambalda in having a much higher proportion of felsic volcanic rocks and their sedimentary derivatives. The areas within the belt that are characterized by mafic and ultramafic volcanism contain a high proportion of masses of adcumulate dunite, some of which contain low-grade disseminated sulfides (see Section 4.1.2) and constitute the deposits of Group 2. Originally the Agnew deposit was regarded as an atypical example of this of group (Billington, 1984). More recently, evidence has accumulated that it belongs with those of Group 1 (Gole et al., 1988), although its large size (45 $\times$ 10^6 tonnes) and lower grade and higher Ni/Cu ratio (2.05 wt % Ni, 0.1 wt % Cu) make it an exception within this group also.

The massive ore of the deposit (Fig. 4.16) occurs within an ultramafic flow that is in contact with and has been deformed by a large mass of adcumulate dunite similar to those found elsewhere in the belt. Disseminated sulfides occur also within the dunite. Gole et al. (1988) have shown that the dunite cuts down through several hundred meters of felsic volcanics to bottom out on the mineralized flow. They pointed to metamorphic temperatures in excess of 1,000°C that

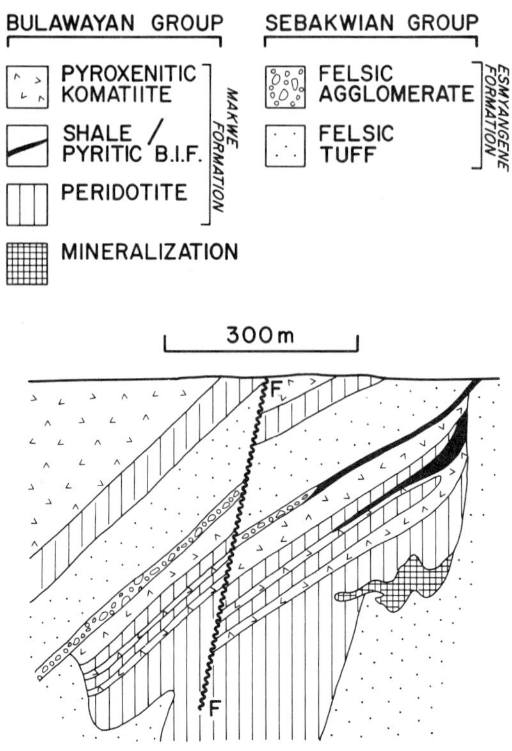

Fig. 4.12. Cross section through the Shangani structure (from Williams. 1979). Line of section is indicated in Fig. 4.11.

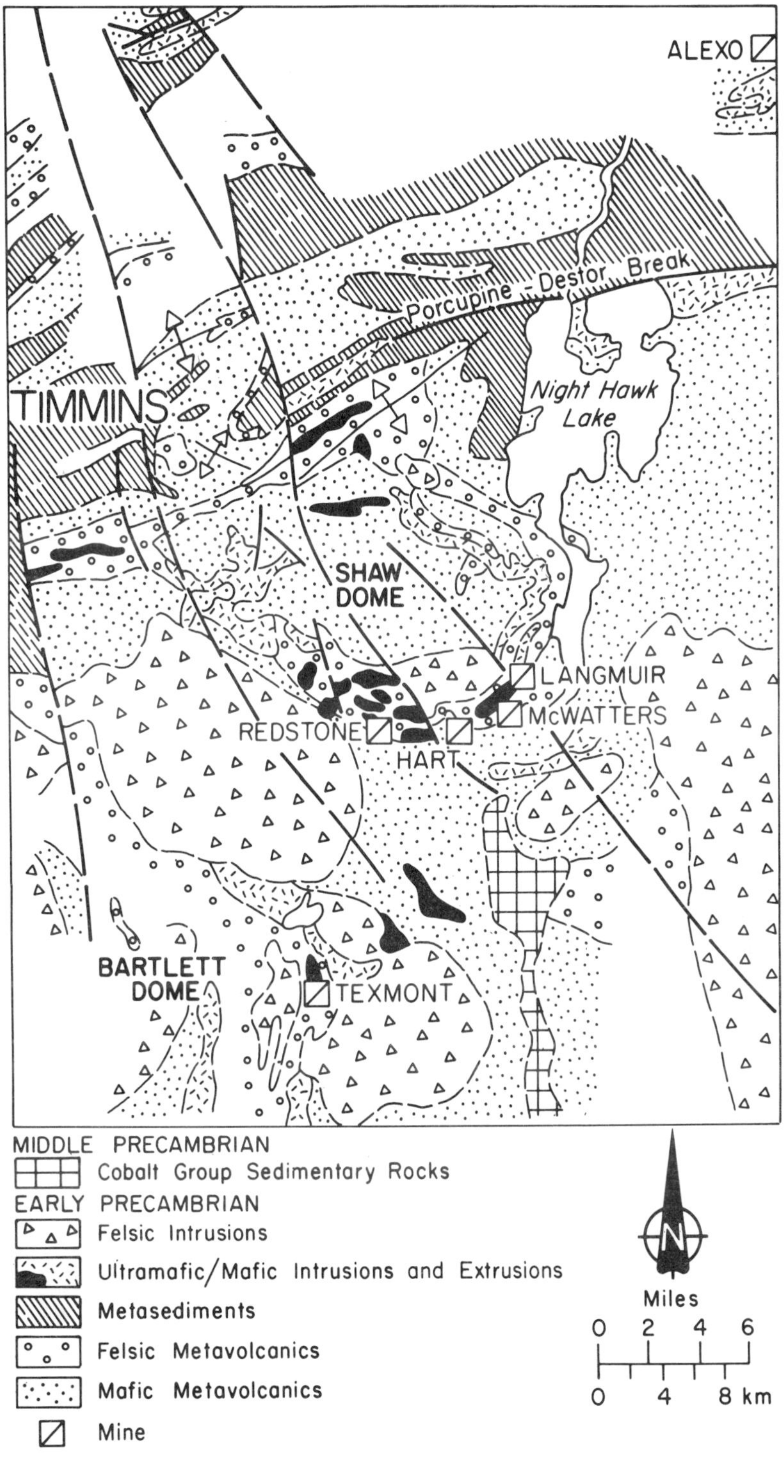

Fig. 4.13. The Timmins area of the Abitibi greenstone belt (After Pyke and Middleton, 1970.)

were produced in a komatiite flow beneath another dunitic mass 50 km to the north. In the light of these extremely high temperatures beneath a body of this type, they suggested (see discussion on thermal erosion by komatiites, Section 4.1.1.5.5) that the Agnew body represents a very large lava river that has thermally eroded its base, and that this thermal erosion was arrested by the

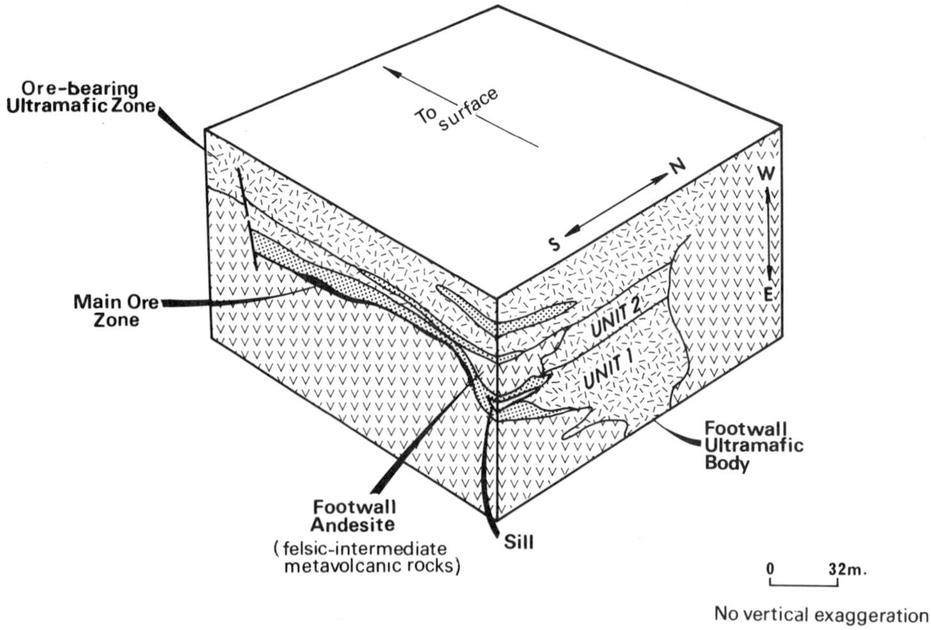

Fig. 4.14. Block diagram of the Langmuir No. 2 deposit, Timmins, Canada, showing the sill-like connection between the footwall ultramafic body (right) and the main ore zone (left). There is no vertical exaggeration. Massive ore shown in black, disseminated ore by dotted stipple, barren ultramafic rock by dashed stipple, and andesitic volcanic rocks by V-shaped stipple. (From Green, 1978.)

higher melting temperature of the underlying komatiitic flow.

4.1.1.5 MODELS OF ORE EMPLACEMENT

Three relationships between ore and komatiite are common to nearly every deposit: (1) the ore is most often found at the bases of the flows; (2) there is a close spatial relationship between massive and net-textured ore, and, when they are found together (as they are in the less structurally deformed areas), the net-textured ore overlies the massive ore; and (3) the contacts between different ore types is sharp and, where shearing is minor, can be seen to parallel the apparent horizontal at the time of extrusion.

4.1.1.5.1 Billiard-ball Model

The three simple relationships outlined previously suggest strongly that the sulfides formed as an immiscible liquid and thus, that the ore had a primary magmatic origin. Naldrett (1973) suggested that the gross features of the ore could be

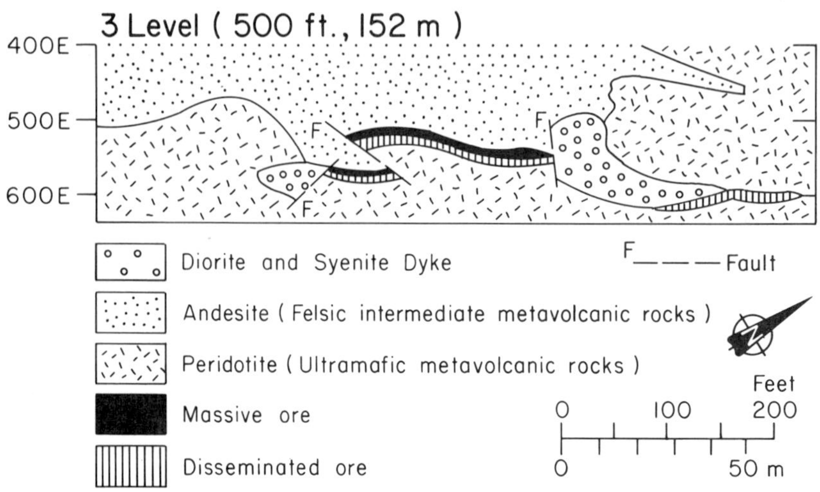

Fig. 4.15. Plan of the 3rd level of the Langmuir No. 2 deposit. (Modified after Green and Naldrett, 1981.)

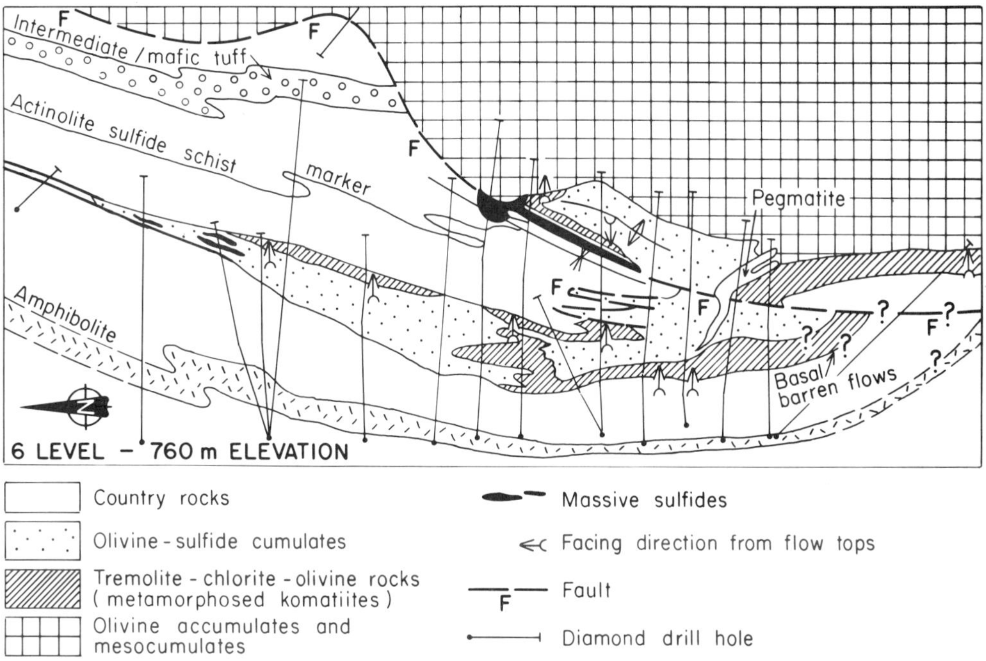

Fig. 4.16. Section through the Agnew deposits, Western Australia. (After Gole et al., 1988.)

explained by analogy with billiard balls in a cylinder of mercury and water. As illustrated in Figure 4.17, the billiard balls are denser than the water but float on the mercury. Some of the balls are forced down by the weight of the overlying balls, to become enclosed in a continuous network of mercury. Naldrett equated the massive ore to the massive mercury, the net-textured ore to the net-textured mercury, and the overlying low grade and barren peridotite to the billiard balls in the water.

Naldrett suggested that as lava flowed over an irregular surface, droplets of liquid sulfide migrated toward the base of the moving lava and hence flowed down into and became trapped within the depressions in the surface, displacing olivine and interstitial liquid to give rise to the present form and fabric of the ore deposits.

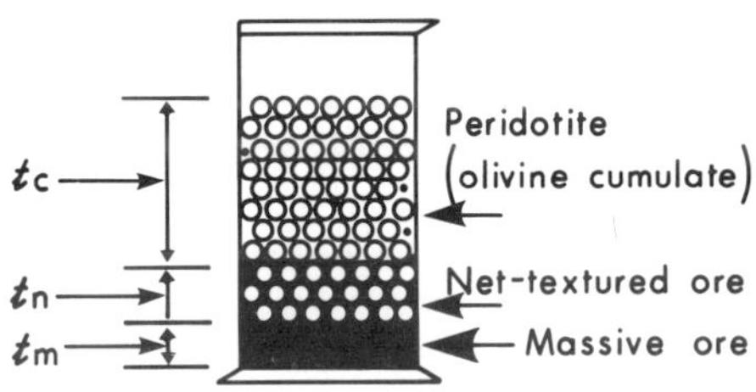

ρs = density of sulfide liquid

ρo = density of olivine

$\rho \ell$ = density of silicate liquid

if porosity of net ore is 50 % and of olivine cumulate is 40 %

then $t_n \times 0.5 \times (\rho s - \rho o) = t_c \times 0.6 \times (\rho o - \rho \ell)$

but $\rho s - \rho o \simeq 0.7$ and $\rho o - \rho \ell \simeq 0.35$

thus $t_c \simeq 1.7 \times t_n$

Fig. 4.17. The billiard-ball model. (Modified from Naldrett, 1973.)

4.1.1.5.2 Exhalative Model

Lusk (1976) proposed that the Kambalda ores are not magmatic but volcanic-exhalative, forming in the same manner as Cu–Zn massive sulfide deposits. He pointed to banding within much of the massive ore, the volcanic association of the Ni–Cu ores, the high pyrite content of the ores themselves, the close proximity of pyrite-rich metasediments to the ores, the presence of millerite within the ore, and the presence of chloritic alteration in the footwall of some deposits. He suggested that melting of the top of the massive sulfide by the overlying komatiite flow would give rise to the net-textured ore. Groves et al. (1976) countered the proposed exhalative origin by pointing out: (1) that the ores are associated with komatiitic lavas of a particularly MgO-rich composition; (2) that, whereas most ores occur at the base, some occur well within individual flow units; (3) that only slight variations in Ni/Cu ratio exist between all of the Kambalda deposits in strong contrast to the wide variation in Cu/Zn ratios observed between different massive sulfide deposits; (4) that, contrary to Lusk's assertion, extensive chloritic alteration and chloritic pipes are not found in the footwall of the Kambalda deposits; (5) that the overall pyrite content of the ores is consistent with a magmatic origin; and (6) that millerite is found within ores showing evidence of oxidation and is thus an alteration product rather than a primary mineral.

Naldrett (1979) pointed out that the elements concentrated by the Kambalda ores are precisely those that have a large value for the distribution coefficient between sulfide and silicate melts ($D_i^{\text{Sul. melt/Sil. magma}}$) and that would, therefore, be expected to be concentrated by a magmatic process. Furthermore, the deposits contain very little zinc (Zn), an element that is characteristic of nearly all Precambrian exhalative deposits but that, with $D_{Zn}^{\text{Sul. melt/Sil. magma}} = 1$, would not be expected to be concentrated in magmatic sulfides. He concluded that the compositional characteristics strongly supported a magmatic model.

4.1.1.5.3 Successive Emplacement Model

The billiard-ball model was criticized by Ross and Hopkins (1975) who found it incapable of accounting for their observation of chromium (Cr) and Zn anomalies both at the interface between net-textured and massive ore and close to the interface between low-grade and net-textured ore. The Cr anomalies are due to the concentration of chromite at these horizons. Ross and Hopkins pointed out that if the billiard model were correct, the chromite should have settled through the underlying, less dense massive sulfide. Accordingly, they proposed that the ore-bearing magma had separated into three magmas with very different viscosities during ascent through the crust and during the early stages of flow: (1) a pure sulfide liquid, (2) a magma consisting of olivine phenocrysts suspended in sulfide liquid, and (3) a magma consisting of olivine phenocrysts suspended in a komatiitic liquid. The emplacement of each occurred consecutively with the pure sulfide liquid first. The top of each separate extrusion became chilled due to contact with seawater, thus preventing the chromite from sinking through it. They also explained the Zn anomalies as resulting from the interaction of chilled sulfide with seawater.

The composition of the chromite is inconsistent with Ross and Hopkins's model. Groves et al. (1977) showed that the chromites in the Ni–Fe sulfide ores are ferrochromites, very rich in Fe and Cr and very low in Al and Mg, with surprisingly high Zn contents. They argued that Al- and Mg-poor chromites could not have crystallized directly from a silicate liquid, but are logical crystallization products of sulfide–oxide liquids (Czamanske et al., 1976; Groves et al., 1977). This conclusion was supported by experiments (Campbell, Naldrett, and Nolan, unpublished data) that showed that sulfide liquids in equilibrium with a komatiitic melt may contain up to 0.8 wt percent Cr_2O_3. Subsequently, Donaldson and Groves (1985) have shown that during metamorphism, chromite in an ultramafic rock may lose Al, Mg, and some Cr, and gain a magnetite component that may cause it to overlap the compositional field of chromites observed in sulfide ores. They regard the interpretation of ferrochromite as crystallizing directly from a sulfide melt as "equivocal." Nevertheless, the presence of Zn-rich chromites in the ultramafic environment would seem to remain strongly indicative of a sulfide melt having played a role in the formation of chromite.

Naldrett and Campbell (1982) suggested that the chromite concentration present at the massive ore–net-textured ore contact at Kambalda may represent either a favorable nucleation site or be the product of reaction between olivines and the sulfide melt. The Zn anomaly noted by Ross and Hopkins (1975) at this contact is clearly due to the Zn contained by the chromite.

4.1.1.5.4 Metamorphic Model

Barrett et al. (1977) also criticized the billiard-ball model, pointing out that the settling of relatively uniform spheres should result in a fabric containing, at the most, 40 percent porosity; they noted that much of the net-textured ore at Kambalda contains more than 60 volume percent sulfide, interstitial to the olivine grains. They also pointed out that the strict application of Archimedes principle to the billiard-ball model requires a precise relationship between (1) the thicknesses of the zones of net-textured ore and overlying barren peridotite and (2) the density difference between olivine and sulfide liquid on one hand and the olivine and intercumulus liquid on the other (Fig. 4.17). This predicted relationship is not upheld in most deposits; there is insufficient net-textured ore to provide flotation for the overlying barren peridotite. According to the billiard-ball model, the overlying mass of cumulus olivine should have forced olivine crystals to the base of the massive sulfide liquid, thereby increasing the thickness of the net-textured ore and eliminating the massive ore. Barrett et al. (1977) suggested instead that the massive ore is the result of plastic flow of original magmatic sulfide to zones of low pressure during metamorphism. They also suggested that the net-textured ore has been upgraded in this way to its present content of more than 60 percent sulfide.

4.1.1.5.5 In situ Crystallization, Dynamic Flow, Thermal Erosion, and Sulfide Assimilation

Usselman et al. (1979) suggested that the preservation of massive ore in a thick flow was due to the sulfides freezing against the base before the flow had crystallized sufficient olivine for the sulfide to be forced right through the sulfide liquid. They demonstrated their argument with a thermal model of the cooling of a flow. They pointed out that their argument was valid as long as the magma contained a restricted proportion of phenocrysts, less than 11 percent in the case of a 60-m flow and less than 23 percent in the case of a 30-m flow. They proposed, therefore, that the bulk of a flow had to crystallize in situ.

Naldrett and Campbell (1982) investigated in situ crystallization from a chemical viewpoint, using a modified version of the program by Duke and Naldrett (1978) to estimate olivine compositions as the flow fractionates. They found that the compositional change is much less than predicted, and concluded that the flows must be "dynamic" (i.e., that magma continued to move through the flows after crystallization at both the bottom and the top had started). The present thickness of any flow might thus be much greater than the thickness when the sulfides were emplaced and started to freeze. Barnes et al. (1983) reached a similar conclusion with respect to flows overlying the Alexo Ni deposit in Ontario.

The narrow, elongate nature of the ore-bearing flows has been emphasized by Lesher et al. (1984), who proposed that flow in the Kambalda district was in the direction of the long axis, comparable to a volcano such as Mauna Loa. They argued that the presence of horizons of crescumulate olivine within the thick, MgO-rich, ore-bearing flows indicates that these flows are not the result of the accumulation of phenocrysts, but that crystallization occurred in situ from an overlying, moving body of magma.

Huppert and Sparks (1985a) showed that the flow of komatiite is likely to have been turbulent. Given this, the transfer of heat to the underlying basalt would have been approximately an order of magnitude more rapid than would have been the case if the flow were laminar. Heat transfer at this speed would inevitably cause severe thermal erosion of the basalt, at a rate varying from several meters per day for a 1,600°C flow within a few kilometers of its source, to several tens of centimeters per day some tens of kilometers away. They proposed that the troughs are almost entirely due to thermal erosion beneath flows traveling many tens of kilometers parallel to the pitch line of the regional slope. Sparks et al. (1984) also suggested that assimilation and incorporation of sulfide from the underlying sediments may have contributed to the sulfide content of the ores.

Lesher and Groves (1986) concurred with the hypothesis that the troughs that channeled the flows may have resulted from thermal erosion of the underlying basalt by the hot komatiitic magma. They argued in favor of a sedimentary source of a large proportion of the S on the grounds that (1) the komatiites would have been unlikely to have reached the surface saturated in sulfide and (2) variations in ore tenor between different deposits and within individual deposits are best explained by variations in the mass ratio of sulfide to magma with which they equilibrated (Campbell and Naldrett, 1979), which are, in turn, best explained by varying degrees of sulfide assimilation.

The assimilation of significant amounts of country rocks, both during ascent through underlying crust and while flowing across the surface, is supported by the data of Arndt and Jenner (1986). They found that the rare earth element (REE) and samarium (Sm)–Nd characteristics of the komatiitic flows could best be explained if komatiitic magma had assimilated up to about 8 wt percent of underlying sediment. On the other hand, basalts belonging to the komatiitic suite, and immediately overlying the komatiites, appeared to have been contaminated primarily by granites of the upper crust.

Gresham (1986) argued against a significant contribution of sedimentary sulfide to the magmatic ores, pointing out that (1) in some deposits (Wannaway, Mt. Windarra), extensive zones of magmatic sulfide are associated with sulfidic sediments with no evidence of assimilation; (2) at Kambalda there is not nearly enough S in the sediments that have been removed to account for the sulfide in the ores; and (3) where there is evidence of sediment assimilation, there is also a higher zinc content in the ores (160 ppm) than in normal ores (40 ppm).

It would seem that the origin of the Kambalda troughs is very probably the consequence of thermal erosion. If so, there is likely to have been some contribution of sedimentary sulfide to that in the ore zones. What is less clear is whether this was a major contribution, which, in essence, accounts for the presence of the ores, or whether the ores would have accumulated in economic amounts without this contribution.

4.1.1.5.6 A Proposed Model

The points discussed previously have been incorporated, together with other information given by Lesher et al. (1984) and ideas of the present author, into the following model, which is designed in particular to explain the differences between the ore-bearing and ore-free environments described by Gresham and Loftus-Hills (1981). The model is illustrated in Figure 4.18. The eight drawings in this figure show a cross section through a lava river that has eroded a trough in the basalt across which it was flowing. The steep walls of the trough are consistent with experiments of Sparks and Huppert (Sparks, personal communication, 1984) involving the flow of hot water across wax. In some of these experiments, the steady-state flow of water resulted in the erosion of channels with re-entrant walls, very similar in configuration to some of those observed at

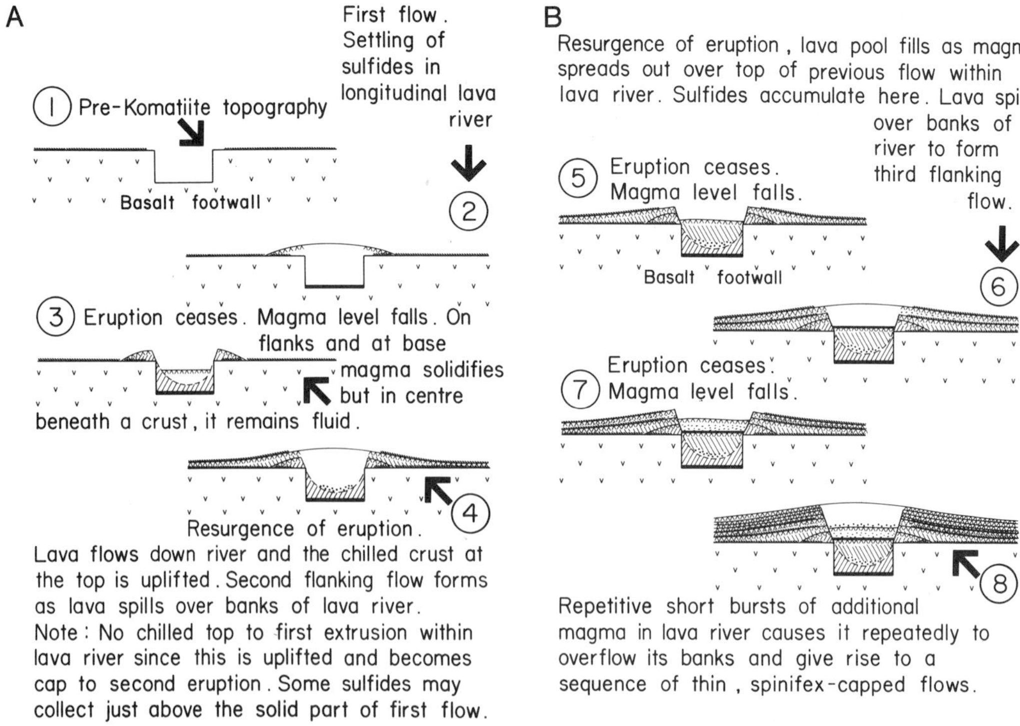

Fig. 4.18. Suggested model for the development of a typical mineralized sequence of komatiites of the Kambalda area.

Kambalda. It is likely that the orientation of the troughs (pronounced at Kambalda) reflects the regional slope.

The first eruption eroded the trough (Fig. 4.18, drawing 1) and, during brief periods of very rapid flow overflowed the banks to a limited extent, to form levees composed of congealed (in part pa hoe hoelike) lava (Fig. 4.18, drawing 2). Sulfides settled rapidly and became trapped by irregularities in the base of the flowing river. As eruption ceased and the magma level fell (Fig. 4.18, drawing 3), levees of congealed lava were left on each bank and a zone of congealed peridotite formed above the now largely solid basal accumulation of sulfide. The peridotite was itself overlain by a zone of fluid peridotite beneath a crust.

Resurgence of the eruption (Fig. 4.18, drawing 4) caused renewed flow along the river, guided by the pre-existing levees; continued thermal erosion of the underlying basalt was prevented by the refractory peridotite at the base of the river. Renewed flooding resulted in additions to the levees; in places, breaching of the banks produced extensive flanking flows that moved away at an angle to the main flow direction of the river, influenced partly by the slope of the levees. Ponding of these flows between levees of adjacent lava rivers and dams composed of their own congealed material, allowed them to crystallize under static conditions. Turner et al. (1986) argued that the turbulent flow of komatiitic lava will prevent all but a thin crust to form while it is in motion, and that well-developed spinifex texture will only form once it has become stationary. It is suggested that the ponding of the flows flanking the main conduits provided the necessary static environment for spinifex development, in contrast with the active environment of the conduit, leading to the best development of spinifex texture on the flanks rather than within the conduit komatiite itself.

It is possible, in some cases, that the renewed flow in the river occurred in a lava tube beneath the chilled crust of the first eruption; therefore, this earlier crust was uplifted to form the cap also of the second flow. Sulfides settling from the second flow would then come to rest on the previously frozen peridotite of the first flow, giving rise to apparent intraflow mineralization as is seen in Figure 4.18, drawing 4. A second cessation in eruption, accompanied by a second fall in the magma level in the river may have been followed by a longer interval of quiescence, giving more time for cooling so that the whole of the second flow solidified (Fig. 4.18, drawing 5).

Further magmatic activity was also guided by the river and its levees, this time flowing above the top of the preceding flow. Sulfides settling from this third eruption gave rise to an ore zone

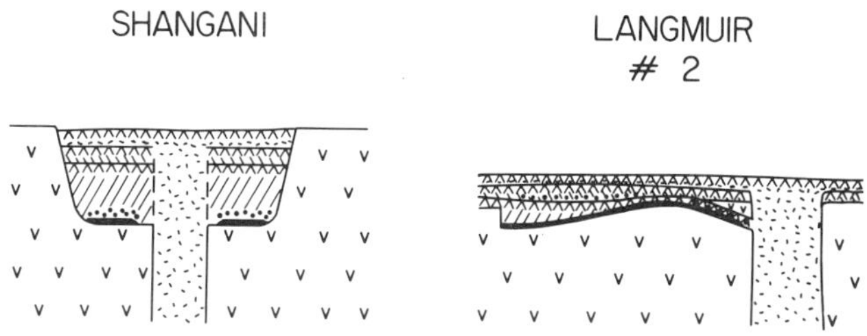

Fig. 4.19. Application of the model shown in Figure 4.18 to the Shangani and Langmuir No. 2 deposits.

located clearly at an interflow contact (Fig. 4.18, drawing 6). The eruption, although broadly following the river, also overflowed its banks to give rise to further lateral flows.

It is likely that away from the hotter zone overlying the lava river itself, chemical and clastic sedimentation could occur, giving rise to the intraflow sediments that characterize the nonore bearing environments. Turbulent motion in the sea overlying the hotter environment of the lava rivers, coupled with thermal erosion by the lavas themselves, probably accounts for the sparsity of sediments in this environment.

A cessation of activity and an accompanying drop in the magma level in the river (Fig. 4.18, drawing 7) was followed by a new phase of activity. In this case, a series of lava floods along the river resulted in repeated overflowing of the banks and gave rise to the thin picritic flows that characterize the upper parts of the nonore environment at Kambalda (Fig. 4.18, drawing 8).

Figure 4.19 illustrates an extension of this model to Shangani where the ultramafic rocks appear to have formed in an elongate trough within flanking felsic volcanics. The "stalk" of the trough may have been the feeder. The sulfides appear to have settled out of the first eruption, to accumulate at the base of the trough. Subsequent flow up the feeder has eliminated any sulfides that may have existed at the center of the trough.

Less is known about the overall shape of the ore-bearing unit at the Langmuir No. 2 deposit. It is known, however, that the deposit lies adjacent to and above a discordant feeder (Fig. 4.19). An arm branches away from the feeder, cutting up through andesitic footwall rocks to connect with the most intensely mineralized flow. The arm itself is heavily mineralized, consisting largely of net-textured ore. Langmuir, although small, is important in that it is the only deposit at which it has been possible to demonstrate a connection between a discordant mineralized feeder and a mineralized flow.

4.1.2 Group 2 Deposits

4.1.2.1 YAKABINDIE AREA

The Six-Mile deposit at Yakabindie, Western Australia, is a typical example of the low-grade disseminated-type Ni mineralization. The deposit is one of a series of lenses emplaced within sediments and felsic volcanics that underlie a sequence of tholeiitic and komatiitic lavas and shallow intrusions. Nickel–sulfide mineralization occurs mainly within a comformable ultramafic lens (Fig. 4.20), which consists of a core of dunite (250–300 m thick) contained within a sheath of peridotite. The peridotite sheath is thicker on the side representing the stratigraphic top, where it exhibits a cyclic variation in the MgO, CaO, and Al_2O_3 contents; Naldrett and Turner (1977) ascribed these variations to fractional crystallization. They postulated the intrusion of a mush of olivine phenocrysts, peridotitic liquid, and sulfide liquid into the sedimentary–volcanic pile. Successive pulses of magma would have flushed out the upper part of the chamber, each pulse undergoing some differentiation and thus giving rise to the cyclic layering before continuing to surface to extrude as komatiitic lavas. In view of the conclusions of Hill et al. (1985) with respect to the extrusive nature of the ultramafic body at the Agnew deposit, a similar interpretation should be entertained for the Six-Mile (and Mt. Keith) bodies, although there is no evidence of an extrusive origin, and the Dumont body (see later) has been established as intrusive.

The sulfides at Six-Mile (consisting of pyrrhotite, pentlandite, heazlewoodite, and godlevskite) occur mostly at triple junctions between the tightly packed olivine grains of the dunite core. This zone grades 0.6 wt percent Ni with a reported reserves of 60×10^6 tonnes. Thinner zones of sulfide with similar grades also occur within the peridotite above the dunite core.

4.1.2.2 THE DUMONT SILL

In his study of the 2.70–2.72 Ga Dumont sill, Duke (1986, 1985) has accumulated evidence for a genetic model that explains the common occurrence of sulfide mineralization containing 0.4 to 0.6 wt percent Ni. The sill (Fig. 4.21) lies conformably within a sequence of Fe-rich tholeiitic basalts of the Abitibi greenstone belt. It comprises a lower ultramafic zone with an average true thickness of 450 m and, in contrast with many of the Australian examples, an upper mafic zone about 250 m thick. The ultramafic zone consists of three subzones, a lower ortho-mesocumulate peridotite subzone, a middle adcumulate dunite

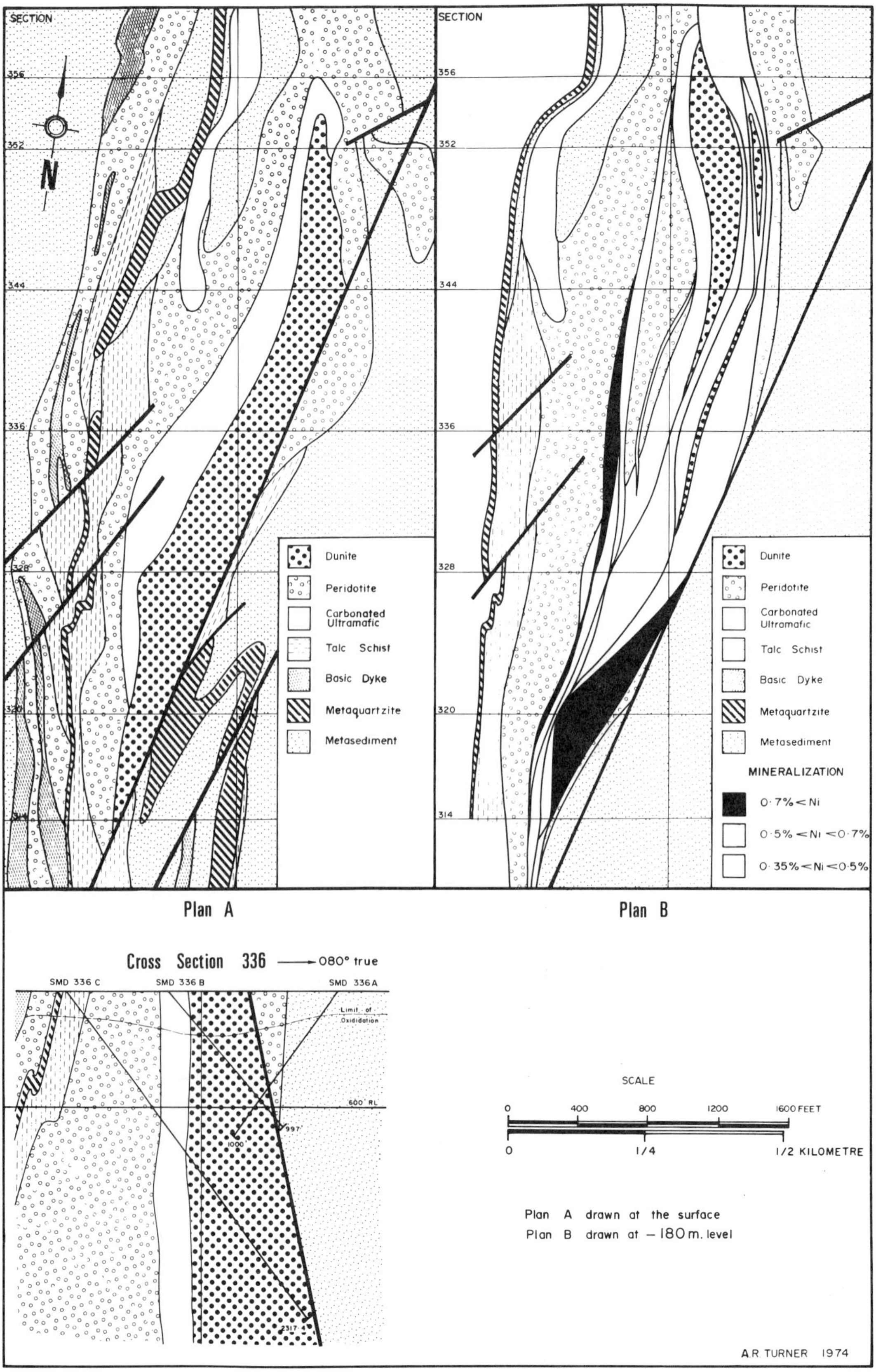

SECTION
356
352
344
336
328
320
314
N
Dunite
Peridotite
Carbonated Ultramafic
Talc Schist
Basic Dyke
Metaquartzite
Metasediment
MINERALIZATION
0·7%<Ni
0·5%<Ni<0·7%
0·35%<Ni<0·5%
Plan A
Plan B
Cross Section 336 ⟶ 080° true
SMD 336 C
SMD 336 B
SMD 336 A
Limit of Oxidation
600' RL
997'
1000'
2317·4'
SCALE
0
400
800
1200
1600 FEET
0
1/4
1/2 KILOMETRE
Plan A drawn at the surface
Plan B drawn at −180 m. level
A.R. TURNER 1974

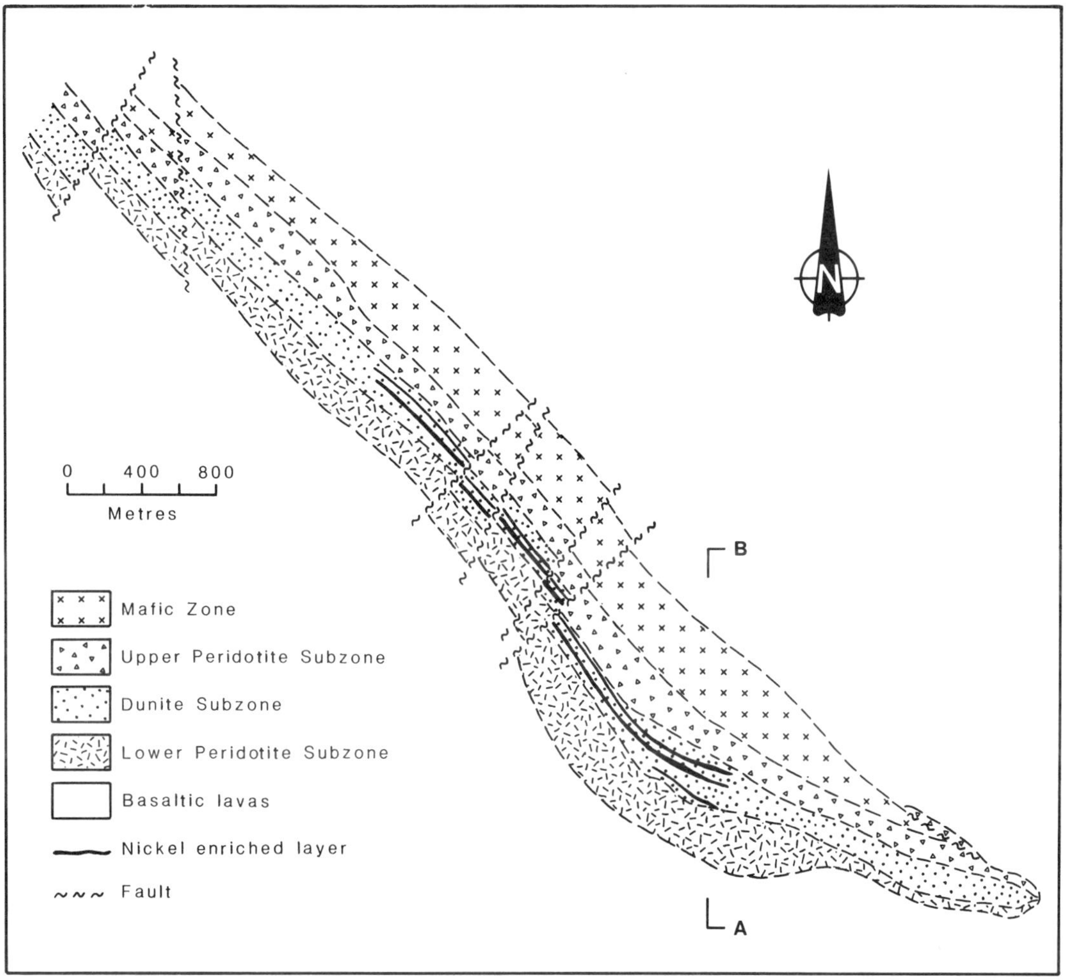

Fig. 4.21. Geologic map of the Dumont sill. (From Duke, 1986.)

subzone, and an upper, meso-orthocumulate peridotite subzone.

The reserves of the mineralization have not been fully determined, but there are three stratiform mineralized zones grading 0.45, 0.50, and 0.41 wt percent Ni, and an enriched section in the central zone, 750 m long by 14 m thick grading 0.71 wt percent Ni. The sulfide zones occur within the dunite, where they form an olivine–sulfide cumulate. Duke observed that the major sulfide zone coincides with an abrupt increase in the forsterite (Fo) content of olivine within the dunite of from 90 to 93 wt percent Fo. Above this level textural features indicate the coprecipitation of olivine plus a sulfide liquid. The very rapid decrease in Ni with Fo content of olivine within this zone provides strong support for the coprecipitation of sulfide and olivine (see Chapter 7). On the basis of trace element data, Duke interpreted the abrupt increase in the Fo content of olivine as due to a change in the regime of crystallization, with the trapping of pore fluid below this level, and the bottom growth of an essentially pore-free adcumulate above it. Where pore fluid was trapped, the olivine became enriched in Fe through reaction with it. He suggested that some of the trapped fractionated pore fluid was squeezed out from the underlying cumulate pile by compaction. It entered the overlying more primitive magma, and, in mixing with it, induced sulfide saturation (Irvine, 1975, 1980), so that sulfide and olivine precipitated together in cotectic propor-

Fig. 4.20. *A.* Surface plan of the Six-Mile dunitic lens. *B.* Plan of the 600 level of the Six-Mile lens showing the distribution of sulfide mineralization. *C.* Vertical cross section through the Six-Mile lens, on section 335N. (From Naldrett and Turner, 1977.)

tions. In the case of primitive ultramafic magma, such as that responsible for the Dumont or Western Australian bodies, he suggested that the cotectic proportions of sulfide plus olivine gave rise to a rock with 0.4 to 0.6 wt percent Ni.

4.2 SETTING IB: DEPOSITS IN PRECAMBRIAN GREENSTONE BELTS ASSOCIATED WITH THOLEIITES

The deposits of setting IB include those of the Pechenga region (the most important representatives), the Lynn Lake deposits in Canada, and numerous other deposits of lesser importance.

4.2.1 Pechenga, Kola Peninsula

Six principal areas of Ni–sulfide mineralization occur within the Kola peninsula, mostly within or very close to rocks of the Pechenga–Varzuga structural and facies zone. This area contains a high proportion of early Proterozoic (Karelian) andesite–basalt, subalkali–basalt, and picrite–basalt extrusive rocks and volcanogenic sediments. Gorbunov et al. (1985a) point out that these deposits fall into two main age groups and magma types. The early Karelian (Sumian) deposits are associated with igneous complexes belonging to the peridotite–pyroxenite–gabbro–norite suite of rocks and include those of Allarechka (Gorbunov et al., 1985b; Yakolev and Yakoleva, 1986), the Monchegorsk area (Gorbunov, et al., 1985b; Kozlov, 1973), and East Pechenga (Gorbunov et al., 1985b). The ultramafic–mafic bodies hosting these deposits occur within faults, either at the margins of greenstone belts (Monchegorsk) or in gneisses forming the basement to the belts (Allarechka and East Pechenga). The late Karelian deposits, associated with wehrlite–pyroxenite–gabbro intrusive rocks and related picritic volcanic suite, are more important, particularly those at Pechenga. They fall into the synvolcanic class of deposit and are described in more detail here.

4.2.1.1 GEOLOGY OF THE PECHENGA AREA

The Pechenga deposits occur at the extreme northwestern end of the Kola Peninsula, very close to the Norwegian frontier. They lie within a tectonic trench (graben–synclinorium), 70 km long by 35 km wide, that contains 10,000 m of volcanic and sedimentary rocks of the upper Karelian Pechenga Group (Fig. 4.22). The following description is drawn from Gorbunov (1968), Gorbunov et al. (1985a,b), and Glaskovsky et al. (1977).

Initial post-Archean activity in the Pechenga–Varzuga zone involved subsidence and submarine basaltic volcanism, gradually giving way to psammitic and pelitic sediments with interstratified carbonates and volcanic rocks. All of these rocks were folded in an early phase of the Karelian orogeny, which was accompanied by development of the 10,000-m Pechenga Group referred to earlier. This consists of basaltic lavas accompanied by lesser amounts of felsic volcanic rocks interstratified with tuffaceous sediments, limestones, and quartzites. Four main phases of mafic volcanism were interrupted by periods of sedimentation. The last phase of volcanism was followed by folding and intrusion of the ore-bearing, stratified mafic–ultramafic bodies into the 1,000-m thick sedimentary zone separating the third and fourth volcanic sequences (Fig. 4.23). The strata have been folded about a northeast trending axis into a broad arc, the center of which defines a distinctive structural trench. The intrusions tend to be concentrated within this trench, with those on the flanks being small and few in number. About one-third of the intrusive bodies are gabbro-diabase. These were intruded at an early stage in the intrusive activity and are unmineralized. The mineralized bodies include peridotite, and are either large bodies showing an upward progression from peridotite to gabbro, or smaller, relatively undifferentiated masses.

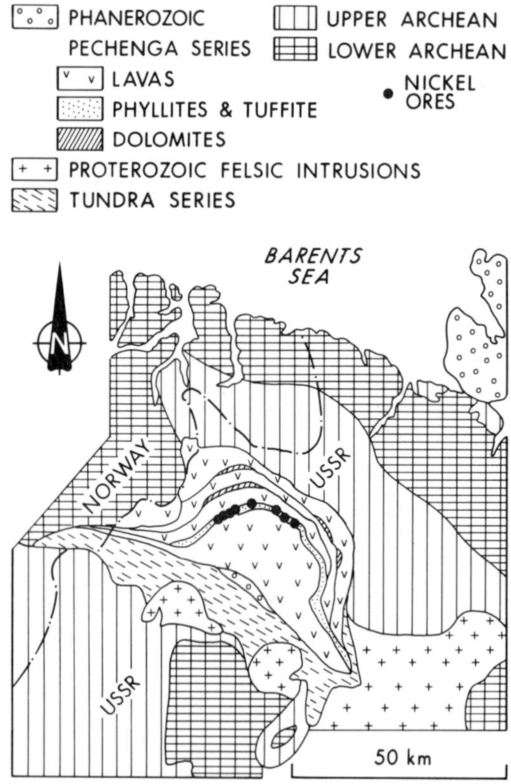

Fig. 4.22. Geologic map of the Pechenga region. (After Gorbunov et al. 1985b.)

4.2.1.2 MINERALIZATION

More than 20 Ni–Cu deposits and prospects have been located within the Pechenga area (Fig. 4.22). Faulting and thrusting have affected many of the

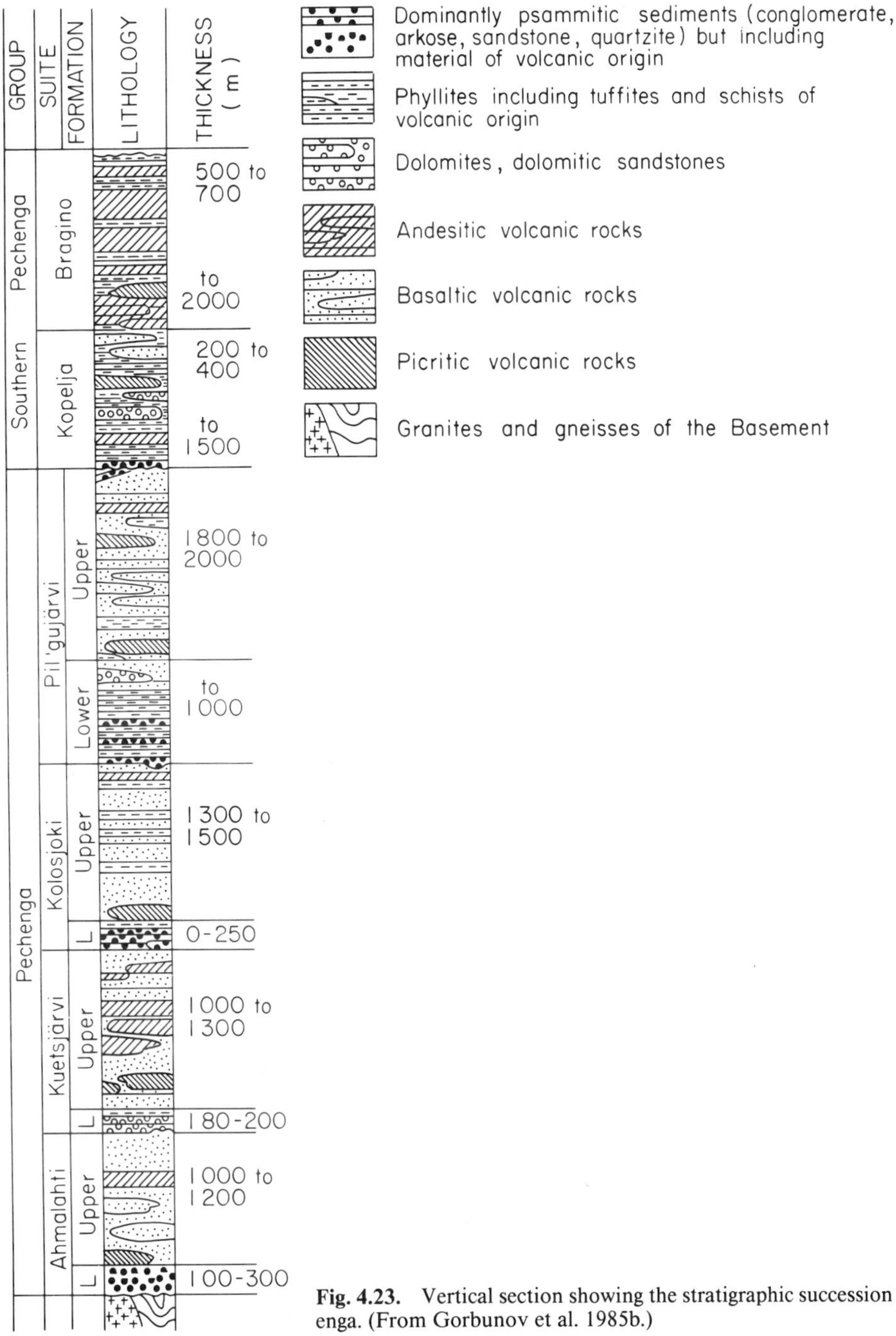

Fig. 4.23. Vertical section showing the stratigraphic succession at Pechenga. (From Gorbunov et al. 1985b.)

ore deposits and resulted in the mobilization of sulfides into tectonic zones. Study of the least deformed bodies has shown that the intrusions are layered, ranging from basal peridotite (now largely serpentinized) through pyroxenite to gabbro. The sulfides are associated with the peridotite or serpentinite and become more abundant where these rocks thicken (particularly in the synclinal troughs) and vice versa.

Ore types include (1) disseminated peridotite-hosted ore in which the sulfide mineralization increases in intensity downward to near the base where it includes massive segregations in small fractures and veinlets, (2) breccia ores located within tectonic zones but passing laterally into the peridotite-hosted ore type, (3) massive ores that are related spatially to the breccia ores and that are economically the most important, and (4)

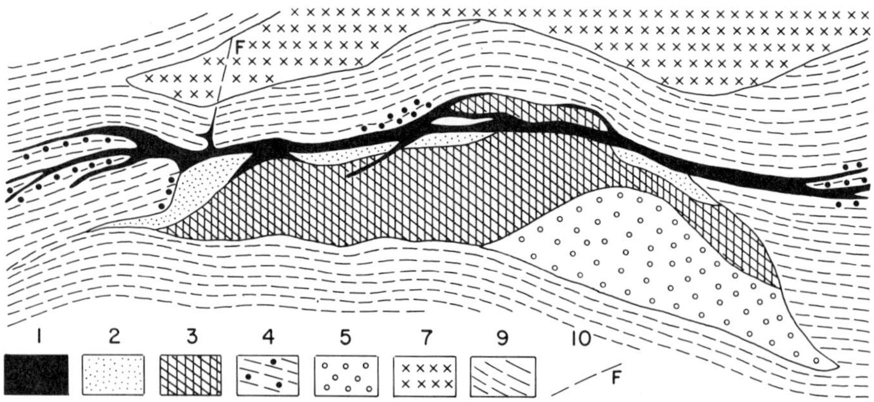

Fig. 4.24. Schematic geologic plan of one of the upper levels of the Kaula deposit. 1 = massive and breccia ore; 2 = high-grade disseminated ore in serpentinite; 3 = low-grade disseminated ore in serpentinite; 4 = mineralized phyllite; 5 = serpentinite; 7 = gabbro; 9 = tuffogenic sediments; 10 = faults. (From Gorbunov et al. 1985b.)

veinlets in the country-rock schist. The Ni content of massive sulfides reaches as high as 10 to 12 wt percent. The Cu content of massive ore is extremely variable, from a few tenths to 13 wt percent. The Cu/(Cu + Ni) ratios of the peridotite-hosted and breccia ore types are fairly constant in most deposits (the Kaula deposit described later is an exception) at about 0.28, but ratios of the veinlets in schist average about 0.51. Nickel/Co ratios average about 50 in all ore types.

The Kaula deposit (Gorbunov, 1968; Gorbunov et al., 1985b) is typical of one of the less deformed deposits (Fig. 4.24, 4.25). Mineralization is confined to the lower part of a differentiated massif consisting mainly of serpentinite. Much of the sulfide ore is confined to a shear zone and extends further down dip than the massif. Where the shear zone leaves the contact of the associated intrusion, the massive, breccia- and phyllite-hosted disseminated ores gradually thin out. Concomitant with this, there is a decrease in the Ni and Cu tenor of the ore to the extent that the place of the ore is eventually taken by barren pyrrhotite and penetrates the surrounding phyllite, both along its foliation and along discordant fractures.

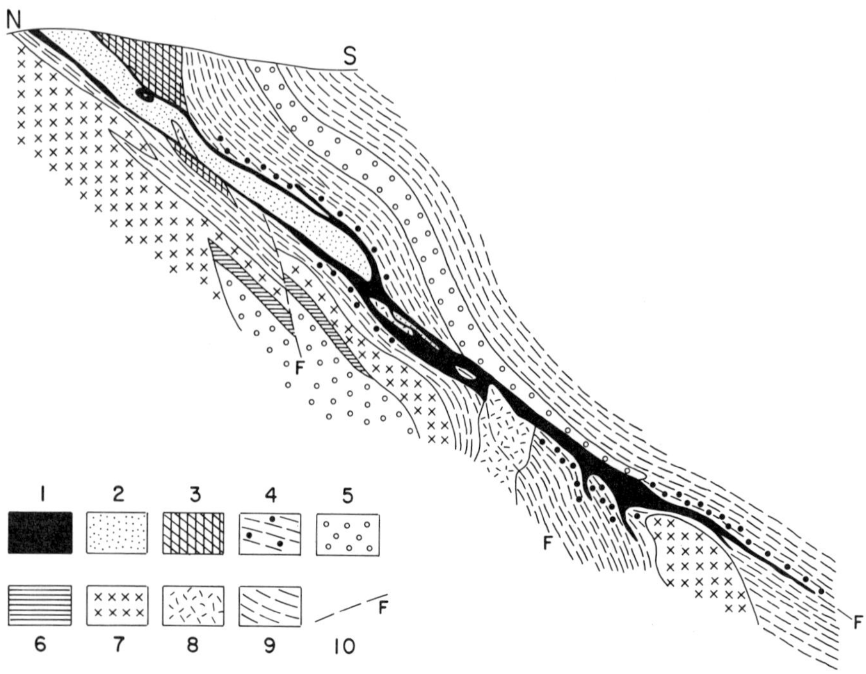

Fig. 4.25. Section through the Kaula deposit, Pechenga region. Key as for Fig. 4.24 plus: 6 = pyroxenite; 8 = diabase. (From Gorbunov et al. 1985b.)

4.3 SETTING II: DEPOSITS RELATED TO KOMATIITES AT RIFTED PLATE MARGINS

4.3.1 The Cape Smith (Ungava) Ni Belt

Wilson et al. (1969), Barnes (1979), Barnes et al. (1982), and Dillon-Leitch et al. (1986) have described the ore deposits of this 1.8×10^9 yr-old belt, which are type examples of the setting II deposits.

As discussed in Chapter 2, Baragar and Scoates (1981) regard this belt as part of the circum–Ungava plate-margin, around much of which continental sedimentation, including red-beds, was followed by mafic volcanism. In the Cape Smith section, the lowermost sediments are dolomites, shales, and quartzites, and pass up into greywacke and shale, which is intruded by numerous mafic and ultramafic sills. The sediments interfinger with and are overlain by shallow water mafic volcanics. These are capped by quartzites, shales, and dolomites, together with some felsic pyroclastics, which are overlain in turn by komatiitic basalts.

The deposits occur at the base of lenses of peridotite, most of which were intruded at the contact between the komatiites and the underlying sediments. The magma associated with these ores was composed of olivine phenocrysts suspended in a liquid that was much less magnesian (16–18 wt % MgO) than Archean peridotitic komatiites. Thin sills and flows containing MgO contents similar to the liquid portion of the magma occur higher in the stratigraphy, and it is probable that the ore-bearing lenses acted as "holding chambers" from which magmas, depleted of much of their olivine and sulfide, then continued to the surface. Consistent with this suggestion, Barnes et al. (1982) have documented zones of varying MgO and olivine content within the lenses hosting the ore that they attribute to the repetitive injection of mixtures of olivine and magma.

A cross section through one of these deposits, the Katiniq body, is shown in Figure 4.26. Barnes et al. (1982) note that sulfides at Katiniq consist of three main textural types: (1) a fine dusting of disseminated grains interstitial to silicate minerals; (2) a net-textured ore, consisting of a continuous network of sulfides enclosing serpentinized olivine grains that comprise 30 to 60 percent of the ore; and (3) massive sulfides. Net-textured ore accounts for about 80 percent of the recoverable reserves. In many places disseminated ore grades downward through net-textured ore into massive ore, with the massive ore showing a sharp contact against the net-textured ore. Massive ore, however, is commonly absent and, in other places, it occurs as zones and stringers within net-textured ore.

Dillon-Leitch et al. (1986) have shown that during serpentinization or regional metamorphism,

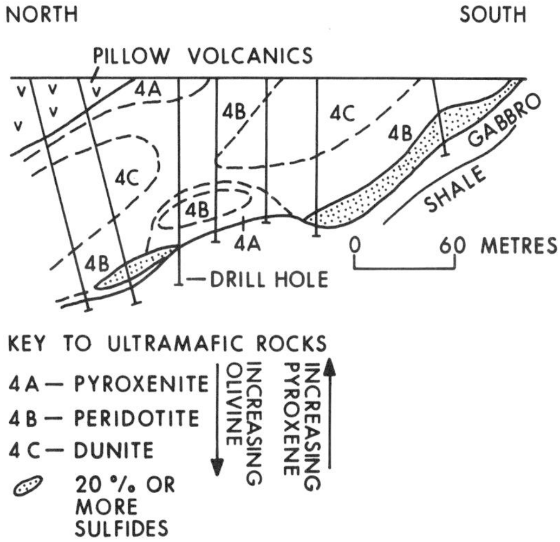

Fig. 4.26. Vertical sections through the Katiniq sill. (After Barnes et al., 1981.)

or both, the relative positions of sulfide and silicate in net-textured ore may become reversed, with sulfides pseudomorphing olivine and serpentine developing in the matrix. They note that this process is accompanied by a loss of Ni and Pt.

4.3.2 Manitoba Ni Belt

The Ni deposits of the Manitoba belt occur within a linear northeast-trending tectonic feature on the boundary between the Superior Province (last regional metamorphism 2.5 Ga) and the Churchill Province (last regional metamorphism 1.8 Ga) of the Canadian Shield. Rocks of the belt are in fault contact to the southeast with granulitic gneisses belonging to the Superior Province, and they are in contact to the northwest with upper amphibolite and granulite gneisses of the Kisseynew Group of the Churchill Province. Within the belt, rocks have been deformed by at least two stages of folding and intruded by late Hudsonian (Baragar and Scoates, 1981) granitoid plutons. The rocks consist of a series of acid- to intermediate-composition gneisses and apparently younger, but still highly deformed, metasediments and coeval metavolcanics. The sediments consist of siltstone, sandstone, quartzite, shale, phyllite, dolomite, and iron formation. Despite the deformation, pillowed and massive sections, and pyroxene–spinifex texture characterize the lavas, which range in composition from ultramafic to basaltic.

The ore is associated with a series of serpentinites that Peredery (1979) regards as deformed differentiated sills ranging in composition from dunite to pyroxenite. The Pipe 2 deposit is the least deformed. The mineralization occurs within and at the margin of a serpentinite lens. The serpentinite becomes less magnesian across the body in the direction in which the sulfides are less abundant. This trend suggests that this direction is stratigraphically upward and that the sulfides exhibit some degree of basal control. The Thompson deposit (Peredery et al., 1982; Zurbrigg, 1963) consists of an apparently conformable horizon of massive and disseminated sulfides within a zone of biotite schist (Fig. 4.27). Numerous inclusions of serpentinized ultramafic rocks occur within the sulfides; some with identifiable layering are interpreted by Peredery et al. (1982) as boudinaged remnants of a sulfide-bearing sill in the ore. One of the complexities at the Thompson deposit is

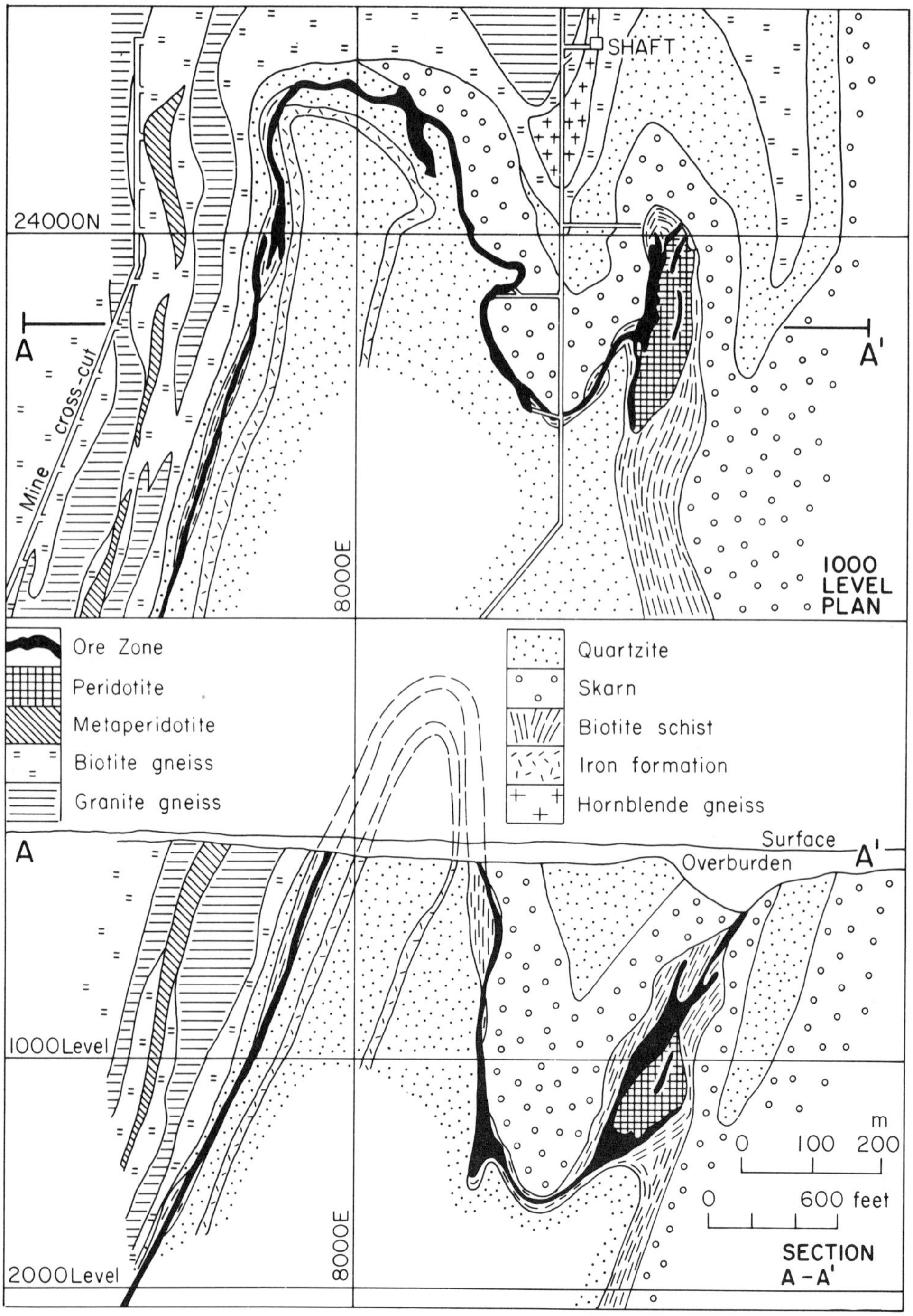

Fig. 4.27. Plan and section of the Thompson Ni–Cu deposit, Manitoba. (From Zurbrigg, 1963.)

the relatively low ultramafic rock/sulfide ratio. There is, however, considerable evidence of deformation along the horizon in which the ore occurs and, if much of the bedding plane slip has been confined to this zone, it is possible that during deformation the ore was mobilized and dragged several kilometers from the main part of the host ultramafic as a result of the slippage.

Many, but not all, ore deposits occur within pelitic phyllites. These ores form discrete horizons within portions of the sediments characterized as well by quartzite, dolomite ("skarn"), and banded iron-formation. There are many similarities between the setting of the Pipe 2 and the Birchtree (another large deposit) deposits on the one hand, and the Thompson deposit on the other, although Peredery et al. (1982) is reluctant to correlate the two horizons. Peredery (1979) pointed to pyroxene–spinifex textures in some of the more MgO-rich members of the basaltic rocks, and to the komatiitic compositional characteristics of the ore-bearing serpentinites and many of the basalts at Thompson, and suggested that much of the sequence is komatiitic. He regards the ore-bearing bodies as a possible intrusive manifestation of komatiitic volcanism.

4.4 SETTING IIIA: ORES ASSOCIATED WITH FLOOD BASALTS EXTRUDED DURING INTRACONTINENTAL RIFTING

Periodically during the earth's history, large-scale rifting of the crust has developed across and within many continents. In some cases this has reached the stage of complete rupture, to produce two separate plates, in which case any related ore-bearing magmas would belong to the category discussed in the preceding section. This section is concerned with less extreme rifting, in which the continent has not ruptured into two.

The rifting has in some places been localized along a single, narrow fracture zone, and in others it has occurred across a zone several 100 km wide along a myriad of parallel fractures. Basaltic magma derived from the underlying mantle has risen along these fractures to spread out across the continental surface as thick, extensive deposits of flood basalt. Two very important Ni camps are associated with intrusive feeders to flood basalt activity: the Lower Triassic ores of the Noril'sk-Talnakh camp, Siberia, and the 1.1 Ga ores of the Duluth Complex, Minnesota.

4.4.1 Noril'sk-Talnakh

4.4.1.1 TECTONIC AND GEOLOGIC SETTING

The Noril'sk-Talnakh Cu–Ni deposits are located on the extreme northwestern margin of the Siberian Platform, which has been a stable craton since the end of the Paleozoic (Fig. 4.28). To the north, the platform is separated by the Khatanga trough from a second platform, the Taimyr Peninsula, which has been stable since the Paleozoic. To the west, a third craton, the East European–Urals Block, is separated from the Siberian Platform by the Yenesei trough, known geographi-

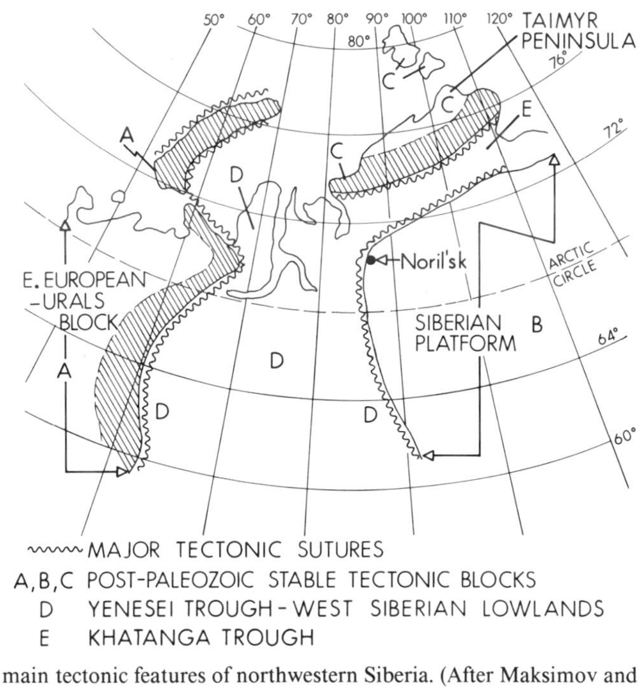

Fig. 4.28. The main tectonic features of northwestern Siberia. (After Maksimov and Rudkevich, 1971.)

as the West Siberian Lowlands. The East European–Urals Block has behaved as a craton since middle Permian time.

Within Western Siberia, Lower Paleozoic dolomites, argillites, and sandstones of marine origin are overlain by extensive Devonian calcareous and dolomitic marls and sulfate-rich evaporites, and Lower Carboniferous shallow-water limestones (Genkin et al., 1981). These are in turn unconformably overlain by Middle Carboniferous lagoonal and continental sediments, including gravels, conglomerates, and coal measures (Smirnov, 1966; Glazkovsky et al., 1977). The emergent sedimentary sequence is covered by an extremely large volume (approximately 1 million km^3) of late Permian and Triassic flood basalt and tuff (Bazunov, 1976; Glazkovsky et al., 1977) known as the Siberian Traps. In the Noril'sk region, these were extruded in four phases, consisting of late Permian quartz and hypersthene normative tholeiites, early Triassic andesite–basalts and olivine–basalts, followed by an early Triassic basaltic, and then an early mid-Triassic basaltic phase. Although the magmatic activity is very widespread, the olivine-rich basalts tend to be concentrated in the Noril'sk region, and individual sheets thin toward the flanks of the region. There is an overall thickening of the flood basalts within the troughs imposed on the underlying sediments. Sill-like tholeiitic intrusions, varying in composition from subalkaline dolerite to gabbro–dolerite (Glaskovsky et al., 1977), were emplaced contemporaneously with, and were, in part, feeders to, the extrusive activity.

The structure of the district is dominated by late Permian to Triassic block faulting, which was coeval with the igneous activity. Individual faults may be over 500 km in length and have throws of up to 1,000 m (Smirnov, 1966). The principal faults trend north–northeast to northeast (Fig. 4.29).

Following extrusion of flood basalt, consider-

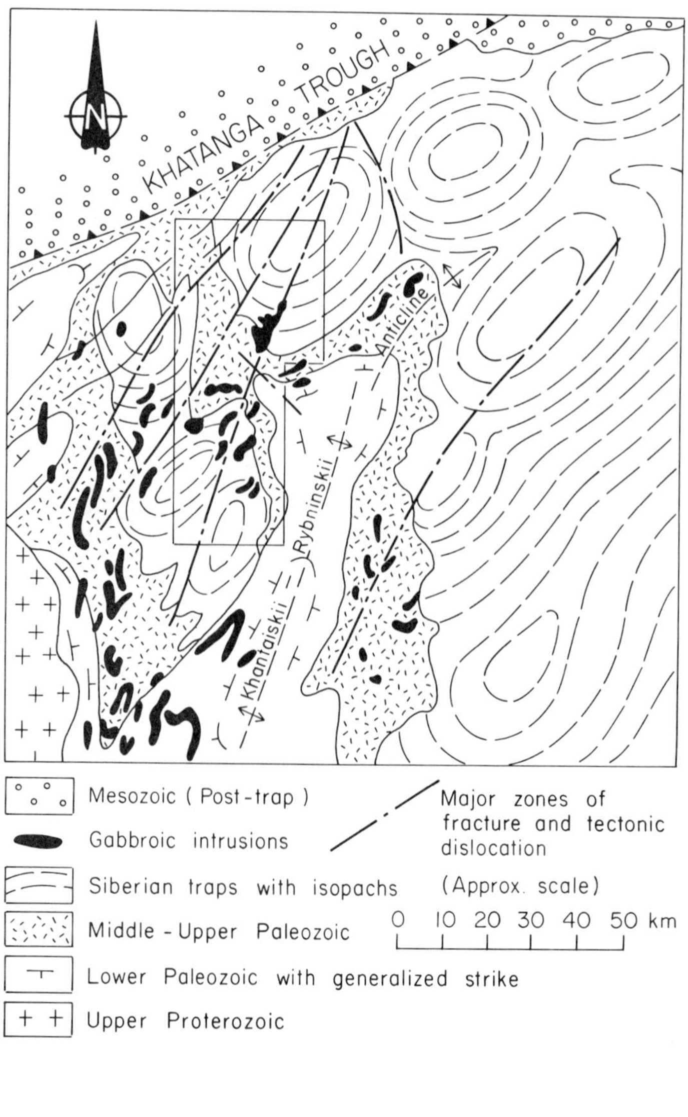

Fig. 4.29. Geologic map of the Noril'sk-Talnakh area showing the main structural elements. (After Genkin et al., 1981.)

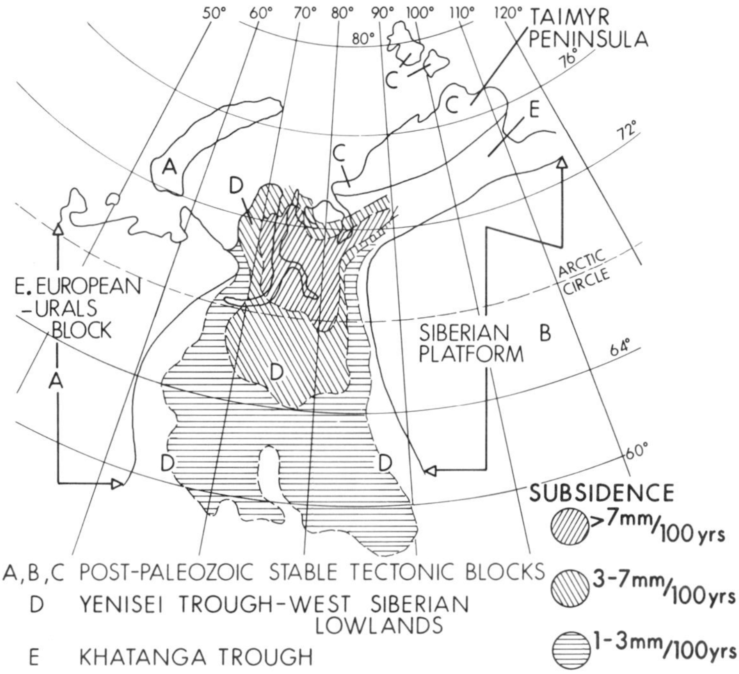

Fig. 4.30. Subsidence during the early and middle Jurassic in northwestern Siberia. (After Tamrazyan, 1971.)

able subsidence took place in the Yenesei and Khatanga troughs from the Early Triassic to the mid-Tertiary; trough-filling sediments attain a thickness in excess of 10 km (Tamrazyan, 1971). The zone of subsidence coincides with the Y-shaped structure (Fig. 4.30), bounded by the block faulting illustrated in Figure 4.28 (Maksimov and Rudkevich, 1971).

Geophysical measurements of depth to the Mohorovicic discontinuity (Tamrazyan, 1971) indicate that, despite the considerable thickness of Mesozoic and early Tertiary sediments in the area, the discontinuity is significantly shallower beneath the Yenesei and Khatanga troughs than beneath the adjacent platforms. The coincidence of considerable subsidence with crustal thinning beneath the troughs is indicative of a rifting environment. In its early stages, the rifting was accompanied by the widespread intrusions and extrusions already described as occurring on the Siberian platform, and which aeromagmatic surveys indicate underlie broad areas of the lowlands themselves. Where they can be mapped on the platform, most of the intrusions are found to be unmineralized (Grinenko [1985] states that only 1–2 % are mineralized), and are tabular to subcordant bodies ranging from gabbro–dolerite to trachy–andesite porphyry.

The emergence during the Carboniferous was accompanied in the Noril'sk region by the development of an anticline (the Khantaiiski–Rybninskii anticline) and two flanking synclinal troughs (Fig. 4.29). These flexures appear to have reflected large-scale fractures that appeared in the basement during the middle and upper Paleozoic, and along which movement persisted until after the Permo–Triassic volcanism. Many of the mineralized intrusive centers of the Noril'sk-Talnakh camp are associated with one of these fractures, the Noril'sk-Kharaelakh fault (Fig. 4.31). It is regarded as a deep-seated fault that acted as the route for upwelling magma (Malich and Tuganova, 1976). Both the Mt. Chernaya deposit to the south and the Talnakh-Oktyabr'sk deposits to the north are located on or adjacent to this fault. The controlling feature for the ore-bearing intrusions in the Noril'sk-Talnakh region appears to be the zones in which this or related faults impinge on the borders of the troughs in the pre-Permian basement which are associated with the thickening of the flood basalts (Genkin et al., 1981).

At Noril'sk, the mineralized Triassic intrusions represent volcanic conduits radiating outward and upward from intrusive centers, cutting the sedimentary sequence and, in places, reaching and apparently acting as feeders to the lavas. Further to the north, at Talnakh, the intrusions do not reach lavas and terminate within underlying sediments.

A further group of gabbro–dolerite-hosted Cu–Ni deposits, the North Kharaelakh ore field, has been described by Dyuzhikov et al., 1976. These deposits, located approximately 375 km north of Noril'sk, have an aerial extent of 1,500 km^2 and lie close to the northerly extension of the Noril'sk–Kharaelakh fault.

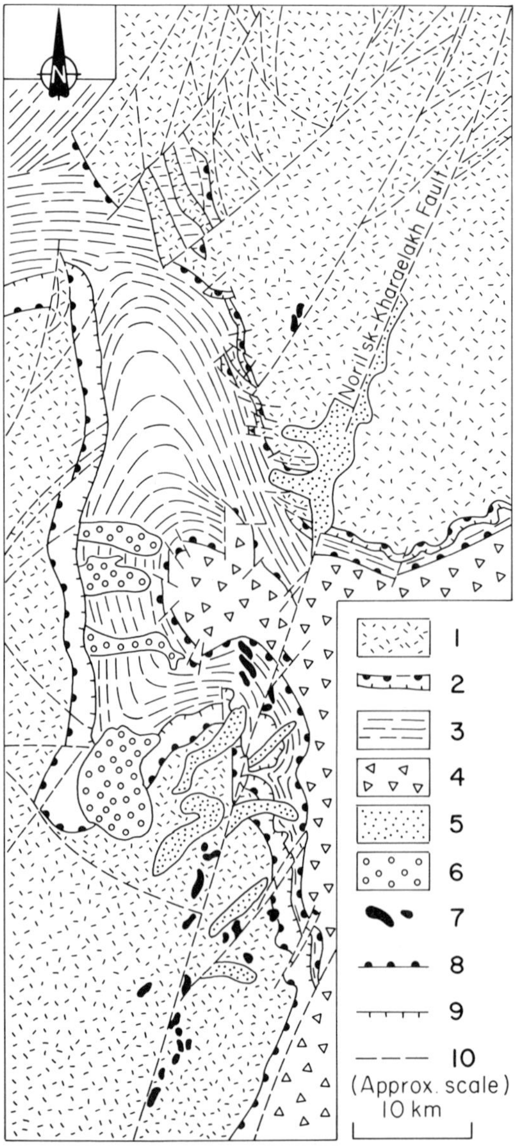

Fig. 4.31. Distribution of different types of differentiated intrusions in the Noril'sk-Talnakh region. 1 = upper Samoedskii and Nadeshdinskii tuffs and lavas; 2 = lower Gudzhikhinskii and Ivakinskii tuffs and lavas; 3 and 4 = sedimentary rocks; 5 = fully differentiated intrusions; 6 = differentiated leucocratic intrusions; 7 = differentiated melanocratic intrusions; 8 = this symbol defines the stratigraphic interval within which differentiated intrusions occur; 9 = sediment − tuff + lava contact; 10 = faults. Enlargement of area outlined in Fig. 4.29. (From Genkin et al., 1981.)

4.4.1.2 HOST INTRUSIONS

The ore-bearing intrusions are differentiated; some have all of the phases to be expected as a result of differentiation of basic magma, in roughly the expected proportions, with picrite and picritic dolerite overlain by more felsic differentiates; other intrusions are particularly rich in melanocratic components and yet others are leucocratic. Cryptic variation is well developed in most intrusions with olivine normally ranging from Fo82 to Fo65 (up to Fo87 in some of the melanocratic intrusions) and the cores of plagioclase grains from An93 to An60 (Genkin et al., 1981; Fig. 4.32).

Genkin et al. (1981) place particular stress on the presence of chrome-bearing spinels, which, at Noril'sk, serve to distinguish mineralized from unmineralized intrusions. Spinels occur primarily in the olivine-bearing rocks, both as sparsely disseminated cumulus grains and as equant agglomerations, ranging from a few centimeters up to 1 m in diameter and containing 25 to 30 volume percent chromite. The agglomerations are restricted mainly to the lower part of the picrite horizons.

The disseminated chromite is of two types; an early, dark colored type containing 15 to 18 wt percent Al_2O_3, 32 to 38 wt percent Cr_2O_3, 6 to 20 wt percent MgO, 8 to 20 wt percent FeO (as Fe^{2+}), and less than 4 wt percent TiO_2, and a later, lighter colored type (often forming overgrowths on the earlier type) containing 5 to 6 wt percent Al_2O_3, 18 to 25 wt percent Cr_2O_3, 2 to 4 wt percent MgO, about 38 wt percent FeO (as Fe^{2+}), and 10 to 15 wt percent TiO_2. Ratios of Fe^{3+}/Fe^{2+} range between 0.5 and 0.8. Chromite of the agglomerations, tends to be richer in Cr (up to 44 wt percent Cr_2O_3), have a lower Fe^{3+}/Fe^{2+} ratio, and the TiO_2-rich variants are absent.

The space interstitial to the chromite in the agglomerations is occupied by pyroxene, plagioclase, and (more rarely) olivine. The interstitial crystals of these minerals are commonly optically continuous with grains forming the fabric of the rock enclosing the agglomerations. One particularly interesting feature of the agglomerations is the presence of clearly defined globules, with a spherulitic internal structure, rich in prehnite, biotite, anhydrite, carbonate, zeolites, chlorite, talc, occasional sulfides, plagioclase, and pyroxene.

Individual sills may attain lengths of up to 12 km and widths of 2 km, whereas thicknesses are typically 200 to 250 m or less. In cross section the intrusions are lenticular, trough-shaped, or, in some cases, tabular. Genkin et al. (1981) note that although the intrusions are generally subconcordant, they may cut down steeply to lower levels in the stratigraphy where they again become subconcordant. They suggest that leucocratic and some normal sills are connected to melanocratic sills that are buried at deeper levels in the stratigraphy, and that together they represent ensembles in which the higher sills represent injections of more fractionated magma that, in fractionating, gave rise to the refractory deposits that remain as the melanocratic sills. They report that magma con-

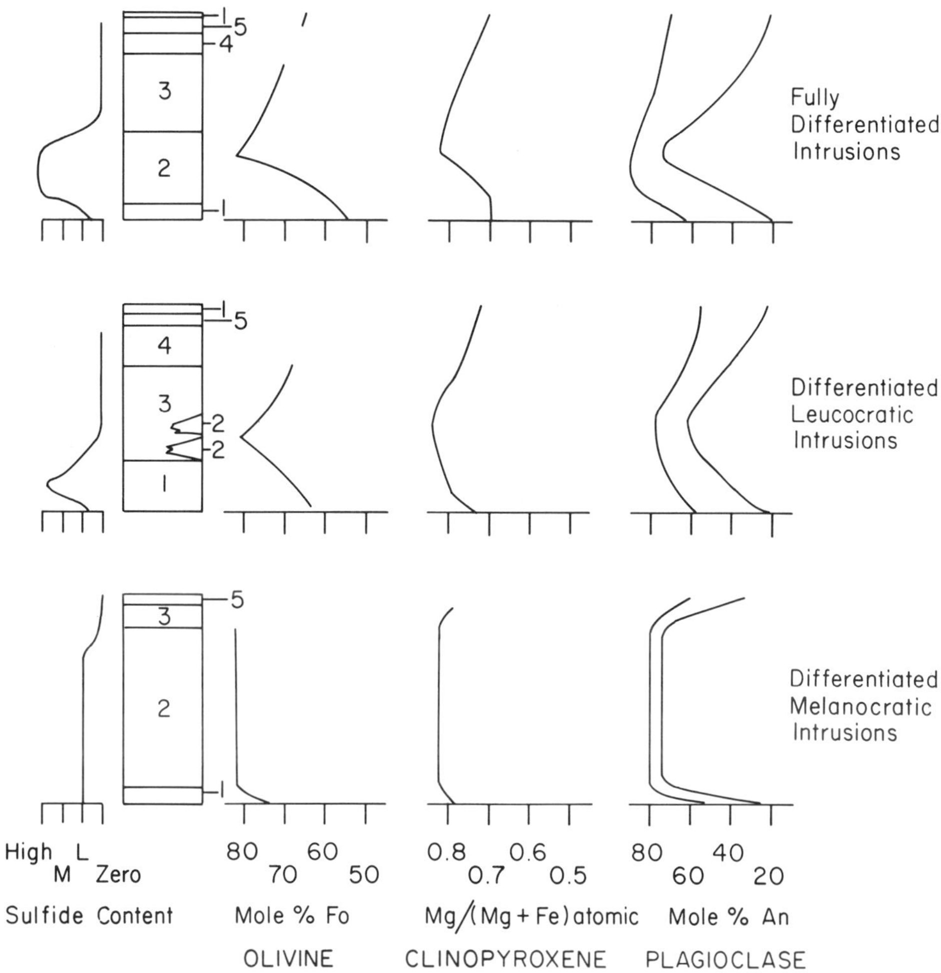

Fig. 4.32. Cryptic variation in the composition of olivine, clinopyroxene, and plagioclase in different types of intrusions of the Noril'sk-Talnakh region. 1 = contact gabbro-dolerite; 2 = picrite and troctolite; 3 = olivine gabbro; 4 = gabbro; 5 = diorite. (After Genkin et al., 1981.)

duits have been established as connecting the normal Talnakh and the melanocratic Nizhne-Talnakh intrusions. They note that the cryptic variation present in the normal and leucocratic sills indicates continuous differentiation in situ, whereas the uniform composition of olivine in the melanocratic bodies indicates that the magma responsible for these was intruded carrying crystals of uniform composition that were presumably derived from elsewhere.

Relationships between an ore-bearing intrusion and both the surrounding rocks and the ore itself have been well described for the Talnakh mineralized body (Fig. 4.33) by Genkin et al. (1981). Much of the following description is taken from their work.

This intrusion can be considered as consisting of three parts, one lying east of the Noril'sk–Kharaelakh fault, one within the fault zone, and one lying to the west of it. The one in the east is the simplest, being an elongate tabular body with few apophyses. It has intruded Permo–Carboniferous rocks, except for a central keel that penetrates Devonian limestones. Its roof comes close to the base of the overlying flood basalts. Within the fault zone, the intrusion has been disturbed by later movements along the fault, but it appears also to have been an elongate, tabular body. The portion of the intrusion occurring west of the fault is quite different and has intruded a highly tectonically disturbed sedimentary sequence. It plunges with the general dip of the country rocks; the stratigraphically lowest part is at the center (where dikes are observed connecting it with the stratigraphically lower, melanocratic Nizhne-Talnakh body), and it rises away from this keel. It is localized within lower and middle Devonian rocks. Where the flanks of the western portion come into contact with sulfate–evaporite and carbonate rocks, they have an intricate structure, breaking

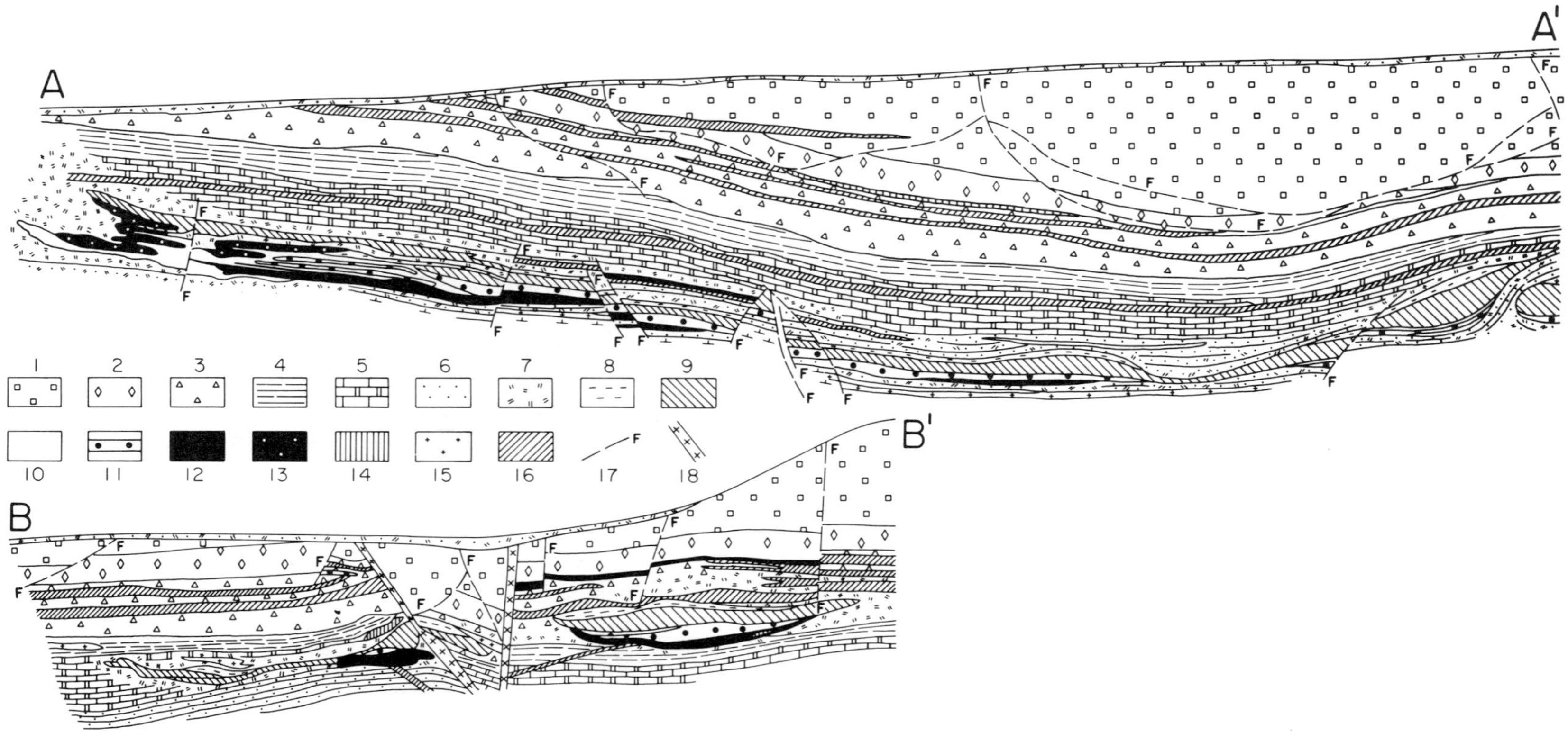

Fig. 4.33. Geologic section through the northwestern part of the Talnakh ore field. 1 = tholeiitic basalts of the Syverminskii suite; 2 = two feldspar titanaugite basalts of the Ivakinskii suite; 3 = sandstones, siltstones, argillites, carbonaceous argillites, and coals of the Tunguska series; 4 = dolomites and limestones of the Kalargonskii suite; 5 = anhydrites and marls of the Manturovskii suite; 6 = argillites of the Razvedochninskii suite; 7 = contact metamorphic and metasomatic rock; 8 = diorite, gabbro–diorite, and gabbro; 9 olivine gabbro; 10 = picrite and taxitic gabbro; 11–14 = ores—11 = disseminated ore in intrusive rocks; 12 = massive ore; 13 = breccia ore; 14 = veinlet ore; 15 = undifferentiated gabbro–dolerite sills; 16 = titanaugite dolerites; 17 = faults. (From Genkin et al., 1981.)

up into a series of apophyses, the larger of which are themselves stratified. In general, the flanking and stratigraphically higher parts of the western portion are marked by a higher proportion of leucocratic differentiates than the central part. A breccia, which the intrusive rocks have penetrated, characterizes the western and southwestern margins of the western body.

The country rocks are extensively metamorphosed adjacent to the body. Typically, zones of metamorphism, which includes both hornfelsic and metasomatic alteration, have thicknesses of several hundred meters, which is two to four times the thickness of the intrusion and its apophyses that they surround.

The products of both the metamorphism and metasomatism are closely related, even on a very fine scale, to the compositional heterogeneities of the country rocks. The hornfelsing has been largely isochemical, ranging from the melilite–monticellite subfacies to pyroxene, then biotite–epidote–amphibolite, and finally to the muscovite chlorite facies.

The metasomatism includes the development of alkali-feldspar and feldspar-quartz rocks in sandstones, siltstones, argillites, extrusive basalts, and olivine-free intrusive rocks. Magnesium and calcium-rich skarns characterize the carbonate-rich rocks, anhydrite evaporites, and olivine gabbros. The magnesian skarns, some of which are postintrusive, include forsterite, fassaite, spinel-forsterite, and spinel–forsterite–pyroxene rocks. Anhydrite-rich and monticellite-bearing skarns are also present. Calcium-rich skarns are more widespread than the magnesium-rich type, and include the assemblages pyroxene–plagioclase, garnet–pyroxene, and wollastonite–vesuvianite.

4.4.1.3 MINERALIZATION

4.4.1.3.1 Ore Types

Mineralization includes the following principal ore types: disseminated ores within the intrusion; massive ores localized at the basal contact, both within the country rocks and, to a lesser extent, within the intrusion; veinlets and associated dissemination related to the massive ores; similar mineralization at the top of the intrusion; and breccia and disseminated ores on the western flanks of the intrusion where it is pinching out.

Within the intrusion, the disseminated ores are restricted to the picrite and the contact phase of the gabbro underlying it. Mineralization in the picrite dies out on the flanks as the intrusion breaks up into a series of apophyses before itself pinching out.

Massive ores are present almost everywhere at the lower contact, although in some places they consist only of thin veinlets. The thickest concentrations of massive ore occur along the axes of the intrusion and are elongated in the down-dip direction. Where the basal contact curves sharply, the massive ore may cut from the intrusion into the country rocks. In some cases massive ore outside the intrusion is continuous with ore localized within stockworks of vertical and horizontal fractures within the intrusion.

The veinlets and their associated dissemination at the base of the intrusion are found primarily in country rocks at places where the massive ores are pinching out laterally. Where this type of mineralization occurs in the roof, it is restricted to that part of the intrusion within the Noril'sk-Kharaelakh fault, and is controlled by fissuring related to the fault. Genkin et al. (1981) note that this ore zone plunges northward, in the direction of basal concentrations of ore of similar type, and suggest that it joins up with and is, therefore, related to these concentrations.

The breccia ores occur only on the western flank of the intrusion, where the country rocks are very brecciated, and where the sulfides have occupied voids in pre-existing breccia. The intrusion is itself very complex in this area, consisting of numerous apophyses and occupying the breccia zone itself, and the ore forms an almost continuous ring around the flank of the intrusion. Genkin et al. emphasize that the intrusion contains little disseminated sulfide itself, although massive sulfides may occur within and adjacent to breccia ore bodies and, in particular, form veins within the intrusion.

4.4.1.3.2 Mineralogy and Mineral Zoning

The primary minerals are pyrrhotite [both monoclinic (po_m) and hexagonal (po_h)], troilite (tr), chalcopyrite (cp), pentlandite (pn), cubanite (cb), talnakhite (tk), mooihoekite (mk), bornite (bo), millerite (ml), and magnetite (mt). The disseminated mineralization within the intrusion is zoned. Horizontal zonation is best demonstrated within the picrite horizon where a (tr + tk + mk) assemblage characterizes the centers of ore bodies, and gives way laterally in an outward direction (Fig. 4.34) to many or all of the following assemblages: (tr + po_h + cp + cb + pn), (po_h + co + pn + cb), (po_h + cp + pn), and (po_m + cp + pn). A similar zonation is also present in both directions vertically, with the (po_m + cp + pn) assemblage characterizing the contact gabbro beneath the picrite layer and the olivine gabbro above it.

Massive ores consist of two types in as far as their mineralogy and zoning is concerned. In the simplest type, a ($po_{m \text{ and } h}$ + pn + cp) characterizes the center of the ore zone and cp increases in amount to dominate at the margins. In the complex type, a (cb + tk + mk) assemblage occurs at the center, giving way progressively to (cb + cp + pn + po_h + tr), (cp + cb + pr + po_h), (cp + pn + $po_{h \text{ and } m}$), and (po_m + pn + cp). Cubanite-rich assemblages are always found separating (tk

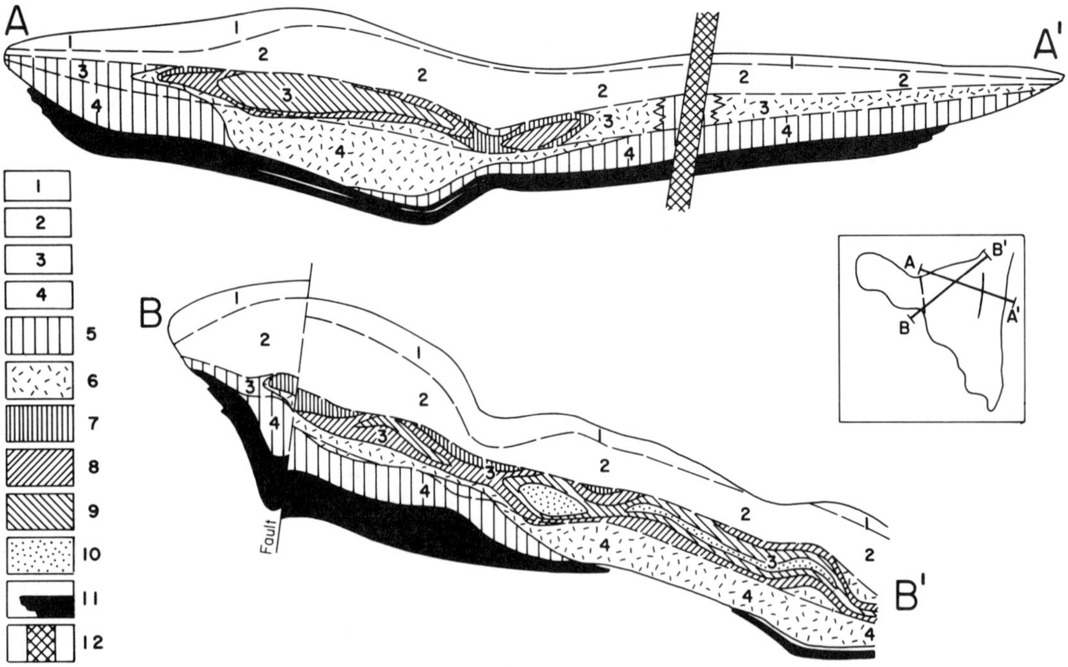

Fig. 4.34. Distribution of sulfide mineral assemblages in two vertical sections through the Talnakh intrusion. 1 = diorite, gabbro; 2 = olivine gabbro; 3 = picrite; 4 = taxitic gabbro; 5–10 = sulfide mineral assemblages—5 = pn + cp + minor po; 6 = pn + cp + po; 7 = pn + cub + cp + po; 8 = pn + cub + cp + tr + po_h; 9 = pn + cub + minor cp + tr; 10 = pn + cub + minor tn (mh, cp) + tr; 11 = massive sulfide ore; 12 = ruptured zones. Note that pn = pentlandite; cp = chalcopyrite; po_m and po_h = monoclinic and hexagonal pyrrhotite, respectively; cub = cubanite; tr = troilite, tn = taenite, mh = mooihoekite. (From Genkin et al., 1981.)

+ mk) assemblages from pyrrohotite-rich assemblages. The S-poor (tk + mk-rich) assemblages in massive ores differ from similar assemblages in disseminated ores in that the latter contain troilite and the former do not.

Veinlet ores are characterized primarily by [po_m(60–80 %) + cp(5–20 %) + pn(5–7 %)] assemblages, but a zoning is sometimes present with cp increasing and the total Fe content decreasing in amount away from the intrusion so that mineral assemblages change to (cp + po_m + pn), (cp + pn), and then (cp + ml). Commonly a secondary alteration to valleriite or pyrite, or both, has been superimposed on the primary mineralization.

The breccia ores are characterized by assemblages of (po_m + cp + pn) with assessory pyrite (py), ml, mt, and valleriite. As in the veins, cp generally increases in proportion away from the intrusion or its apophyses, laterally and in both directions vertically, with assemblages changing to (cp + po_m + pn), (cp + pn), and then (cp + ml). The pyrrhotite of the veinlet and breccia ores is exclusively monoclinic.

The massive sulfides have their own associated metamorphism, which they appear to have imposed on previously metamorphosed and metasomatized country rocks. These effects include the recrystallization of these rocks to a coarser grain size, and the melting of sandstones displaying potash feldspar metasomatism to produce pegamatoidal potash feldspar–quartz aggregates.

4.4.1.3.3 *Sulfur Isotopes*

Grinenko (1967, 1985), Godlevski and Grinenko (1963), and Gorbachev and Grinenko (1973) have studied the isotopic composition of sulfides in unmineralized and mineralized intrusions of the northwest Siberian platform. The majority of the intrusions are not mineralized and are barren or carry only a sparse dissemination of sulfide. Among the mineralized intrusions, Grinenko (1985) distinguishes between those that contain economic mineralization and those that do not. Her data are illustrated in Figure 4.35.

Representative samples were selected from drill core and subjected to whole-rock S and S isotope analysis. Because, in most cases, less than 20 samples were taken, it is unlikely that the average S values are representative of the whole intrusion, but merely of the samples analyzed. Most unmineralized intrusions contain an average of less than 0.1 wt percent total S (equal to less than 0.3 % sulfide) with average $\delta^{34}S$ values ranging from +0.1 to +4.6. Average S contents for the mineralized noneconomic intrusions range from 0.17 to

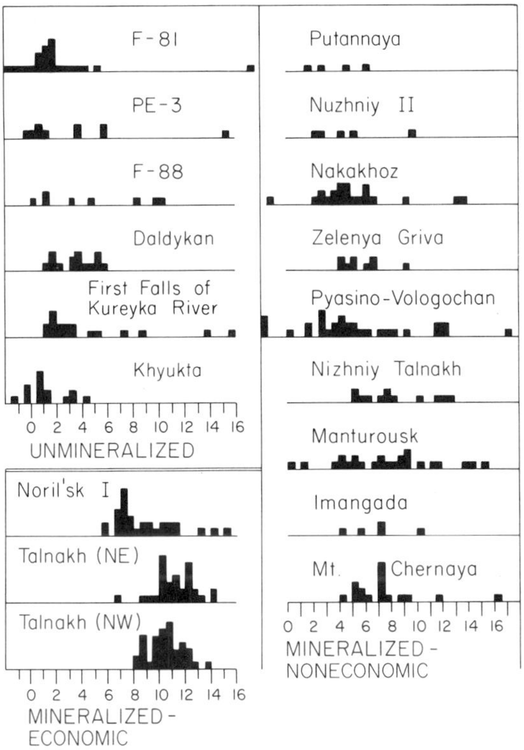

Fig. 4.35. Isotopic composition of S in unmineralized, mineralized–economic, and mineralized–subeconomic intrusions of the Noril'sk-Talnakh region. (After Grinenko, 1985.)

0.34 wt percent, with average $\delta^{34}S$ values from +5.5 to +8.4. The samples from the economic mineralized intrusions contain on average 0.95 to 2.2 wt percent sulfides, with average $\delta^{34}S$ values from +8.9 to +11.4.

The sulfides from individual intrusions with economic mineralization show less spread in $\delta^{34}S$ than those of barren or noneconomic intrusions, suggesting that those of the first group have been better homogenized than those of the two latter groups.

4.4.1.4 GENESIS OF MINERALIZATION

Grinenko (1985) noted that there is a direct correlation between the S content and the average $\delta^{34}S$ value of the sulfides of the barren and noneconomic mineralized intrusions. She also noted that for these two groups there is a relationship between the nature of the wall rocks and the amount and isotopic composition of S. Barren intrusions tend to occur in S-free clastic sediments of the Tunguska supergroup and in the overlying flood basalts. Among the noneconomic mineralized group, most of those intruded into sulfate-bearing sediments have higher S contents than those in S-free sediments. She suggested that much of S has come from the wall rocks, and that the lack of isotopic homogenization shown by the sulfides in these bodies is due to the S contamination having occurred in situ.

Assuming that any original magmatic S had an isotopic composition of zero, adopting the measured value for the $\delta^{34}S$ of the evaporite S of +20, assuming that any contamination was not accompanied by isotopic fractionation, and assuming that the measured average S content of her samples represented that of the magmas responsible for the intrusions, Grinenko (1985) calculated from mass–balance considerations that 20 to 36 percent of the S now present in the intrusions has been added to the noneconomic mineralized bodies, and that their initial (before contamination) S content was between 0.12 and 0.23 wt percent. This, she argued, is consistent with the solubility of S in mafic magma (this author would regard these amounts, equivalent to 0.5–1.0 wt % sulfide, as distinctly high for dry basaltic magma, see Chapter 3).

Grinenko also noted that if a similar calculation is made for the economic intrusions, also assuming the evaporites to be the source of S, the initial S content is much higher than could have been dissolved in the initial magma. Making an assumption as to the amount that could be transported in solution into the magma, she calculates that the $\delta^{34}S$ value of the contaminant was +10.7 to +12. She interpreted the homogeneous isotopic composition of these sulfides to be due to the contamination having occurred before final emplacement of the igneous bodies, and concluded that the contaminant was sour gas S, which is known to be common in the region and has an isotopic composition of about +10.

This author is in agreement with the interpretation that the economic intrusions were contaminated before their final emplacement, and that the sulfides have become homogenized during this emplacement, but believes that the conclusion as to contamination by sour gas is not proved by her evidence. Among other debatable assumptions, her conclusion is based on the assumption that S added as a contaminant has to account for all sulfide in excess of the amount that could have been dissolved by the basic magma responsible for the intrusions. This assumption is open to doubt, indeed is probably not true. The very high PGE content of the sulfide ores indicates a high "R" factor for the sulfides (see Chapter 6), which is to say that they have equilibrated with and concentrated PGE from many times their own mass of magma. Values of $10^{3.5}$ to 10^4 are suggested in Chapter 6 for the mass ratio of magma to sulfide for the Noril'sk-Talnakh ores, in comparison with the value of 16 or so that arises from the premise behind Grinenko's argument.

Grinenko places the Nizhniy-Talnakh (Lower Talnakh) noneconomic mineralized intrusion in a separate category to the Talnakh body, implying a different source for S. As already seen, Genkin et al. (1981) have argued on geologic and petrologic grounds that the two are related and connected, with Nizhniy-Talnakh representing a lower chamber in which the magma differentiated, and thus gave rise to ultramafic cumulates, before rising farther in the stratigraphy to form the Talnakh body. Unless S was added to the magma between the two horizons (which from the descriptions of Genkin et al. seems very unlikely), the source of the S must have been the same for the two bodies.

The present author favors the view of Genkin et al. (1981) that Nizhniy-Talnakh was an intermediate holding and differentiation chamber, and that the contamination occurred either in this, or before the magma intruding to form this. A limited proportion of the magma that gave rise to this body, enriched in sulfide that had been present within the body and that had had, therefore, the opportunity to scavenge ore metals from a much larger volume of magma than itself, then rose farther in the stratigraphy to form the Talnakh intrusion.

The way in which such a sulfide-enriched magma was derived from the Nizhniy-Talnakh body is not entirely clear at present, although it seems likely that not all of the sulfide was necessarily introduced at the same time as the magma. Distler et al. (1986) have argued that the introduction of much of the massive and veinlet sulfide at Talnakh postdates injection of the main intrusion. They base their arguments on the observations (1) that the intrusion at Talnakh had reacted in response to tectonic stress as a solid mass, and that it, therefore, gave rise to dilatent zones into which massive sulfides were then introduced (some of these are even present in the roof of the intrusion as outlined above) and (2) that massive sulfides were accompanied by their own metamorphic effects that were superimposed on the metamorphism and metasomatism due to the intrusion.

Distler et al. (1986) use their conclusion of the later introduction of massive sulfide to support their hypothesis that the separation of sulfide liquid from silicate magma occurred deep in the crust and not within the supercrustal sulfate-rich rocks. Godlevsky and Likachev (1986) believe that these magmas have carried 3 to 10 wt percent sulfide all of the way from the mantle. Thus, they do not accept contamination by country-rock S as a significant ore-forming process, ascribing the heavy S of the ores to the presence of heavy S in some parts of the mantle (A. D. Genkin, personal communication, 1982). The present author is impressed by Grinenko's (1985) S isotope evidence as to the involvement of crustal S. He suggests that, because of their lower melting temperature, sulfides in and beneath the Nizhniy-Talnakh intrusion remained liquid after the ultramafic cumulates of this and parts of the Talnakh body had solidified enough to fracture, and that the liquid sulfides were then squeezed up higher in the stratigraphy into these fractures in response to the ongoing tectonic movements.

The present author suggests that the prehnite + biotite + anhydrite + carbonate + zeolite + chlorite ± sulfide globules that occur within the chromite agglomerations in the picrite of the intrusions could represent melted remnants of partially assimilated country rock. The assimilation of anhydrite-rich rocks, coupled with the reduction of sulfate to sulfide would have introduced considerable oxygen into the silicate melt, which could have been the cause of the precipitation of chrome spinel. Inclusions of anhydrite-rich material, floating in the magma, could have served as locii for chromite crystallization, thus giving rise to the association between the agglomerations and the globules.

Genkin et al. (1981) noted that the normal progression of zoning from pyrrhotite-rich to chalcopyrite-rich assemblages is readily explicable in terms of the fractional crystallization of sulfide magma, as, for example, Naldrett et al. (1982) have suggested for the Sudbury ores. Genkin et al. remarked that the existence of large masses of sulfide with talnakite and mooihoeckite assemblages is unique to the Noril'sk-Talnakh camp, and attributed the zoning from pyrrhotite + chalcopyrite to mooihoeckite + talnakhite to the separation of the sulfide melt into two liquids, one Fe- and S-rich and the other Cu-rich and S-poor. They suggested that the cubanite ores are the result of reaction between the two adjacent liquids as they crystallized. Although there is no experimental evidence for liquid immiscibility in the relevant parts of the Cu–Fe–S and Cu–Fe–Ni–S systems, Distler et al. (1986) have described globular masses of sulfide from oceanic basalts in which the sulfides have segregated into two halves, one with the composition of nickeliferous Mss and the other close in composition to cubanite. Similar globules have been described from the Insizwa deposit by Lightfoot et al. (1984) where the Cu-rich part is observed to be at the top and to consist largely of chalcopyrite. An alternative explanation is that the compositional separation is due to the crystallization and settling of a refractory, dense Mss phase from a sulfide melt, leaving a Cu-rich residual liquid. The separation between the two domains in the globules is very sharp for the latter explanation to be entirely satisfactory. On the other hand, the progressive zoning toward the center of the disseminated ore zone in the picrite, or the center of zones of massive ore is also difficult to explain on the basis of either theory.

Perhaps one of the most perplexing questions remaining at Talnakh is the size of the metamorphic and metasomatic aureoles, several times larger than the intrusive bodies themselves. Although most of the intrusive bodies have an aureole, Genkin et al. (1981) stressed that the aureoles are much larger about well-differentiated bodies than about others, concluding that the former must have been much richer in volatiles. The aureoles would be easier to explain if the intrusive bodies were all conduits for the overlying volcanism, as seems to be the case at Noril'sk itself, because in this case one could call on a much larger mass of magma than is present in the intrusions. The Talnakh intrusions, however, appear to be blind and not linked physically to the overlying volcanics. The resolution to this problem may come from a quantitative evaluation of the heat and mass transfer that has occurred between the Noril'sk-Talnakh intrusions and their surroundings.

4.4.2 Mineralization of the Duluth Complex

The Duluth Complex crops out as an arcuate mass extending about 150 miles northeast from Duluth, Minnesota, to the Canadian border. It is the host to more than 4×10^9 tonnes of mineralization averaging 0.66 wt percent Cu and 0.2 wt percent Ni (Listerud and Meineke, 1977).

Sims and Morey (1972) have presented a generalized discussion of the stratigraphic setting of the Complex. It is closely associated with Keewanawan (Late Proterozoic) basalts of the Lake Superior area and was intruded 1.12 Ga ago (Faure et al., 1969), close to the contact of these rocks with older Precambrian basement rocks that are now exposed to the northwest (Fig. 4.36). These basement rocks consist, in the north, of a complex of Archean volcanics and felsic intrusions (primarily the Giants Range Granite), which are overlain unconformably to the south by the south-dipping Biwabik Iron Formation. The iron formation is itself overlain by black argillites, greywackes, siltstones, graphitic slates, and sulfide facies iron-formation comprising the Virginia Formation.

The Duluth Complex itself is a composite intrusion that the studies of Grout (1918), Green et al. (1966), Bonnichsen (1970, 1972), Phinney (1969, 1972), and Weiblen and Morey (1975, 1980) have shown to consist of two series, an older series of anorthositic rocks that is cut to-

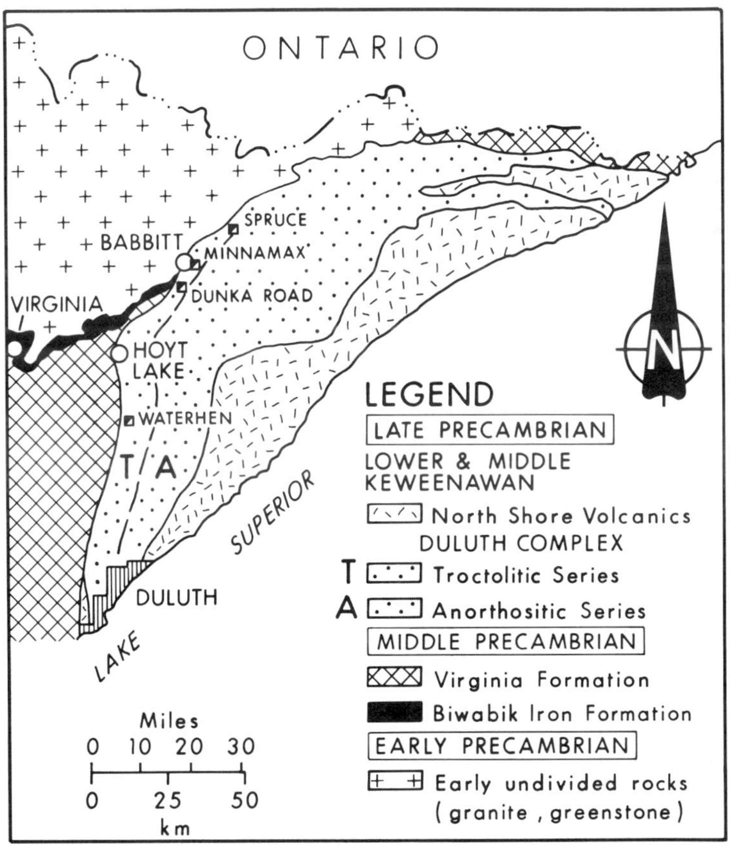

Fig. 4.36. Map of the western end of Lake Superior showing the location of the Duluth Complex and its relationship to the country rocks.

ward its western margin by a series of later intrusions consisting predominantly of troctolite. In the central part of the Complex, Foose and Weiblen (1986) describe the troctolitic rocks as comprising three separate intrusions, Bald Eagle, South Kawishiwi, and Partridge River, although the distinction between the latter two is made on the basis of subtle variations in texture.

Foose and Weiblen (1986) report that the South Kawishiwi intrusion consists of an upper portion, 1,000 to 1,200 m thick, in which troctolite and minor anorthosite are interlayered. Pegmatoidal layers of plagioclase are commonly in sharp contact with underlying troctolite, and form the bottoms of packages of rocks that grade upward into medium-grained plagioclase cumulates and then plagioclase–olivine cumulates. Constituent minerals show normal and reverse cryptic variation within individual rock packages, but no systematic overall compositional variation, leading Foose and Weiblen to suggest that the chamber was repetitively recharged with new magma, possibly some of it from adjacent intrusions.

The layered troctolites are underlain by a heterogenous zone, 70 to 300 m thick, in which medium- to fine-grained troctolites, together with picrites, norites, anorthosites, and oxide cumulates are the host to disseminated sulfides and numerous hornfels inclusions.

The mineralization occurs along the western margins (stratigraphic bases) of these troctolitic intrusions. Wager et al. (1969) described the geology of the Spruce deposit of INCO Metals Ltd; Matlock and Watowich (1980), the geology of AMAX Exploration Ltd's Minnamax deposit; Rao and Ripley (1983) and Ripley (1981), the petrology and geochemistry of U.S. Steel Corporation's Dunka Road deposit; and Mainwaring and Naldrett (1977), the mineralization of the Waterhen Complex. The locations of these deposits are shown in Figure 4.37. In the first three deposits, the mineralization consists of pyrrhotite, chalcopyrite, pentlandite, and cubanite weakly disseminated in troctolite and norite within 300 m of the base of the Complex. The mineralized zones are characterized by numerous inclusions of country rocks consisting of hornfelsed Virginia and Biwabik Formations, together with barren gabbro and peridotite. Inclusions of Virginia and Biwabik rocks occur even at the Spruce deposit, where granite forms the footwall and the Virginia and Biwabik do not appear to be present in situ at surface within several kilometers of the deposit (Fig. 4.38). At both the Minnamax and Dunka Road deposits, norite is more common and troctolite less common in the vicinity of the sulfide horizons, attesting to reactions between the troctolite and country-rock inclusions.

The Waterhen Complex is somewhat different in that the sulfides are associated with a zone of dunite and also with peridotite layers within the

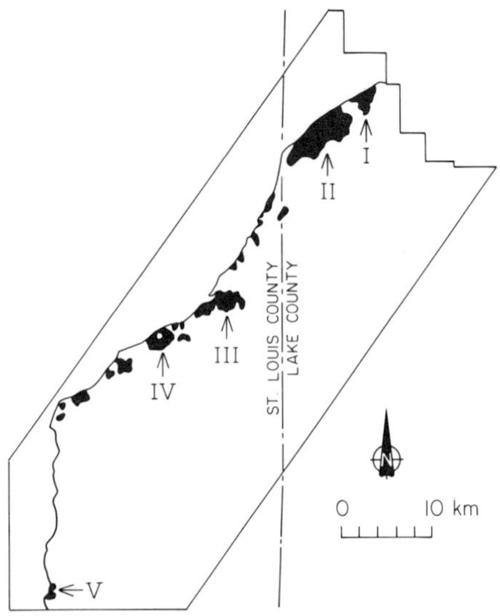

Fig. 4.37. Map of parts of St. Louis and Lake counties, Minnesota, showing the main areas of mineralization outlined to date. I, Spruce Pit area, 635 × 10^6 tonnes > 0.5 wt % Cu; II, 2100 × 10^6 tonnes > 0.5 wt % Cu; III, Minnamax Area, 725 × 10^6 tonnes > 0.5 wt % Cu; IV, Dunka Road Area; V, Waterhen Area.

overlying troctolite. Although the mineralization consists of a normal assemblage of pyrrhotite, pentlandite, and chalcopyrite, occurring interstitial to silicate grains in a characteristic magmatic "net texture," an uncommon feature is the occurrence of as much as 10 percent graphite in certain zones. Other unusual features of the sulfide environment include the presence of green (Mg–Al) spinel in some of the troctolites, and the local occurrence of cordierite. As in other deposits, numerous partially resorbed remnants of what appear to be hornfelsic Virginia slate occur in the troctolite. These observations led Mainwaring and Naldrett (1977) to question whether the graphite and sulfide might not have been derived from the Virginia Formation. The results of their sulfur isotope study are included in Figure 4.39. Barren sulfides from the Virginia Formation average about 18 $\delta^{34}S$ per mil, whereas the Waterhen sulfides range from 11 to 16 $\delta^{34}S$ per mil. They noted that these results are consistent with a model in which perhaps as much as 75 percent of S was derived from the country rocks.

The S isotope data from Ripley (1981) for the ore of the Dunka Road deposit (Fig. 4.39) ranged from 0.2 to 15.3 per mil with a mean of 7.5 per mil. Pyrrhotite in the underlying Virginia Formation shows a similar range in isotopic composition, also supporting a country-rock source for much of the sulfur. The heterogenous distribution of the S isotope values throughout the deposit, coupled with a wide range in the Ni contents of olivine in the mineralized zone lead Ripley (1986)

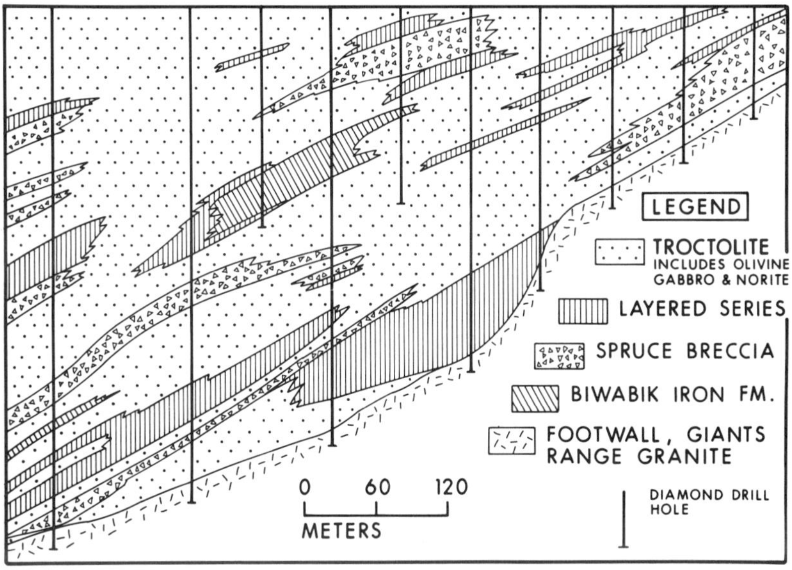

Fig. 4.38. Vertical section through the Spruce deposit, Duluth Complex, Minnesota. The layered series refers to zones of well-banded picrite, mafic troctolite, and troctolite. The Spruce breccia refers to troctolite with more than 30 percent of inclusions of clinopyroxene–plagioclase–olivine-bearing rock, Biwabik Iron Formation, and recrystallized mafic rocks. (From Wager et al., 1969.)

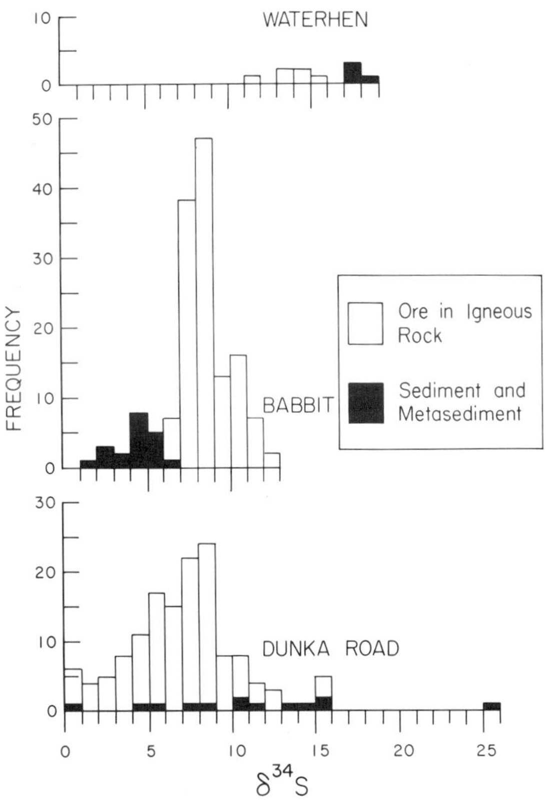

Fig. 4.39. Isotopic composition of S in Ni–Cu and adjacent sedimentary sulfides of deposits in the Duluth complex. (Data for Dunka Road [Ripley, 1981], for Babbit [Ripley, 1986], and for Waterhen [Mainwaring and Naldrett, 1977].)

to suggest that S introduction had occurred essentially in situ, and that widespread equilibration between the sulfides and their enclosing silicates had not occurred.

Ripley (1986) found that the $\delta^{18}O$ values at Dunka Road ranged from 5.8 to 9.6, but that anomalous values (>8 per mil) only occurred near (within 3 m) the margins of country-rock xenoliths (which themselves have a range of from 8.6–11.1 per mil). The olivine of the troctolite gives way to orthopyroxene in these zones, suggesting that SiO_2 contamination has accompanied the change in $\delta^{18}O$. The localized nature of the O isotope contamination lead Ripley to suggest that whole-rock assimilation was not a major cause of S contamination. He proposed that S had been introduced in a volatile phase that was released from the footwall rocks as they were metamorphosed by the intrusion. This is supported by the common association of hydrous minerals, such as biotite and amphibole, and patches of pegmatite with the mineralization.

At Babbit, a much higher proportion of the igneous rock is isotopically anomalous with respect to O, the sulfides are more abundant, and their PGE contents are higher (indicative of higher R factors, as discussed in Chapters 3 and 6). The sulfides are more homogenous isotopically, particularly those of the better grade ores, and significantly higher than those of the adjacent country rocks (Fig. 4.39). These factors lead Ripley (1986) to suggest that contamination had occurred at an earlier stage here than at Dunka Road, possibly during intrusion.

As stated previously, the Duluth Complex is an intrusive part of the Keewanawan flood basalt province of Lake Superior. A midcontinental gravity high extends this province southward through Minnesota to Kansas, and also southeastward into central Michigan (Fig. 4.40). The zone of anomalous gravity is also marked by numerous strong positive and negative magnetic anomalies. Pre-Paleozoic basement rocks intersected in boreholes within the zone include a high proportion of Keewanawan basalts and mafic intrusions. These and the related gravity and magnetic anomalies are expressions of magmatic activity associated with intracontinental rifting, which has its greatest development in the Lake Superior area with the extrusion of the Keewanawan flood basalts and the intrusion of the Duluth Complex.

4.5 SETTING IIIB: SUDBURY

The Ni–Cu ores of the Sudbury district are associated with the Sudbury Igneous Complex (SIC), a layered intrusion ranging from quartz norite at the base, through gabbro, to a granophyric cap. The purpose of this section is to describe the geologic setting at Sudbury, to discuss the composition of the Complex, to show how the composition accounts for the profusion of magmatic sulfide ores, and to propose a model to account for the geology of the Complex and some aspects of its setting.

4.5.1 Regional Geologic Setting

The SIC was intruded beneath the Onaping Formation (Fig. 4.41), a breccia composed of fragments of country rocks and recrystallized glassy material set in a matrix of glassy shards (Muir and Peredery, 1984). The Onaping Formation is variably interpreted as an ignimbrite or the "fall-back" breccia from the impact of a meteorite. The SIC and strata overlying it are exposed as a series of concentric, crudely elliptical rings that dip toward the center of the Complex and suggest the structure as a basin. The Sudbury basin is located at the contact between (1) gneisses, remnants of greenstone belts and intrusive tonalites all of Archean age to the north and (2) rocks of the Proterozoic Southern Province, which overlie the Archean basement unconformably and thicken to the south.

In a recent study of regional gravity and magnetic data, Gupta et al. (1984) analyzed the combined residual gravity and magnetic anomaly that marks the Sudbury region itself. After subtracting large-scale "super regional" gravity trends, they conclude that the broad +20 to 30 mGal anomaly characterizing the Sudbury region cannot be explained by the rocks of the Complex itself. Their modeling indicates that a large (60 × 40-km) mass of mafic rock, with a density similar to gabbro or gabbro–anorthsite (3.02 ± 0.03 g/cc) underlies the Complex at a depth of at least 5 km, and extends beyond it as indicated in Figure 4.42. Their analysis of the regional magnetic data indicates that some of the rocks of this body are partially serpentinized.

4.5.2 Local Geologic Setting

A detailed description of the local setting at Sudbury is beyond the scope of this book; it is well described by a number of authors in a book on Sudbury geology (Pye et al., 1984). Many aspects of the local setting suggest that an explosion of unusually large intensity gave rise to a crater at Sudbury. The evidence includes:

1. The basinal shape of the structure as interpreted from surface and underground mapping and drilling.
2. An upturned collar around the basin, as seen particularly in the Huronian rocks along the southern margin (Dressler, 1984).
3. Shock metamorphic features in the country rocks around the structure (Dressler, 1984).

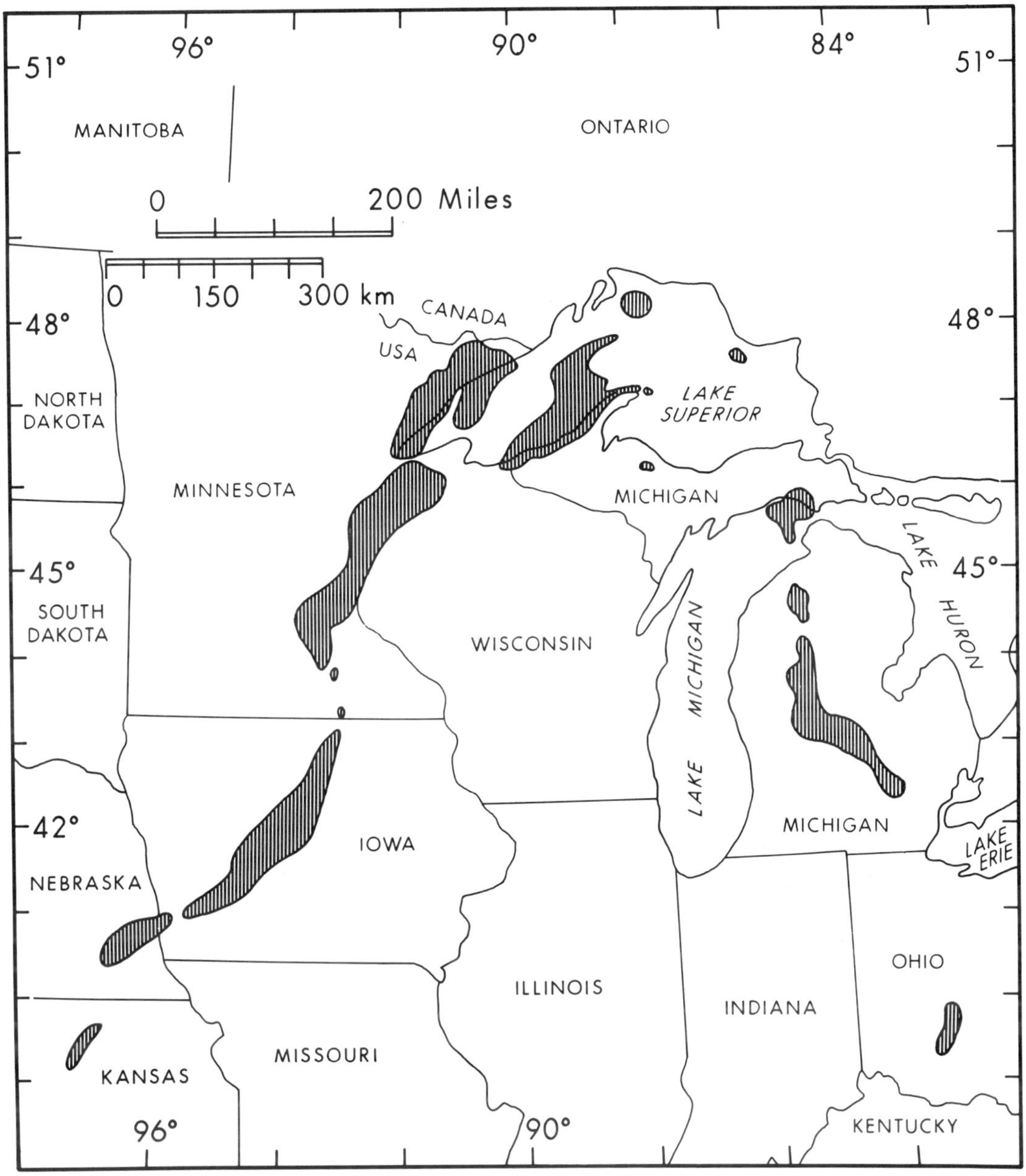

Fig. 4.40. The midcontinent gravity anomaly. Shading indicates areas of positive Bouguer anomaly (>0 mGal). (After Naldrett and MacDonald, 1980.)

4. The Sudbury Breccia (compared with the pseudotachylite of the Vredefort and Ries structures, Dressler, 1984) in the country rocks around the structure and the Footwall breccia beneath the Complex.
5. Evidence of shock metamorphism in country rock inclusions in the Onaping Formation (Muir and Peredery, 1984).
6. The 1,800 m of Onaping Formation itself, the lower part of which is variably interpreted as a meteorite fall-back breccia or an ignimbrite (Peredery and Morrison, 1984; Muir, 1984).

Opinions are divided between an extraterrestrial and endogenic origin for the structure. This author believes that meteorite impact is the more likely origin, primarily because so many of the features observed at Sudbury are also found at known impact sites. There are, however, many difficulties with such an origin and these are summarized by Muir (1984).

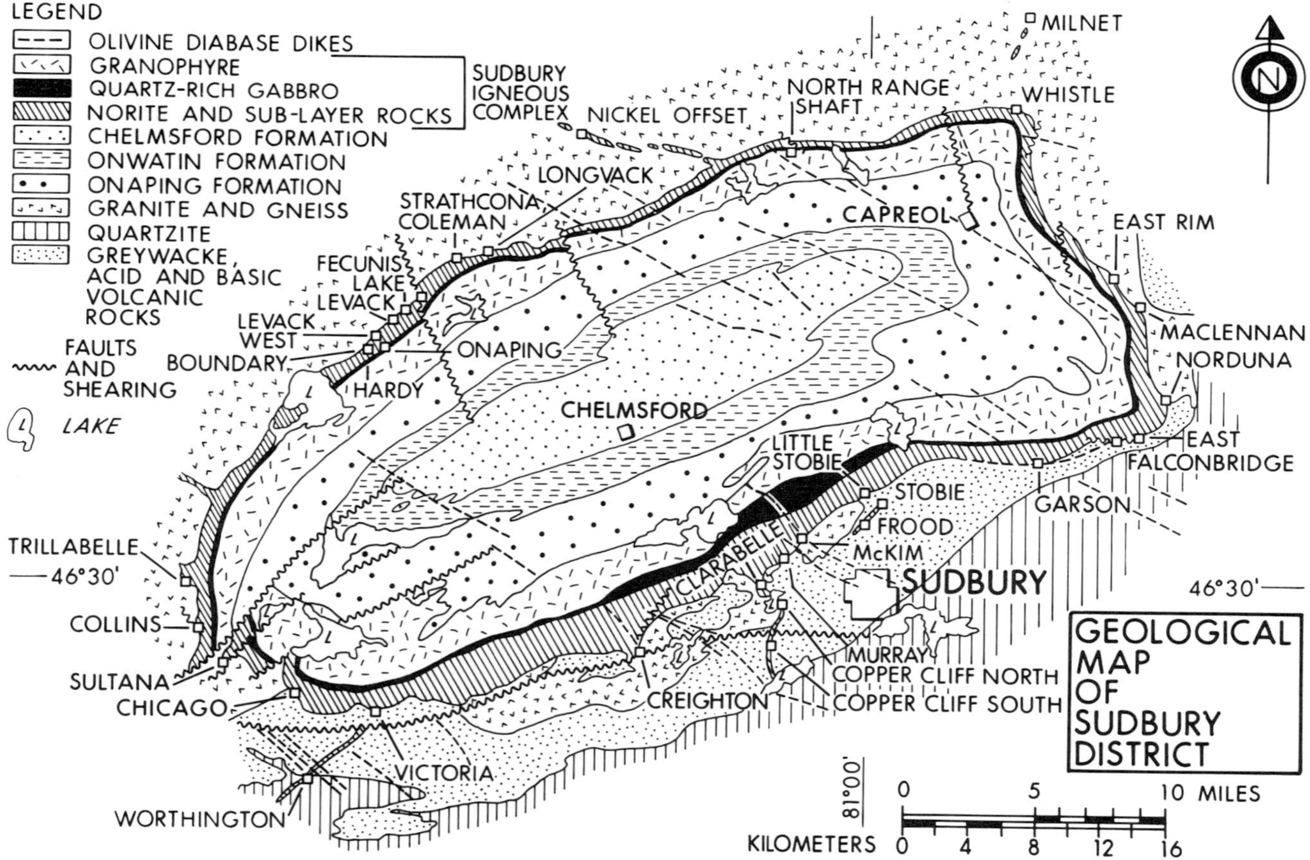

Fig. 4.41. Geologic map of the Sudbury Region. (Drawn after Pye et al., 1984.)

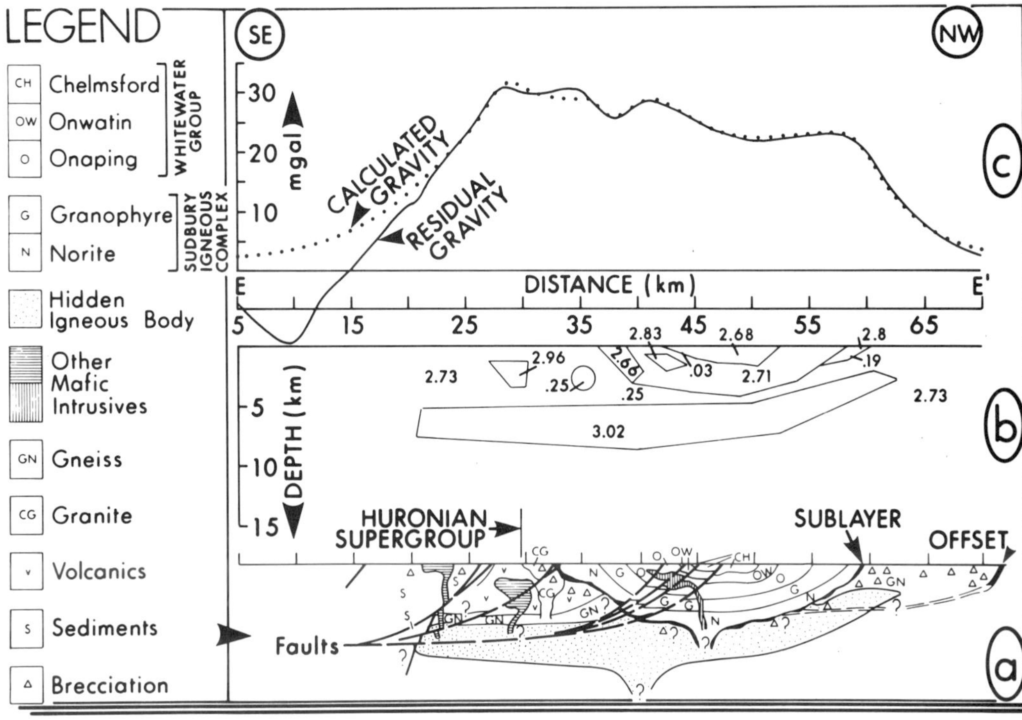

Fig. 4.42. A northwest–southeast section across the Sudbury district showing the residual and calculated gravity profiles, a two-dimensional reconstruction of the model used to calculate the gravity profile, and a geologic interpretation consistent with the reconstruction. (After Gupta et al., 1984.)

4.5.3 Petrology of the Sudbury Igneous Complex

The main units of the Complex include (Fig. 4.43): (1) the Sublayer, (2) the marginal Quartz-rich Norite of the South Range and Mafic Norite of the North Range, (3) the South Range Norite and Felsic Norite, (4) the Quartz Gabbro, and (5) the Granophyre- and Plagioclase-rich Granophyre. All units except the Sublayer are included within the Main Mass of the Complex.

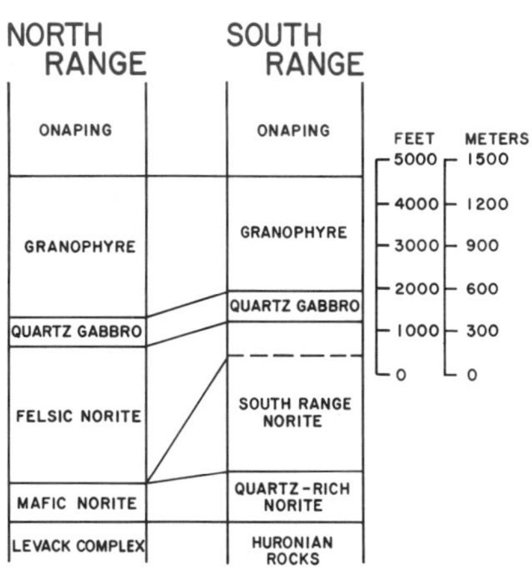

Fig. 4.43. Principal units of the Sudbury Igneous Complex.

4.5.3.1 MAIN MASS

The Felsic (found on the North Range) and South Range Norites are plagioclase–orthopyroxene–clinopyroxene cumulates; they show cryptic variation in the Mg/(Mg + Fe) ratio of the pyroxenes and An content of plagioclase, with these variables changing upward in a manner consistent with fractional crystallization (Naldrett et al., 1970). Orthopyroxene disappears and titaniferous magnetite and apatite appear as cumulus phases in the overlying Quartz Gabbro. The cryptic variation characteristic of the underlying norites has been traced into the lower part of the Quartz Gabbro, but cannot be traced across it because of the intense hydrous alteration that has affected the upper part. The upper part is characterized by a rapid increase in a granophyric intergrowth of plagioclase and quartz at the expense of cumulus phases as the gabbro grades into the overlying Granophyre. Most of the Granophyre is a uniform rock consisting of 75 modal percent granophyric intergrowth and 25 modal percent idiomorphic plagioclase plus clinopyroxene, although zones containing up to 50 modal percent idiomorphic plagioclase are present. Peredery and Naldrett (1975) point to the continuity of modal and compositional trends between the Quartz Gabbro and the Plagioclase-rich Granophyre, suggesting that the bulk of the Granophyre has been intruded subsequently, laterally from the center of the Complex into its present position.

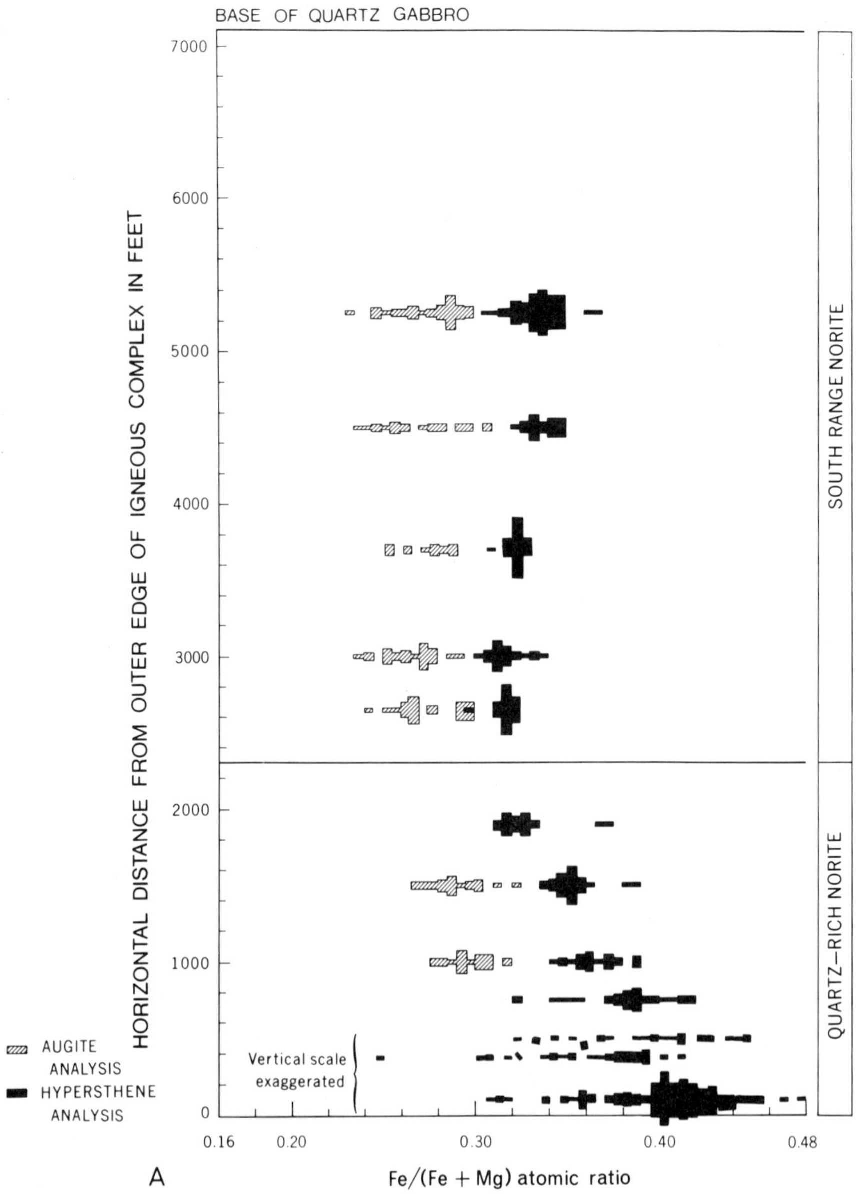

Fig. 4.44. Variation in the FeO/(FeO + MgO) ratio of orthopyroxene and clinopyroxene along a section through the South Range of the Sudbury Nickel Complex. (From Naldrett et al., 1970.)

The marginal unit of the Main Mass on the South Range is the Quartz-rich Norite. In this case, the quartz content increases progressively toward the contact over the outer 300 m (Naldrett et al., 1970). This increase in silica is unlikely to be due to contamination because the increase in quartz occurs as much where the footwall is composed of SiO_2-deficient greenstone as where it is composed of granite. This observation indicates that if contamination is involved, it is not in situ contamination. The increase in quartz is accompanied by an equally progressive decrease in the average Mg/(Mg + Fe) ratio of the pyroxenes (Fig. 4.44) (Naldrett et al., 1970). This decrease is due to progressively more strongly zoned pyroxenes, with the Mg/(Mg + Fe) ratio decreasing toward the edges of grains, whereas the cores retain a constant composition. These observations, coupled with a decrease in grain size toward the margins (Naldrett et al., 1970), indicate that the outer part of the Quartz-rich Norite is a noncumulus rock that crystallized essentially in situ.

4.5.3.2 SUBLAYER

The Sublayer occurs discontinuously around the Complex. It can be divided logically into those variants that occur close to the outer contact and those that occur as dikes that radiate outward from the Complex and are known locally as "offsets."

The Contact Sublayer consists of a suite of fine- to medium-grained norites and gabbros that are distinguished from the Main Mass Felsic Norite and Quartz Gabbro by their lower quartz content in relation to pyroxene (Naldrett et al., 1972).

Sublayer rocks within the offsets generally have the composition of quartz diorite and are referred to as such. Grant and Bite (1984) recognize a number of variants of Quartz Diorite that form a continuum grading from Hypersthene to Biotite Quartz Diorite, a change that is attributed to varying degrees of contamination.

Geologic relationships between the Sublayer and the Main Mass give conflicting evidence on the relative ages of the two units. Inclusions of marginal Quartz-rich Norite have been observed in Sublayer and inclusions of Sublayer have been observed in norite of the Main Mass of the Complex. Intrusive contacts are never marked by fine-grained chill zones, suggesting that whichever unit was the older at any particular location, it was still warm at the time of intrusion of the younger unit. On the North Range, the distinction between Main Mass and Sublayer is consistently clear, with the Sublayer having the finer grain size and lower quartz content. This distinction is not necessarily the case on the South Range, where a number of researchers have commented on gradations between the two units (Slaught, 1951; Cochrane, 1984). It would seem, therefore, that the introduction of the Sublayer and the Main Mass was a complicated process with each preceding the other in different localities.

4.5.3.3 INCLUSIONS IN THE SUBLAYER

In some areas the Sublayer is characterized by inclusions. These can be divided into two groups, those of obviously local derivation, and those composed of mafic and ultramafic rocks, many of which do not outcrop in the Sudbury area. The mafic–ultramafic groups are of particular interest here. Scribbins et al. (1984) described them as ranging from peridotite, through clinopyroxenites and orthopyroxenites, to olivine gabbro and norite. Most of the inclusions either display cataclastic or cumulus textures. Olivines range in composition from Fo86 to Fo72, with the Fo content decreasing as the plagioclase content increases (Fig. 4.45). Scribbins et al. conclude that the inclusions are derived from layered intrusions that have fractionated at moderate depths in the crust.

4.5.4 Geochemistry of the Sudbury Igneous Complex

4.5.4.1 MAJOR ELEMENTS

A selection of average and individual major element analyses of representative material from the Complex are listed in Table 4.2.

4.5.4.1.1 Main Mass

Judging from field and petrographic criteria, the Quartz-rich Norite was close in composition to the SIC magma when it was emplaced along the South Range. The Mg/(Mg + Fe) atomic ratio (henceforth referred to as MgNo) of 0.61 indicates that this is a reasonably primitive rock.

The SIC was intruded into a near-surface environment in a continental setting, thus Naldrett

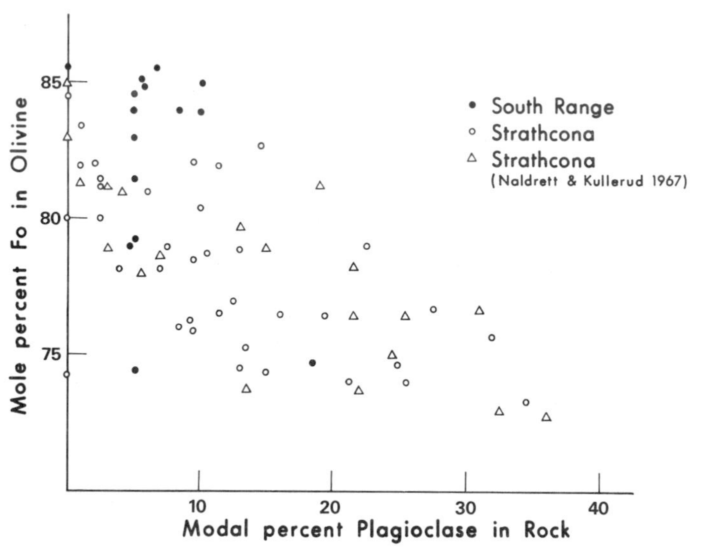

Fig. 4.45. Plot of Fo content of olivine versus the modal content of plagioclase for samples of mafic and ultramafic inclusions in the sublayer. (From Scribbins et al., 1984.)

TABLE 4.2. *Comparison of Major Elements of Rocks of Sudbury Igneous Complex*

	1 Quartz-poor Offset(SR)	2 Typical Offset(SR)	3 Offset(NR)	4 Quartz-rich Norite	5	6 South Range Norite	7 Felsic Norite	8 Quartz Gabbro	9 Granophyre
SiO_2	50.7	55.3	61.5	56.7	57.0	55.4	60.4	58.9	68.1
TiO_2	1.56	0.94	0.73	0.57	1.34	0.49	0.65	1.62	0.65
Al_2O_3	16.3	14.3	14.9	16.3	16.4	17.8	15.6	13.5	13.7
FeO	11.4	10.8	6.84	7.44	7.33	7.35	7.19	10.5	5.21
MnO	0.17	0.15	0.11	0.12	0.13	0.12	0.13	0.14	0.09
MgO	4.91	4.43	3.82	6.10	6.40	6.68	3.81	2.66	1.08
CaO	7.92	6.71	4.70	6.66	7.28	7.75	6.05	5.48	2.15
Na_2O	2.76	2.61	3.15	—	2.41	2.50	3.12	3.13	3.31
K_2O	1.66	1.70	2.60	1.40	1.55	1.10	1.96	2.13	5.00
Total	97.28	96.9	98.4	—	99.84	99.2	98.90	98.06	99.29
MgNo	0.43	0.42	0.50	0.59	0.61	0.62	0.47	0.31	0.27
Na_2O/K_2O wt %	1.76	1.54	1.21	—	1.56	2.27	1.59	1.47	0.66

Explanation to Table 4.2

1. Grant and Bite (in Pye et al. 1984, Table 12.2, No. 4) (average of 4)
2. Grant and Bite (in Pye et al. 1984, Table 12.2, No. 2) (average of 14)
3. Grant and Bite (in Pye et al. 1984, Table 12.2, No. 5) (average of 17)
4. Grant and Bite (in Pye et al. 1984, Table 12.2, No. 6) (average of 31)
5. Collins (1934) P.140 No.92 W.
6. Collins (1934) P.140 No.89 W.
7. Grant and Bite (in Pye et al. 1984, Table 11, No. 9) (average of 31)
8. Grant and Bite (in Pye et al. 1984, Table 11, No. 10) (average of 17)
9. Collins (1934) P.139 No.56 W.

(1984) argued that it is logical to compare its magma composition with that of continental flood basalts. He pointed to the high SiO_2 and K_2O, low CaO and Na_2O/K_2O ratio of the Quartz-rich Norite when it is compared with Keewanawan and Columbia River flood basalts on the basis of MgNo, and showed that contamination of a relatively unfractionated Columbia River basalt with 50 percent of a 1:2 mixture of Archean quartz monzonite and tonalitic gneiss, respectively, gives rise to a major element composition similar to that of the Quartz-rich Norite.

4.5.4.1.2 Sublayer

The texture and field relations of the constituent units of the Sublayer indicate that these units are also not cumulates but are rocks that have solidified essentially in situ, and thus represent magma compositions. MgNos of North Range samples of Contact (as opposed to Offset) Sublayer given by Rao et al. (1983) range from 0.37 to 0.58 and average 0.48 for the North Range and 0.51 for the South Range. Thus the magmas are relatively fractionated, considerably more so than the Main Mass as represented by the Quartz-rich Norite. At the same time, the MgNos indicate a variable degree of fractionation for the Sublayer.

Comparison of the major element compositions (Fig. 4.46) of Sublayer with those of flood basalts indicates that the Sublayer is also enriched in SiO_2 and K_2O, and low in CaO and Na_2O/K_2O ratio for a given degree of fractionation as indicated by the MgNo. These data, therefore, are also consistent with contamination by country rocks.

4.5.4.2 TRACE ELEMENTS

Data on trace elements in the SIC are much less abundant in the literature than are those on major elements. Data exist for REE elements on most units of the Complex, but those of Rao et al. (1983) on the Sublayer are the only modern data on other incompatible trace elements.

In Figure 4.47, the REE pattern by Kuo and Crocket (1979) for the Quartz-rich Norite is compared with those for the two flood basalts samples of similar MgNo. The much steeper slope of the Sudbury sample (La/Yb = 17) as compared with the flood basalts (La/Yb = 3) is obvious.

The range of REE patterns for the Sublayer of the SIC, together with the averages for the North and South Ranges are compared with flood basalts of similar MgNo in Figure 4.48. The Sublayer patterns, with their La/Yb ratios of 16, are essentially identical in slope and magnitude to the average pattern for the Quartz-rich Norite and, again, are very much steeper than those for the flood basalts (La/Yb = 6).

Naldrett (1984) discussed the possibility that the light rare earth element (LREE) enrichment of the Sudbury rocks is attributable to contamination by LREE-enriched country rocks. He concluded that combination of equal parts of a primitive flood basalt and a 1:2 mixture of quartz monzonite and typical tonalitic gneiss, respec-

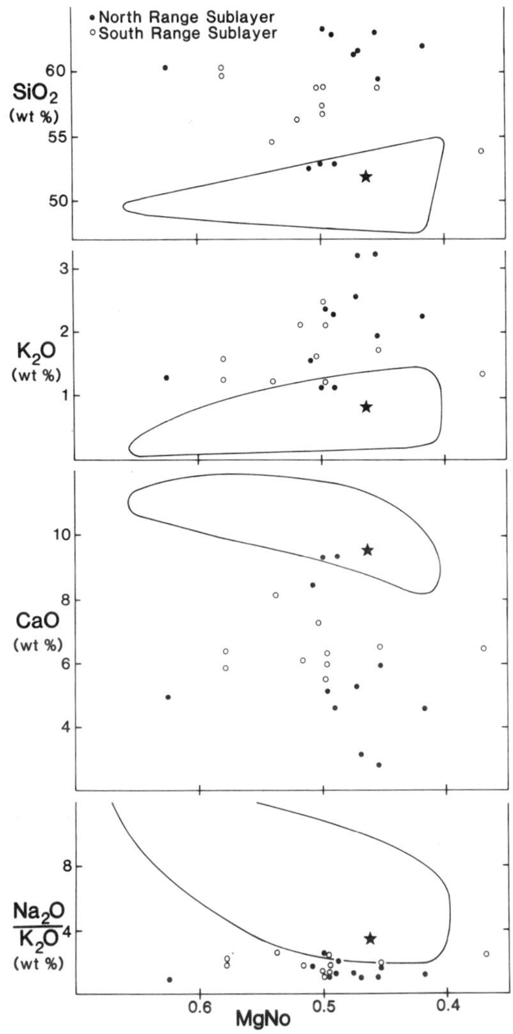

Fig. 4.46. Comparison of some major element concentrations in the sublayer with those in typical continental flood basalts. The solid line encloses the compositions of the flood basalts used as a comparison by Naldrett (1984, Figure 25.3). (After Naldrett, 1984.)

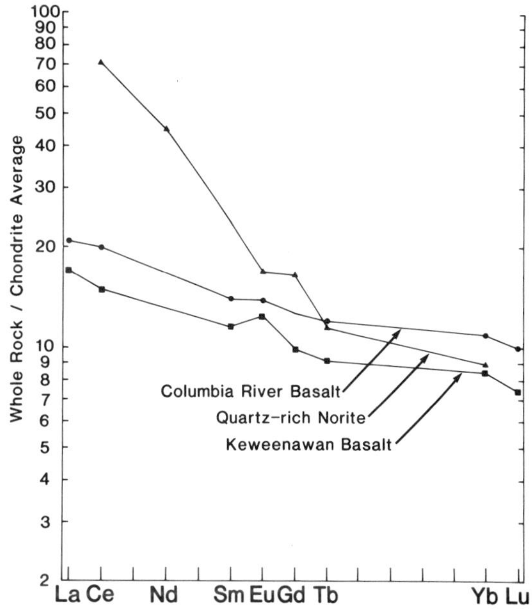

Fig. 4.47. Comparison of REE data for the quartz-rich norite with that for a Keewanawan and a Columbia River basalt of similar MgNo. Data for flood basalts from Basaltic Volcanism Study Project (1981). Data for quartz-rich norite from Kuo and Crocket (1979). (After Naldrett, 1984.)

tively, can account for the REE data. This conclusion is consistent with the close match that this mixture also produces for the major elements. Similarly, with regard to the Sublayer, he showed that the REE profile of one of the least contaminated Contact Sublayer samples can be modeled by mixing 25 percent of the postulated granite-gneiss contaminant with 74 percent of a relatively fractionated Keewanawan basalt. Rao et al. (1983) emphasized that the majority of their Sublayer samples are best modeled as the consequence of the fractionation of the more primitive variants, coupled with assimilation of country rocks. From the previous discussion, particularly the observations on REE profiles, even the more primitive Sublayer material seems to have been

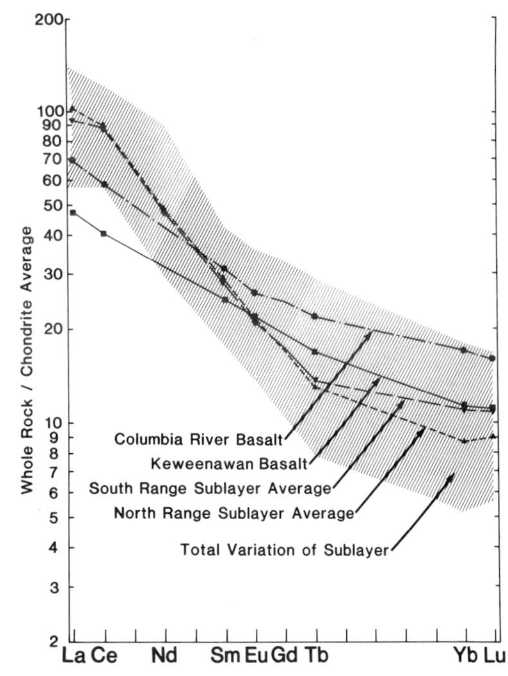

Fig. 4.48. Comparison of REE data for the sublayer with that of relatively fractioned Keewanawan and Columbia River basalts. (Data as for Fig. 4.46).

contaminated to some extent, although this contamination has been less intense than the one that affected the initial magma of the Main Mass before the fractionation process started as described by Rao et al. (1983).

4.5.4.3 ISOTOPES

Krogh et al. (1984) show from lead (Pb)–U radiometric dating of zircons and baddeleyite that the SIC has an age of 1.85 Ga. Using this as a model age, Rao et al. (1984) show from their studies of Nd–Sm and Sr–rubidium (Rb) isotopes that the Sublayer contains less radiogenic Nd and more radiogenic Sr than the bulk earth (Fig. 4.49). This observation is consistent with contamination of these magmas by rocks that had resided in the crust (with its low Sm/Nd and high Rb/Sr) for a considerable period before contamination occurred. Typically, continental flood basalts also have compositions within the southeast quadrant of Figure 4.49, but they normally show a lower degree of crustal interactions than the one shown by the Sublayer.

The Levack gneisses, which have the mineralogy and texture of granulites, plot in the southwest quadrant, which is also typically where lower crustal granulites plot. It is interesting to note that although the Sublayer samples from the South Range plot close to a mixing line between primitive mantle values and the average Huronian crust, the North Range samples plot well to the left of this line. Levack gneiss is not present on the South Range. Rao et al. conclude from this correspondence between local geology and isotopic composition of the Sublayer magmas that these magmas have been modified in composition as a result of interaction with local rocks.

4.5.5 Ore Deposits

The Ni–Cu ore deposits of the Sudbury camp can be divided into four categories: marginal South Range deposits, marginal North Range deposits, Offset deposits, and a miscellaneous group. Detailed descriptions of many of the deposits have been given by Souch et al. (1969), Naldrett and Kullerud (1967), Cowan (1968), Pattison (1979), and by a number of authors in Pye et al. (1984).

The marginal deposits of the South Range, of which the Murray mine is a typical example (Fig. 4.50), are generally zoned from massive ore at the footwall to disseminated sulfide ore toward the hanging-wall. The massive ores rest directly on the footwall rocks and contain inclusions of footwall material as well as fragments of gabbro and peridotite. The host rock of the disseminated ores is sublayer norite, which is in sharp contact with the overlying Quartz-rich Norite of the Main Mass.

The Strathcona deposit (Fig. 4.51) is typical of

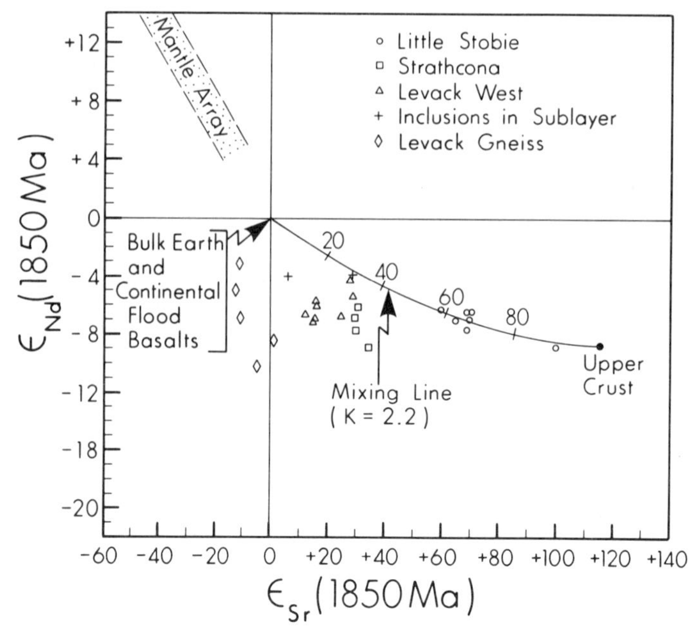

Fig. 4.49. ϵ_{Nd} versus ϵ_{Sr} diagram for samples of mineralized sublayer. The measured values were recalculated to 1.850 Ga, the age of the Sudbury Igneous Complex (Krogh et al., 1984). Mixing curves calculated between the bulk earth and the upper crust are shown. The upper crustal composition (at 1.85 Ga) was calculated using the data from McCullogh and Wasserburg (1978) and Taylor (1964). The Sr and Nd contents of the initial magma were assumed to be 400 and 15 ppm, respectively. The tick marks on the mixing curves indicate the proportion of contaminant in the mixture. (After Naldrett et al., 1986b.)

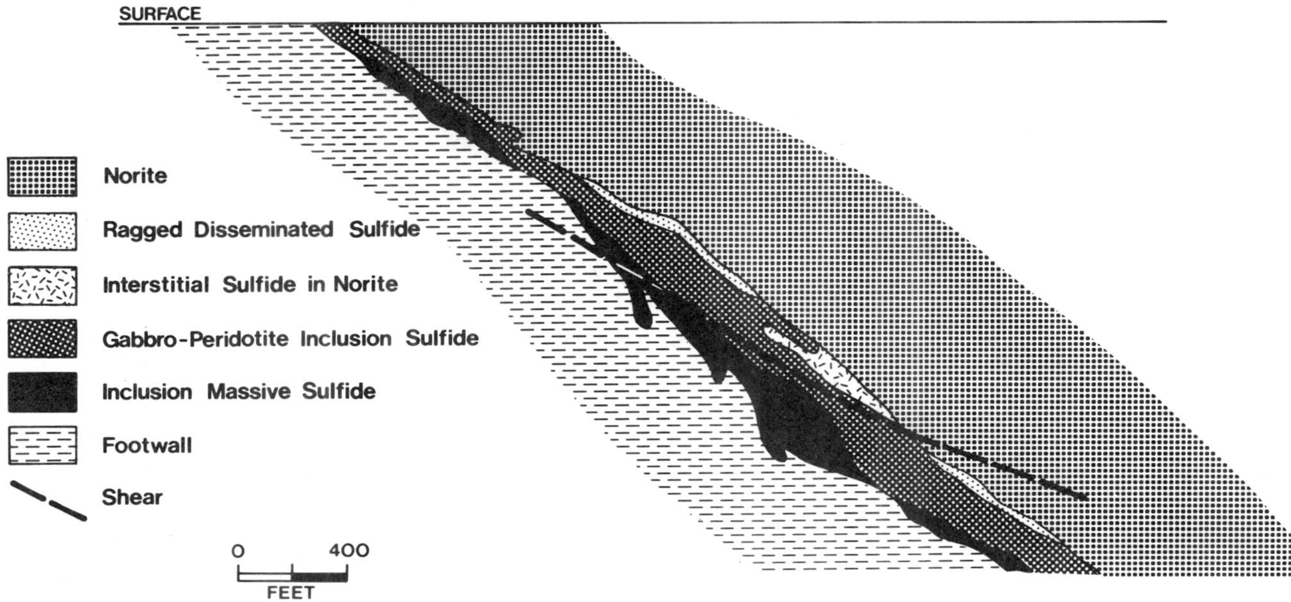

Fig. 4.50. Generalized section through the Murray Mine, Sudbury, Ontario, looking west. (After Souch et al., 1969.)

Fig. 4.51. 2625 ft Level plan of the Strathcona Mine. (After Coats and Snajdr, 1984.)

the North Range marginal deposits. In these deposits, the mineralization occurs primarily within brecciated country rocks at the basal contact of the SIC and in fractures in country rock underlying the breccias. The footwall breccias consist of fragments of country rock, ultramafic inclusions, and rare norite in a quartzo–feldspathic matrix. It has been postulated that the breccia resulted from and lined the base of the impact crater, although postore brecciation can be demonstrated in some places. The sulfides occur (1) as fine and blebby disseminations and as massive stringers within the footwall breccias, (2) as stringers in the footwall fractures, and (3) more rarely, as disseminations within overlying sublayer norite (hangingwall ore). An uncommon Cu-rich ore type consisting of nearly massive chalcopyrite enclosing a small percentage of pentlandite has recently been discovered at the Strathcona mine (Abel et al., 1979; Abel, 1981). This Cu ore occurs in fractures within the footwall gneiss up to several 100 m from the basal contact of the Complex.

The offset deposits occur in the dikelike offsets of sublayer norite and gabbros that extend several kilometers away from the Complex into the footwall. In many cases, the sulfides form steeply plunging, lenslike pods of massive and interstitial disseminated ore associated with high proportions of inclusions in the offset dikes. The Frood–Stobie, the largest offset deposit, is unusual in that it lies parallel to the southern contact of the Irruptive (Fig. 4.52). The offset resembles a dike in plan and a downward-pointing wedge in cross section, and dips steeply to the north. Massive, inclusion-bearing sulfide ore is concentrated at the margins of the wedge and toward the base of the deposit, whereas the upper part of the ore body consists of disseminated sulfides in inclusion-bearing sublayer norite.

The Falconbridge deposit belongs to the miscellaneous class, and is unusual in that it is localized along a fault that parallels the contact of the Complex. The fault forms the contact between rocks of the Complex and the country-rock greenstones over much of its length, although to the west it enters and dies out within the Complex and to the east it enters the greenstones.

4.5.6 Relationship of Ore Deposits to the Rocks of the Complex

Although, as outlined above, there is considerable variation in the characteristics of different ore deposits, there are a number of features common to all of them. These include:

1. *Embayments or other irregularities at the base of the SIC.* An increase in sulfide content is typically observed at the lower contact throughout the Complex, but it is where irregularities exist that the zone of sulfide thickens and increases in intensity sufficiently to form ore.
2. *The presence of the Sublayer.* The spatial relationship of ore to Sublayer rocks is such that the sulfides constituting the ore bodies appear

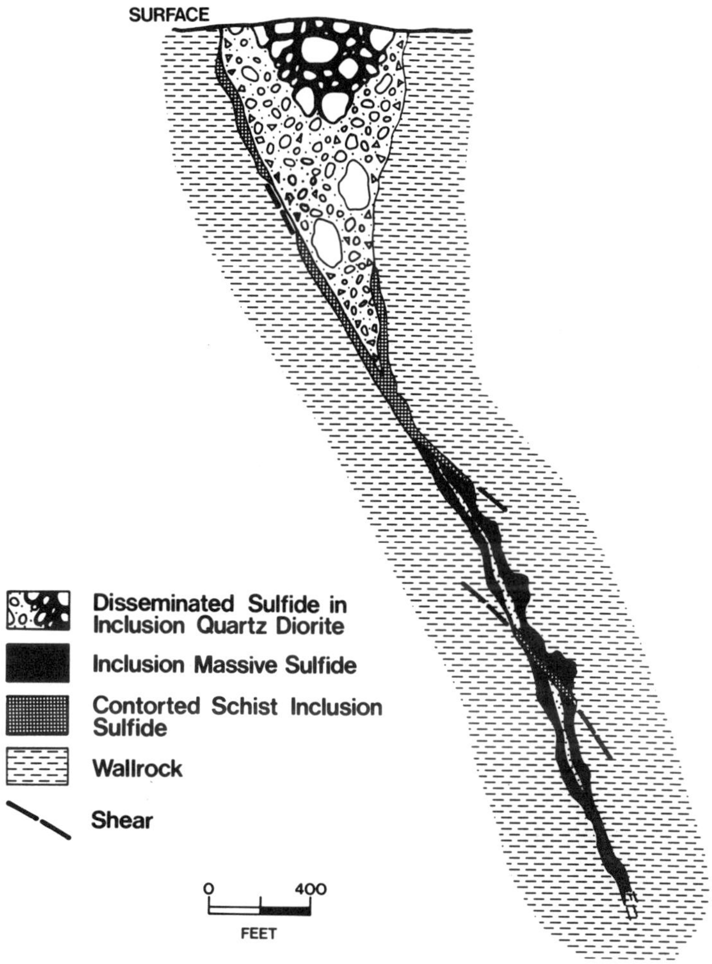

Fig. 4.52. Generalized section through the Frood ore body, looking southwest. (After Souch et al., 1969.)

to have settled out of bodies of the Sublayer. Sulfides also occur within Quartz-rich and Mafic Norities; but, except where they have been moved from their original position by faulting, ore bodies are invariably associated with sublayer rocks.

3. *Ultramafic inclusions within Sublayer host-rocks.* Some of the Sublayer is mineralized, other varieties seem to be devoid of significant sulfide. There is little to distinguish mineralized from unmineralized Sublayer as far as texture, mineralogy, and chemical composition of the rocks are concerned, except for the obvious presence or absence of sulfides. All types of Sublayer contain inclusions, but both in the contact and in the offset environments, the mineralization is associated only with Sublayer host-rocks that carry members of a suite of mafic and ultramafic inclusions that Scribbins et al. (1984) conclude are derived from one or more deep-seated layered intrusions.

4.5.6.1 RELATIONSHIP BETWEEN INCLUSIONS, SUBLAYER, AND MAIN MASS OF THE SIC

As discussed previously, the REE profiles of the Sudbury rocks differ from those of continental flood basalts and suggest that contamination by country rocks has occurred on a large scale. The similarity in the profiles of the Sublayer and the Main Mass rocks suggests that the source magmas have a common heritage. REE profiles for inclusions have the same steep slopes (Fig. 4.53), although the abundances are lower. Thus, there is some evidence that the inclusions were derived from a magma that had suffered contamination similar to the remainder of the SIC rocks.

Using the relationship of Roeder and Emslie (1970) between the Mg/Fe ratio of olivine and the basaltic liquid in equilibrium with the olivine, Naldrett (1984) shows that the MgNos of liquids in equilibrium with the olivine of the inclusions

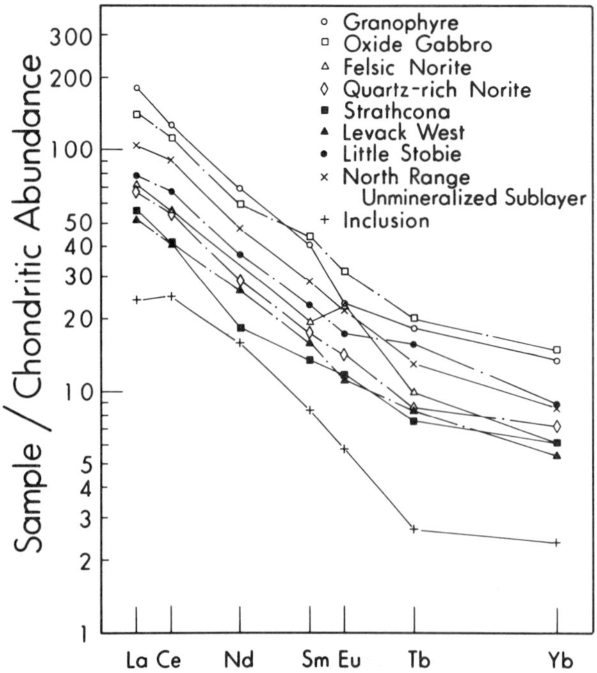

Fig. 4.53. Comparison of REE data for a peridotite inclusion from the sublayer with similar data for main mass and sublayer rocks. (Data from Rao et al., 1984.)

span the range of MgNos exhibited by the Sublayer samples. He argues that the range of olivine compositions observed in the inclusions is precisely the one that would be expected in cumulus rocks crystallizing from a magma evolving along a compositional trend such as that recorded within the Sublayer.

Thus, two lines of evidence (REE profiles and olivine compositions) suggest that inclusions, Sublayer, and Main Mass norite are all genetically related.

4.5.6.2 SEGREGATION OF SULFIDES

The SIC differs from other layered complexes in a number of significant ways; these include:

1. Evidence that the area into which it was intruded had been involved in a catastrophic explosion, probably the consequence of the impact of a meteorite in the view of many workers.
2. Rocks of the Complex are very siliceous in comparison with other intrusions judged on the basis of the Mg/(Mg + Fe) ratios of their pyroxenes; this observation implies that the source magma was unusually siliceous for its state of fractionation.
3. Other compositional data (major and trace elements and isotopic analyses) indicate that the high SiO_2 content is the consequence of country-rock assimilation.
4. The presence of an unusually large number of occurrences of very concentrated sulfide mineralization.

Irvine (1975) pointed out that assimilation of SiO_2-rich material by a mafic magma could lower the solubility of S within it, and suggested that this effect had occurred at Sudbury. Naldrett and MacDonald (1980) used the 1,200°C isothermal section of the FeO–FeS–SiO_2 system, reproduced here as Figure 4.54, to illustrate the principle. Composition A lies in the field of homogenous liquid, but addition of SiO_2 to change its composition to B moves it into a two-liquid field, and

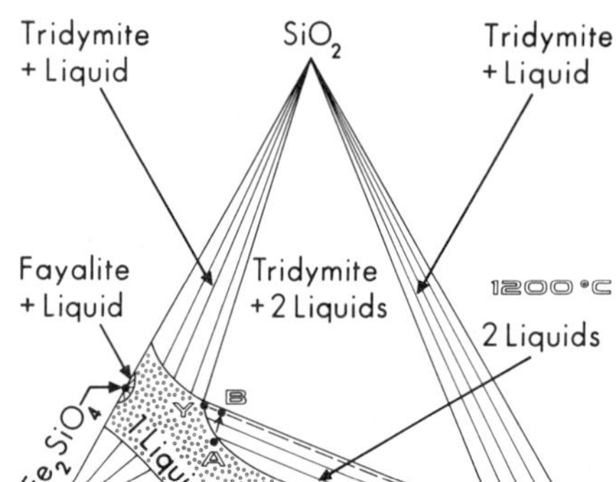

Fig. 4.54. The 1,200°C isotherm of the Fe–S–O system illustrating how the addition of SiO_2 to a homogeneous, sulfide-rich liquid of composition A will change its composition to point B, where it will consist of two liquids, one silicate-rich (Y) and the other sulfide-rich (X). (After Irvine, 1975.)

results in the segregation of two immiscible liquids, one silicate-rich (Y) and the other sulfide-rich (X). The SiO_2-rich nature of the SIC lends support to this hypothesis. Evidence indicating that extensive country-rock assimilation has occurred points to how the SiO_2-rich composition has been achieved. The highly fractured ground underlying an explosion site, coupled with the heating and melting resulting from the explosion would have provided an ideal environment for country-rock assimilation.

4.5.7 A Model for the Sudbury Igneous Complex

It is suggested that a catastrophic explosion occurred in the Sudbury region 1.85 Ga ago (this author believes that the weight of available evidence points to an extraterrestrial impact, although impact is not essential to the development of this model). The explosion gave rise to the Onaping Formation, fractured the underlying crust, and triggered the ascent of a magma similar in its initial composition to continental flood basalt. Magma probably rose through a centralized plumbing system in the lower crust, but some of it spread out through numerous subsidiary channels as it entered the highly fractured ground beneath the crater (Fig. 4.55). It is suggested that these hidden, flanking intrusions account in part for the buried mass of mafic and ultramafic rock that the analysis of gravity and magnetic data by Gupta et al. (1984) indicate underlies and extends beyond the present Sudbury structure.

On its initial ascent, the magma fractionated to some degree and also became contaminated. Batches of magma that infiltrated flanking subsidiary channels cooled, became contaminated further through reaction with their surroundings, and fractionated further, giving rise to olivine and pyroxene-rich cumulates. The contamination, coupled with cooling, caused sulfides to precipitate at an early stage; these sulfides collected as massive bodies beneath the cumulates in the flanking chambers of the plumbing system. The main body of silicate magma, contained within a large, vertically oriented chamber, lost heat less quickly than in the flanking chambers, and thus

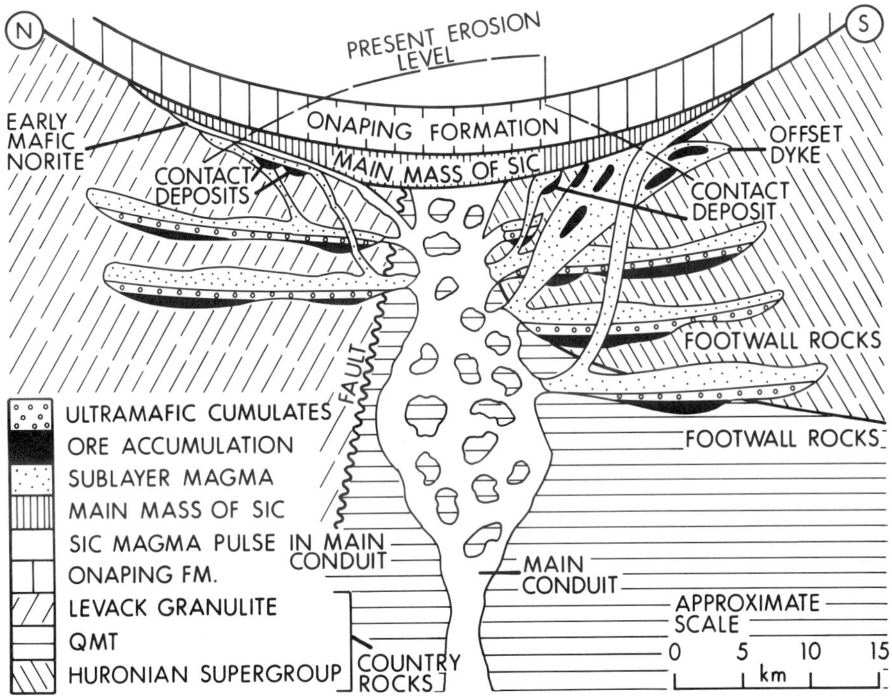

Fig. 4.55. Schematic cross section illustrating the emplacement of the SIC. Magma rises up a vertical conduit into a highly fractured zone beneath an explosion (possibly impact) crater. The central body of the magma cools somewhat slowly, assimilating much of the fractured country rocks. Subsequently it is emplaced into the crater, beneath the low density Onaping Formation, in response to structural readjustments in the area. It is probable that this is a multistage process. Magma has also spread out as a series of sills peripheral to the central conduit. Assimilation also occurs within these, giving rise to an imprint of localized contamination. The magma of the sills deposits sulfide and fractionates, giving rise to olivine and pyroxene-rich cumulates. Fractionated magma from the flanking sills also works its way up to the crater floor, intruding to form the sublayer of the contact and offsets. Where magma from a lower sill has cut and disrupted an overlying one, it has picked up sulfides and inclusions. As it rises to the base of the Complex, it carries these to the sites of the ore deposits. (After Naldrett et al., 1986b.)

fractionated less rapidly; because it cooled more slowly, it reacted with country rocks to an even greater extent than in the flanking bodies.

Successive injections of magma occurred along the unconformity between the crater floor and the overlying, relatively unconsolidated (and thus low density) Onaping Formation to give rise to the Main Mass of the Complex.

The faster cooling, and thus more fractionated (consistent with the MgNos) magma of the flanking bodies was also squeezed up to intersect the crater floor, where it spread out and exploited "troughs" and "embayments" in the floor filled with Footwall Breccia; the magma also followed fractures in the basement rocks to form the offset dikes. This magma is thus responsible for the Sublayer. In some instances (Fig. 4.55), as the magma rose up through the basement, it intersected and disrupted overlying bodies, picking up sulfides and ultramafic cumulates in the process, thus accounting for the very important association between the mineralization and the xenoliths. Other batches of magma did not intersect overlying bodies and were emplaced carrying only country-rock inclusions, thus giving rise to the unmineralized sublayer. The derivation of sublayer intrusions from a number of bodies following parallel, but distinct, evolutionary trends accounts for the greater variability of the Sublayer in MgNo and Sr initial ratio in comparison with the Main Mass. Probably some of the Offset dikes represent channel-ways along which ore-bearing magma moved upward, to spread out along the floor of the crater, although in other dikes, emplacement has been from the top downward.

5. PLATINUM GROUP ELEMENT DEPOSITS

Despite the fact that platinum group elements (PGE) are found concentrated in a variety of geologic settings, ranging from high temperature magmatic to late diagenetic or sedimentary (Kucha, 1982), the only deposits from which substantial quantities of PGE are produced are those associated with mafic or ultramafic rocks. Identified resources within these are summarized in Table 2.2.

The Merensky Reef and the Upper Group-2 (UG-2) chromitite of the Bushveld and the J–M Reef of the Stillwater Complex are stratiform mineralized intervals in very similar layered complexes, which together account for 56 percent of the total reserves of Pd, 69 percent of Pt, and 80 percent of Rh, and form the main focus of this chapter.

In this chapter, cumulate rocks are referred to in two ways, to be consistent with the terminology normally used in the literature. Most papers on the Bushveld Complex use standard petrologic names, such as dunite for an olivine cumulate, harzburgite for an olivine–bronzite cumulate or an olivine cumulate with abundant intercumulus bronzite, bronzitite for a bronzite cumulate, and norite for a plagioclase bronzite cumulate. On the other hand, most recent papers on the Stillwater Complex use a notation in which oC indicates olivine cumulate, ocC olivine chromite cumulate, bC bronzite cumulate, pbC plagioclase bronzite cumulate, and other similar combinations, in which the most common cumulate mineral is indicated first, the second most abundant second, and the capital letter C indicates a cumulate rock. This notation is also used in the following.

5.1 THE BUSHVELD COMPLEX

5.1.1 Stratigraphy

The Bushveld Complex is divided into four zones (Fig. 5.1). The lower zone is composed mainly of bronzitites, harzburgites, and dunites (Cameron, 1978). It is overlain by the critical zone, which hosts the economically interesting LG-6 chromitite layer, and the PGE-bearing Merensky Reef and UG-2 chromitite (Cameron, 1980, 1982). The critical zone is overlain, in turn, by the norites, gabbros, and anorthosites of the main zone, which are themselves capped by the ferrogabbros and ferrodiorites of the upper zone (von Gruenewaldt, 1973; Molyneaux, 1974).

The variation in the MgNo of bronzite and olivine (Fig. 5.1) indicates an overall upward Fe-enrichment in the Complex, consistent with fractional crystallization, although in detail there are numerous reversals and large thicknesses across which no systematic variation is present.

The critical zone (Fig. 5.2) is divided into two parts (Cameron, 1980, 1982). The lower critical zone consists primarily of bronzitites, chromitites, and some harzburgites and dunites. The upper critical zone, marked by the incoming of cumulus plagioclase, can itself be subdivided into two parts: (1) a lower part consisting of anorthosites, norites, and minor bronzites, which occur in no systematic order and define no classic cyclic units; and (2) an upper part (starting with base of the UG-1 chromitite unit) that, in contrast, comprises units made up of a regular succession of rock types, consisting of some or all of chromitite, harzburgite, bronzitite, norite, and anorthosite, in most cases in this order.

The different zones and units vary greatly in thickness throughout the Complex, and are absent in some areas. The lower zones vary in thickness due in part to their transgressive nature and appear to have had the most restricted development. In many places, the critical zone transgresses over a wider area of the Bushveld floor than the lower zone, and the main zone a still wider area. Von Gruenewaldt (1979) has attributed the transgression to the progressive addition of the new magma as the Complex crystallized.

5.1.2 Mineralization

PGE mineralization occurs in four distinct settings in the Complex: (1) the Merensky Reef, (2) the UG-2 chromitite layer, (3) the Platreef of the Potgietersrus area, and (4) the discordant dunite pipes of Onverwacht, Mooihoek, Driekop, and Maandagshoek. Mining of the pipes ceased in the late 1920s and early 1930s; production was insignificant in comparison with the reserves for the other three environments.

5.1.2.1 SETTING

The geology of the Bushveld Complex is shown in Figure 5.3. Apart from the Platreef, which lies at

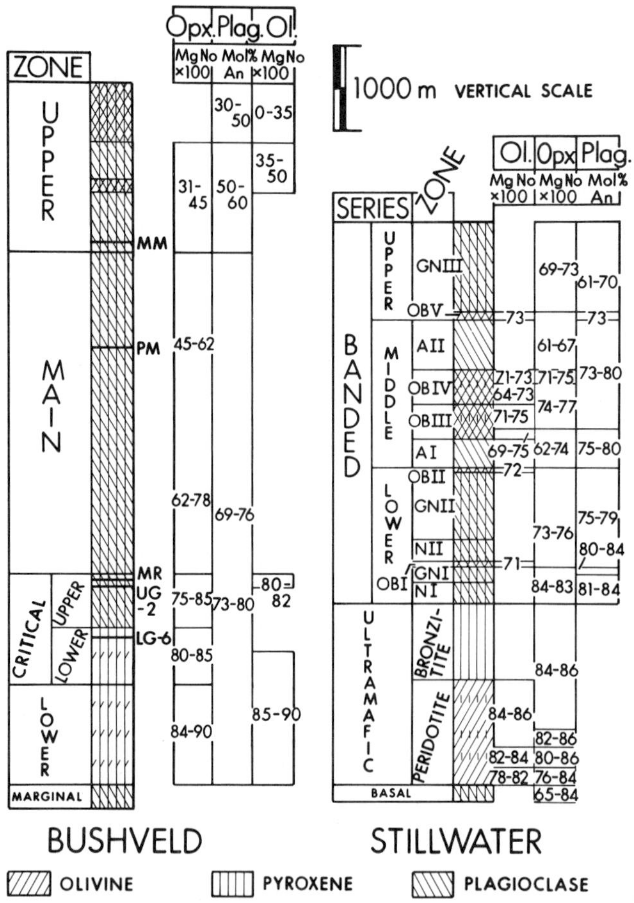

Fig. 5.1. Generalized stratigraphy of the Bushveld and Stillwater Complexes, showing compositional variation in olivine, bronzite, and plagioclase. MM = main magnetitite layer; PM = pyroxenite marker horizon; MR = Merensky Reef; UG-2 = UG-2 chromite layers; LG-6 = LG-6 chromitite layer. Data from Cameron (1978, 1980, 1982), Naldrett et al. (1986), Von Gruenewaldt et al. (1985), Wager and Brown (1968). (From Naldrett et al., 1987.)

the base of the Complex in the northernmost limb, the principal PGE mineralization in the Bushveld Complex occurs in the upper part of the upper critical zone. Five or six well-defined cyclic units occur in this part of the stratigraphy (Fig. 5.4); these are the UG-1, UG-2, UG-3 (in some areas only), Pseudoreef (in some areas only), Merensky Reef, and Bastard Reef. The units are composed of some of all of the following rock types, in most cases in this order; chromitite, harzburgite, bronzitite, norite, and anorthosite. Concentrations of PGE are associated with the lower parts of most of the cyclic units. Von Gruenewaldt et al. (1986) report values of 3.43, 9.64, and 4.71 ppm (total PGE + Au) for the UG-1, UG-2, and UG-3 chromitites, respectively, in the Maandagshoek area of the eastern Bushveld. PGE-rich sulfides occur scattered near the base of the Pseudoreef unit (see following section), the Merensky Reef is reported by Buchanan and Nolan (1979) to grade 8.1 ppm total PGE, and PGE are reported in more modest concentrations at the base of the Bastard unit (see following section).

Where the norites are parts of cyclic units, they are mostly transitions from bronzitite to anorthosite. They do not represent cotectic norites, which should consist of a nearly constant proportion of about ⅓ bronzite to ⅔ plagioclase. The principal norites that have cotectic compositions are the large thicknesses that intervene between the Boulder and UG-2 units at Rustenburg and between the Merensky and UG-3 units at Jagdlust.

Bronzite in most of the cyclic units shows an overall iron enrichment trend (Fig. 5.5). At Union, however, where individual rock sequences are thicker and variations within them are more easily seen, the upward iron enrichment is much stronger in the norites than it is in the bronzite cumulates.

Both in Figure 5.5 and in the diagrams of Kruger and Marsh (1985), plagioclase shows marked zoning in the bronzite cumulates, consistent with an orthocumulate origin for these rocks. In the plagioclase cumulates of the Merensky unit, plagioclase shows relatively little upward albite enrichment, and only slightly more upward enrichment across the anorthosites of the Bastard unit.

Naldrett et al. (1984) and Kruger and Marsh (1985) have shown that the Sr content of plagio-

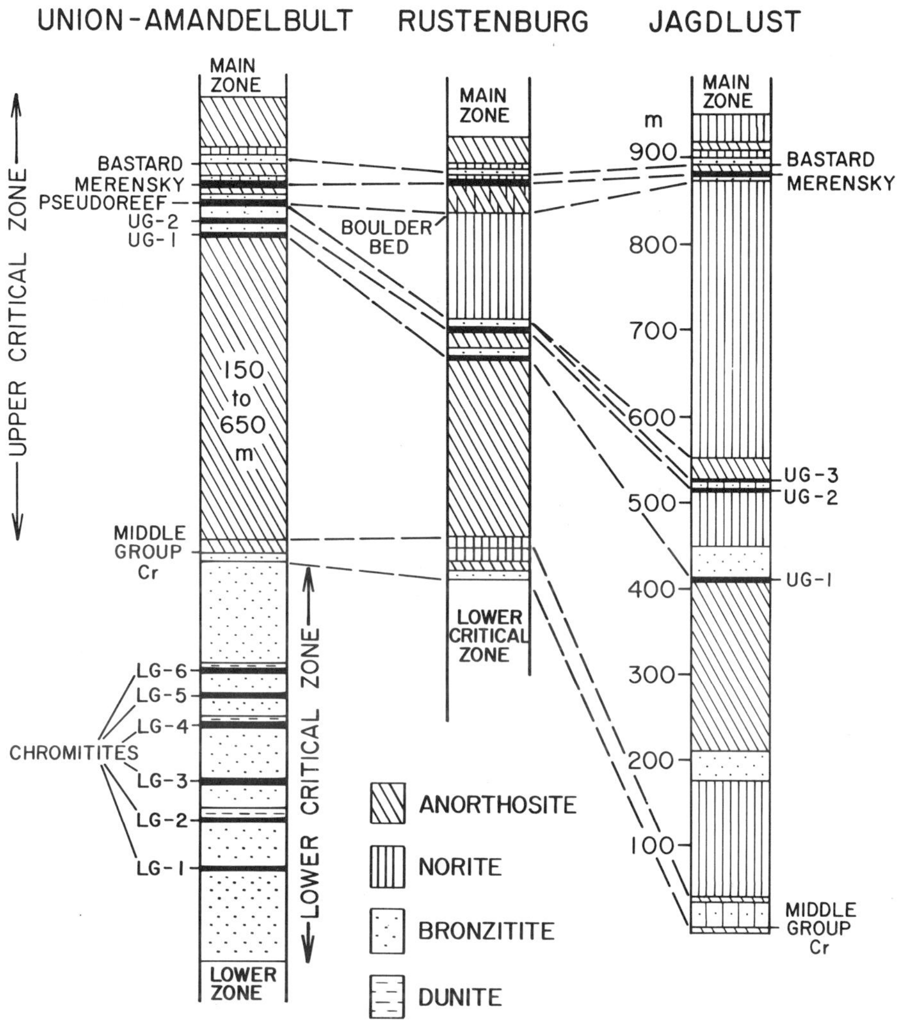

Fig. 5.2. Stratigraphy of the critical zone with special reference to the upper critical zone in the northwestern (Union-Amandelbult), southwestern (Rustenburg), and eastern (Jagdlust) portions of the Bushveld Complex. Including data from Cameron (1980, 1982), von Gruenewaldt et al. (1985). (From Naldrett et al., 1987)

clase in the cyclic units shows some systematic changes. In the profiles from Union and Rustenburg (Fig. 5.6), there is relatively little change below the Merensky Reef, and plagioclase contains close to 500 ppm Sr. Even the lowermost norites of the Merensky unit have substantially less Sr than the rocks below them, and there is a decrease in Sr across the norites and the overlying anorthosites of this unit. This decrease continues in the norites of the overlying Bastard unit, although in the anorthosites of the Bastard the Sr contents are relatively constant.

In the eastern Bushveld, the major decrease in the Sr content of plagioclase occurs in the anorthosite of the Bastard unit and not in the Merensky unit. In the western Bushveld the decrease in Sr content in plagioclase, which is proportional to the Sr content of the magma from which the plagioclase is crystallizing, appears to coincide with a major change in Sr isotopic composition of the rocks of the Complex. Figure 5.7 incorporates data from Hamilton (1977) (from throughout the Bushveld) and from Kruger and Marsh (1982) (from the western Bushveld). The change in isotopic composition occurs within the norites and anorthosites of the Merensky cyclic unit, precisely where changes occur in the Sr content of the plagioclase and where the bronzite shows strong upward Fe-enrichment.

Sharpe's (1985) data for the eastern Bushveld Complex indicate that there is a similar jump at approximately the same position. Sharpe's analyses, however, are not spaced sufficiently close to indicate whether this occurs in the Merensky unit, or just above in the anorthosite of the Bastard unit, where it has been shown (see previous dis-

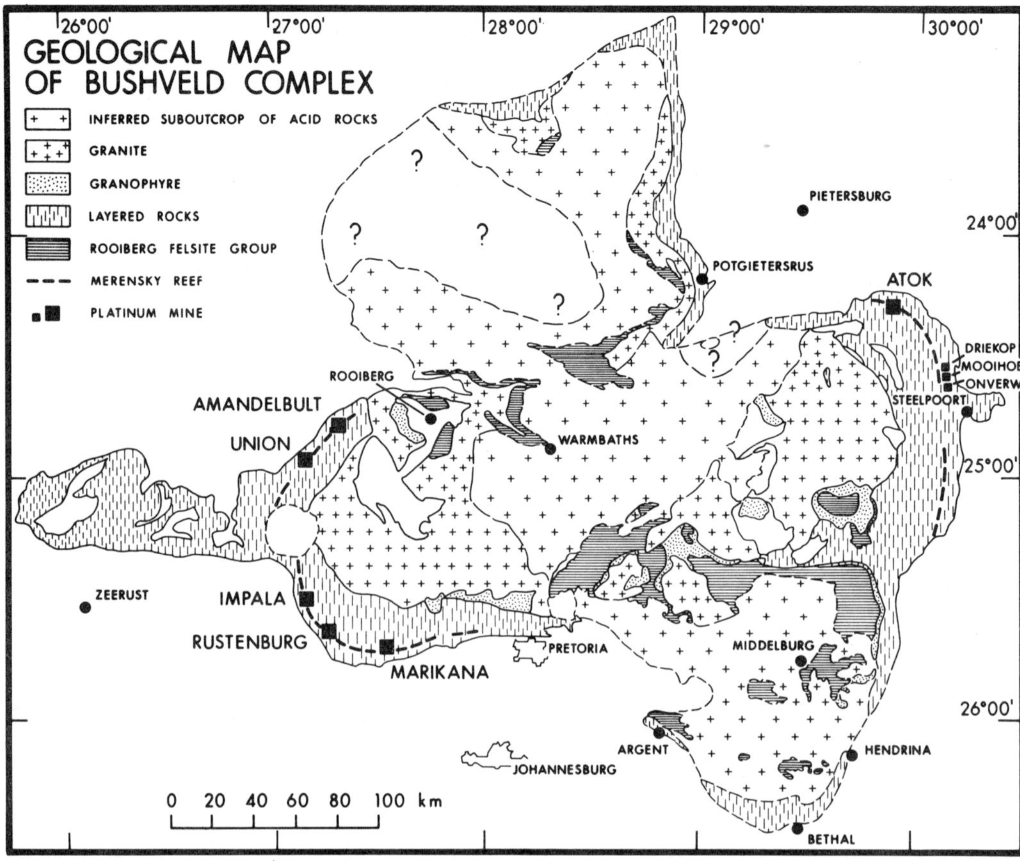

Fig. 5.3. Generalized suboutcrop geologic map of the Bushveld Complex, showing the main PGE mines in the area. The farm Jagdlust is very close to the Atok mine.

cussion) that the Sr content of the plagioclase changes abruptly. It is likely that, as in the west, the isotopic and bulk Sr changes occur at the same horizon.

To understand the development of the cyclic units, Naldrett et al. (1987) attempted to model variations in the MgNo of bronzite and the An and Sr contents of plagioclase in the Merensky and Bastard units. They found that the rate of upward change in these compositional variables was inconsistent with the different minerals (or even the different elements in any single mineral) coming from a single body of magma and concluded that fractional crystallization cannot, by itself, account for the observed compositional variation.

A number of authors, including Kruger and Marsh (1982, 1985) and Sharpe (1985) have attributed the isotopic shift to an influx of isotopically distinct magma. Naldrett et al. (1987) conclude that the mixing of two somewhat dissimilar magmas can also account for the observed major and trace element compositional variations.

Both Kruger and Marsh (1985) and Sharpe (1985) have speculated that the presence of the Merensky Reef is related to the influx of fresh magma, although Naldrett et al. (1987) point out that if the change in isotopic composition is coincident with the change in Sr content in the eastern Bushveld, it implies that, in the east, the major influence of the new influx on the cumulate record was felt during formation of the Bastard unit, not the Merensky unit. They raise the question as to whether the onset of crystallization of this isotopically distinct magma is as important to the origin of the Merensky Reef as has been suggested.

5.1.2.2 MERENSKY REEF

In the Rustenberg area of the western Bushveld Complex, the ore zone comprising the Merensky Reef is confined to a pegmatoid that occurs at the base of the Merensky cyclic unit (Fig. 5.8). PGE are concentrated in the vicinity of two chromite layers that occur at the top and the bottom of the pegmatoid, and are particularly concentrated near the upper chromite layer.

In the Union–Amandelbult area, the ore is also associated with chromite layers within pegmatoid, but here olivine is an important component of the pegmatoid, particularly toward its base.

In the Marikana area, east of Rustenburg Mine,

the pegmatoid is less common, occurring as isolated patches within porphyritic bronzitite, and the chromite layers are much wider apart. The upper chromite occurs approximately 50 cm below the contact of the bronzitite with the overlying anorthosite, whereas the lower chromite occurs at the base of the bronzitite (Fig. 5.8). The most important metal values are associated with the upper chromite layer.

The Merensky Reef is also different in the vicinity of the Atok Mine near Jagdlust in the eastern section of the Complex. Here a pegmatoidal pyroxenite is present within bronzitite, but the chromite layers and PGE values occur above the pegmatoid, within overlying bronzitite (Fig. 5.8).

In the Rustenburg area the Merensky pegmatoid consists of coarse bronzite crystals (1–4 cm in length) enclosed in a matrix of plagioclase. Biotite, apatite, and tremolite are common constituents. The lower bounding chromite layer varies from a few millimeters to a few centimeters in thickness. The upper chromite layer is only a few millimeters thick and is discontinuously developed in some areas. In many places some pegmatoid extends above the upper chromite layer into the overlying bronzite. Elsewhere medium-grained, nonpegmatoidal bronzitite masses occur enclosed within the reef pegmatoid.

Naldrett et al. (1986) pointed out that the composition of the pegmatoidal reef bronzite is more magnesian than the bronzite immediately overlying the top chromite layer of the pegmatoid.

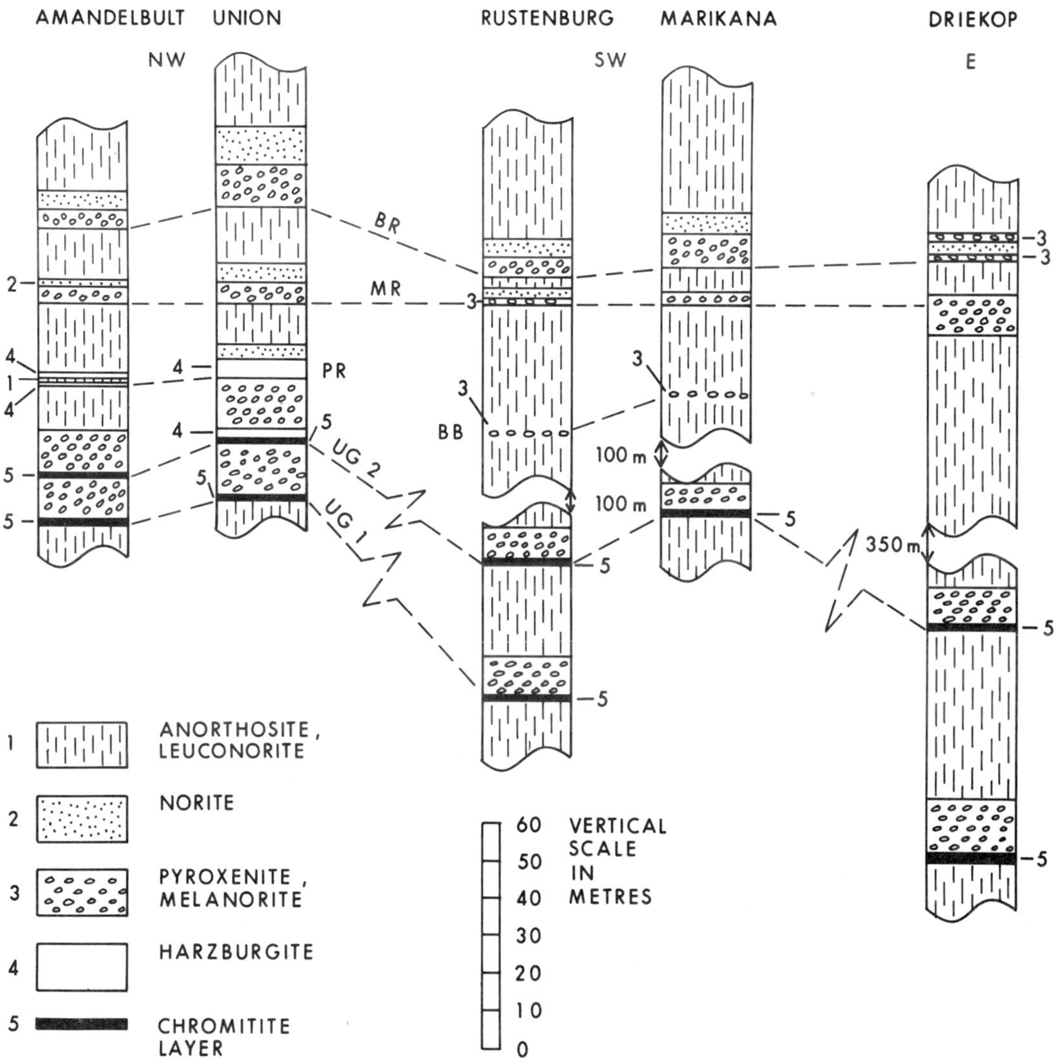

Fig. 5.4. Cyclic units forming the upper part of the upper critical zone of the Bushveld Complex. Data from Vermaak, 1976a; Farquhar, 1985; and Gain, 1980.) BR, MR, PR, and BB refer to Bastard Reef, Merensky Reef, Pseudoreef, and Boulder Bed, respectively. (From Naldrett, 1981b)

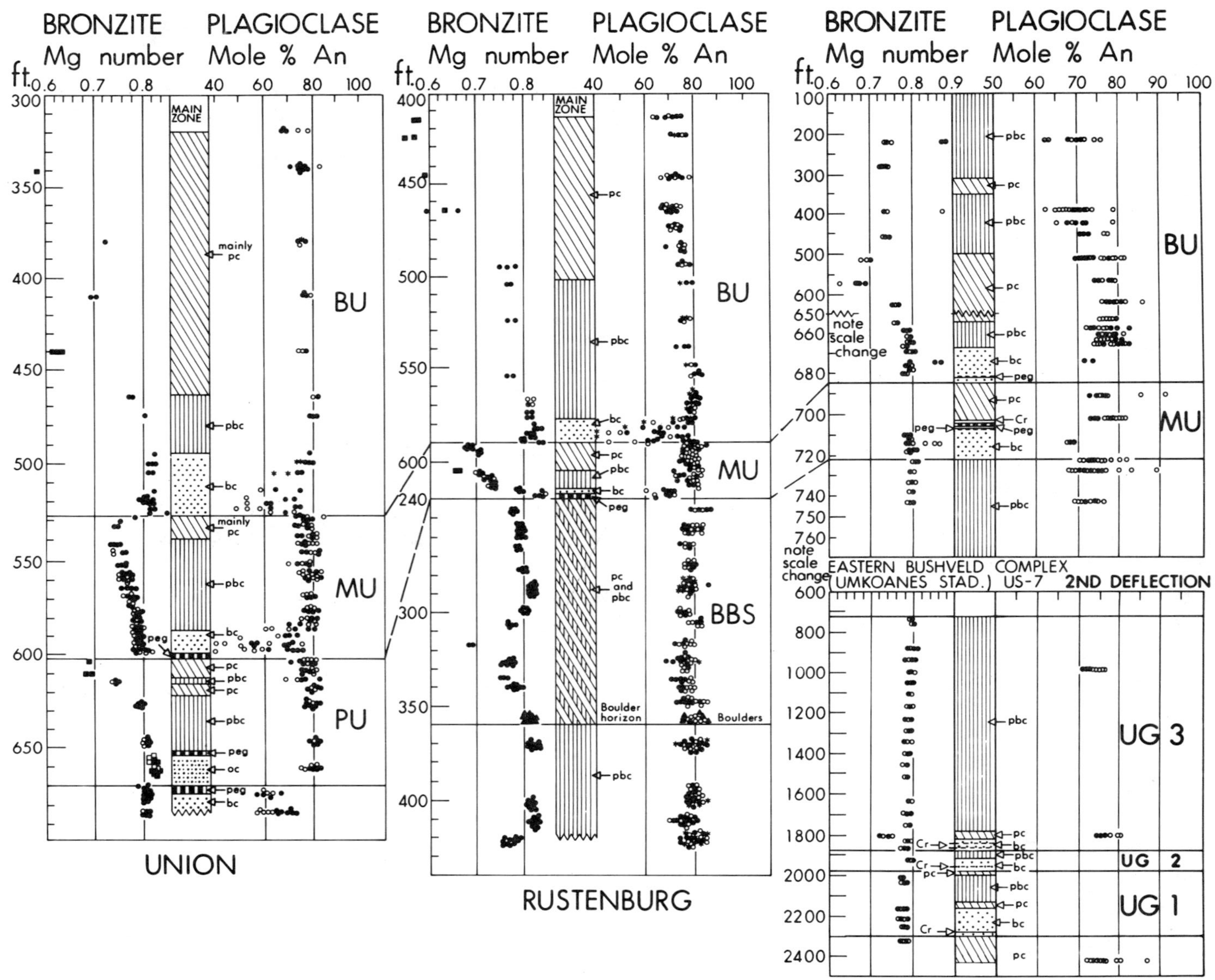

BRONZITE
Mg number
PLAGIOCLASE
Mole % An
ft.
MAIN ZONE
BU
MU
PU
BBS
mainly pc
pbc
bc
pc
peg
oc
pc and pbc
Boulder horizon
Boulders
Cr
note scale change
UG 3
UG 2
UG 1
UNION
RUSTENBURG
EASTERN BUSHVELD COMPLEX
(UMKOANES STAD.) US-7
2ND DEFLECTION

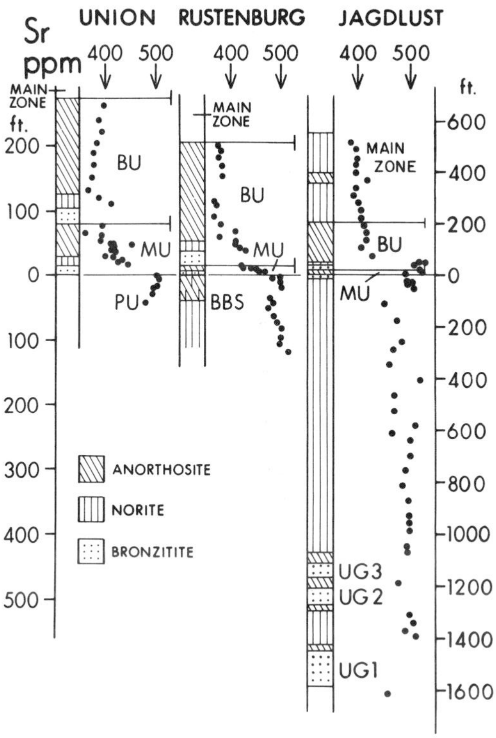

Fig. 5.6. Change in Sr content of plagioclase over the Union, Rustenburg, and Jagdlust sections. PU = Pseudoreef unit; MU = Merensky unit; BBS = Boulder bed; BU = Bastard unit. Data from Naldrett et al. (1986a) and Cameron (1988). (From Naldrett et al., 1987.)

They used this as supporting evidence for their argument that the Reef was derived from magma more primitive than the one that gave rise to the overlying Merensky unit. This argument is discussed more fully later.

Ballhaus and Stumpfl (1985a) have studied fluid inclusions in postcumulus quartz from the Merensky and Bastard Reefs. In the Merensky Reef they observed (1) rare primary gas inclusions filled with CO_2, CH_4, and, in some cases, traces of hypersaline brine, that homogenized above 580°C; (2) complex polyphase fluid–melt inclusions that contained very variable proportions of solids (including chlorides, silicates, and, very rarely, oxides and sulfides); and (3) abundant secondary three-phase H_2O–NaCl inclusions with more than 25 volume percent NaCl. The latter homogenized at 230 ± 15°C. Finally, they found (4) secondary two-phase inclusions with equivalent NaCl of less than 25 volume percent. Balhaus and Stumpfl (1985b) have drawn attention to the widespread occurrence of graphite in both the Merensky Reef and in the platiniferous dunite pipes. The graphite has $\delta^{13}C$ in the range of −19 to −21 per mil, which they suggest indicates that it was derived from a crustal source, probably the footwall sediments of the Transvaal System. They found graphite and hydrous silicates to contain significant chlorine. Boudreau et al. (1986) have found that apatites of the PGE-rich zones of the Bushveld and Stillwater Complexes have Cl/(Cl + OH + F) ratios of between 0.45 and 1.0 in comparison with all other noncumulus and cumulus apatites from the Bushveld, Stillwater, Skaergaard, Kiglapait, and Great Dyke intrusions, which have Cl/(Cl + OH + F) ratios <0.2. Boudreau et al. (1986) also report that the Cl, F, and OH contents of phlogopites and amphiboles from all of these intrusions are more variable, but that Stillwater and Bushveld phlogopites are richer in Cl in comparison with those of the Skaergaard and Kiglapait intrusions. These observations are in agreement with those of Johan and Watkinson (1985), who found up to 1.0 wt percent Cl and 1.7 wt percent F in mica just above the Merensky Reef.

Boudreau et al. (1986) have argued that, in view of the lower solubility of Cl with respect to F in silicate melts, and the fact that melts with high Cl/F ratios are unknown, their data indicate that the apatites of the PGE-rich zones have equilibrated with Cl-enriched fluids that separated during the solidification of the cumulate sequence.

5.1.2.3 COMPOSITION OF SULFIDES IN THE MERENSKY AND ADJACENT UNITS

Naldrett et al. (1986) have studied the bulk composition of sulfides in the Reef itself, in the overlying Merensky Unit, and in the flanking Pseudoreef and Bastard Reef and their associated units. The Ni/Cu and Pt/Cu ratios of their samples are illustrated in Figure 5.9, augmented by data from Lee (1983). Considering first the data from Naldrett et al. for the Merensky unit, the Ni/Cu and Pt/Cu ratios of the sulfides disseminated throughout the rocks of the unit are appreciably less than those of the Reef (in the case of Pt/Cu ratios, a factor of 40 to 13 less). The data from Lee (1983) of Pt/Cu ratios show similar differences, although his data for Ni/Cu ratios do not coincide with those of Nadrett et al.

Most of the sulfides in the Pseudoreef and Bastard units have Pt/Cu ratios and, where they have been determined, Ni/Cu ratios, very similar to those in the Merensky unit. Both the Bastard and Pseudoreef units have patches of rock enriched in disseminated sulfides close to their basal contacts. Sulfide-enriched zones occur within the Pseudo-

Fig. 5.5. Variation in the composition of bronzite and plagioclase at three locations in the Bushveld Complex. Note the scale changes over the Jagdlust section. Depths indicated are depths in the borehole. Data on Union and Rustenburg from Naldrett et al. (1986a) and on Jagdlust from Cameron (1988).

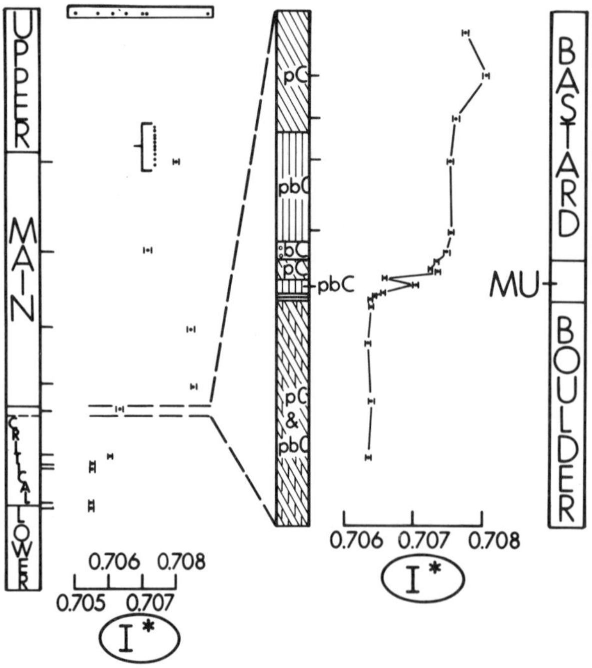

Fig. 5.7. Variation in initial $^{86}Sr/^{86}Sr$ ratios (I*) in the Bushveld Complex. Data from Hamilton (1977) for the whole of the Complex and from Kruger and Marsh (1982) for the interval around the Merensky Reef in the western Bushveld. The horizontal bar containing dots at the top of the left-hand diagram is the data of Hamilton on the upper zone, which could not be located accurately within the stratigraphic column.

reef unit in the Pseudoreef itself at the base of the unit and in a harzburgite immediately overlying the area known as the *tarentaal;* the zones are characterized by sulfides with very much higher Pt/Cu ratios than the rest of the unit. The same is true of the patches at the base of the Bastard but here the difference is not nearly as great.

The difference in metal ratios is also reflected in the Pt tenor (Pt content in 100% sulfides) of the sulfides. For example, sulfides contain about 20 ppm Pt in the Merensky unit, in comparison with 250 to 600 ppm in the sulfides of the Merensky Reef itself. Sulfides in the Bastard unit contain 3 to 20 ppm Pt (one highly anomalous value of 50 ppm has been omitted here) in contrast with 15 to 25 ppm in the sulfides of the sulfide-rich patches at the base of the Bastard pyroxenite. Thus the differences in Ni/Cu and Pt/Cu ratios between sulfides in the Reef and the overlying Merensky unit and between the basal patches of sulfide and sulfides in the overlying rocks of the other two units represent differences in the Ni and Pt tenor of the sulfides. In the case of the Merensky Reef, the upper chromite layer marks a major break in sulfide composition, with sulfides that are extremely enriched in Pt below and sulfides with a high but more normal tenor above. Naldrett et al. (1986) stress that the change occurs over a very few centimeters; samples 1 to 2 cm below the chromite layer have a high tenor, those 2 to 10 cm above have much lower tenor. Any genetic model for the reef must account for these features.

5.1.2.4 UPPER GROUP-2 CHROMITITE LAYER

Recent studies of the UG-2 chromitite include those of McLaren and de Villiers (1982), Gain (1985), and Hiemstra (1985, 1986). The chromitite layer occurs within a bronzite cumulate at the base of the UG-2 cyclic unit. It is underlain by a zone of pegmatoid, consisting of coarse bronzite and interstitial plagioclase, and containing irregular layers of harzburgite and chromitite. The pegmatoid penetrates the overlying chromitite in places for 10 to 20 cm.

The main chromitite normally ranges in thickness from 75 to 250 cm in the southwestern Bushveld, 90 to 150 cm in the northwest, and from 60 to 140 cm in the east. It is commonly separated by 10 to 15 cm of bronzitite from an overlying 0.5- to 1.0-cm thick "leader" layer. In certain areas, particularly in the southeastern Bushveld, 2.5 to 4 m of bronzitite split the main chromitite layer.

The chromitite consists of 60 to 90 modal percent chromite, 5 to 25 percent bronzite, and 5 to 15 percent plagioclase, with accessory clinopyroxene, base metal sulfides, PGE minerals (PGM), ilmenite, magnetite, rutile, and biotite.

Gain (1985) has shown that the UG-2 cyclic unit is characterized by bronzite with a slightly higher MgNo (0.79–0.80) than the underlying rocks and that this composition shows little change across the unit. The Cr content of the

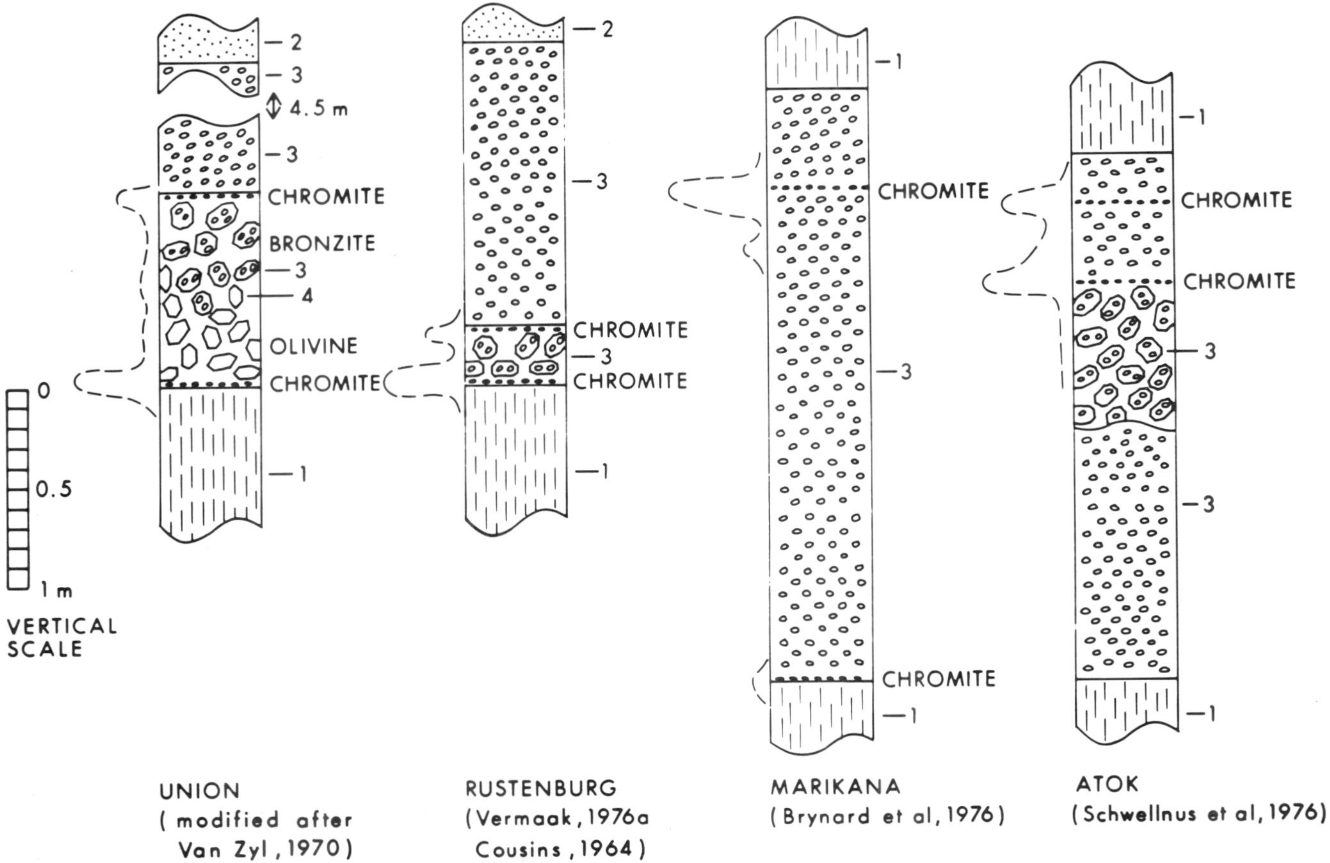

Fig. 5.8. Variations in the relationship between PGE-rich zones (shown schematically by the dashed lines to the left of each column) and the pegmatoid (olivine shown as 6-sided symbols; bronzite as 8-sided symbols that contain 2 open circles), underlying anorthosite and chromitite, and overlying bronzitite layers in the vicinity of the Merensky Reef in different areas of the Bushveld Complex. Vertical shading = anorthosite; open circles = bronzitite, dotted pattern = norite. (From Naldrett, 1981b).

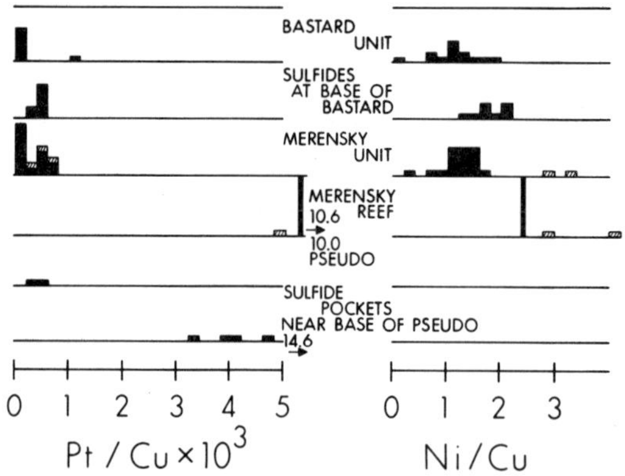

Fig. 5.9. Variations in Ni/Cu and Pt/Cu ratios in sulfides from the Pseudoreef, Merensky, and Bastard reefs and respective cyclic units. The data shown as solid symbols are from Naldrett et al., 1986; those shown with shading, and also the Pt/Cu ratios lying to the right of the bar for the Merensky Reef are from Lee, 1983. (Modified after Naldrett et al., 1986a.)

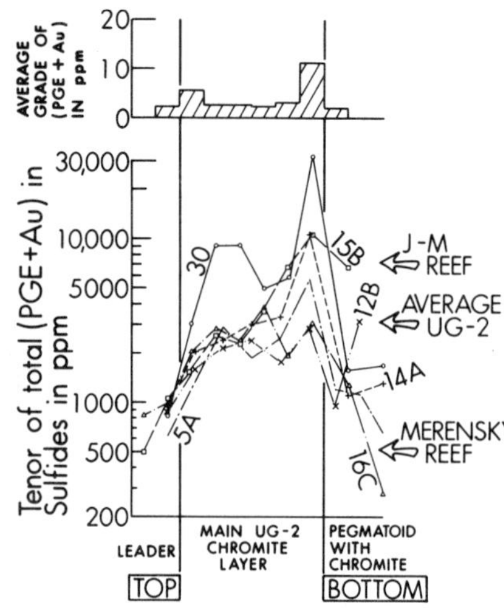

Fig. 5.10. Average (PGE + Au) grade from six intersections and individual tenors in sulfides of (PGE + Au) in each intersection, across the UG-2 chromitite. To compensate for varying thicknesses of the chromitite the positions of the samples have been adjusted so that they bear the same relative position to each other as they do to the top and bottom of the chromitite. Note that grade refers to concentration in the rock, tenor to concentration in the sulfide phase. Data from Gain (1985). (From Naldrett et al., 1987.)

bronzite increases slightly above the level of the UG-2 from 3,100 to 3,600 ppm, before decreasing sharply to 2,000 ppm at the top of the UG-2 unit.

Both Gain (1985) and McLaren and de Villiers (1982) have observed a very close correlation between base metal sulfides and PGE mineralization. Gain notes that the Ni, Cu, and PGE contents of bulk rocks increase slightly in the bronzitite beneath the chromitite layer, and then increase sharply within the layer, before decreasing abruptly to background values above it.

Average grades for the UG-2 range from 4.3 to 4.9 ppm total PGE at the Western Platinum Mine, to 8.2 at Maandagshoek in the eastern Bushveld, and to 9.3 ppm in the northwestern Bushveld Complex (McLaren and de Villiers, 1982).

Hiemstra (1985) found that in a core from Western Platinum Mine, Pt, Pd, and Rh concentrations define two coincident maxima, one at the base of the unit and a second half-way up. Both maxima start abruptly, decay steeply upward, and are followed by intervals over which the metals show little variation. McLaren and de Villiers (1982) found that a similar distribution characterized three of the five cores that they studied, but that in the other two the maxima were the top and the bottom of the chromitite. An average for the six intersections studied by Gain (1985) from Maandagshoek is shown in Figure 5.10, where it can be seen that maxima also occur at the base and the top.

The data of Gain (1985) have been used to calculate the PGE tenor (PGE content in 100% sulfide) of the sulfides. These are also shown with respect to position within the chromitite in Figure 5.10. It has been assumed that all PGE are concentrated in sulfide. This may not be entirely true for Ru, Ir, and Os, as some of these metals may have been concentrated by chromite. These metals, however, constitute a relatively small proportion of the total PGE and their inclusion in sulfide does not affect the validity of the following argument. It is seen that the PGE tenor of the sulfide is relatively low in the pegmatoid (1,300 ppm), highest at the base of the chromitite layer (3,000–30,000 ppm), decreases across the layer (to 2,000 ppm at the top), and is lowest in the leader (800 ppm). There is an indication in some of the profiles that there is a slight increase in PGE tenor one-half to two-thirds up through the chromitite.

Both McLaren and de Villiers (1982) and Gain (1985) consider it important to account for the average tenor of the UG-2 (3,300 ppm total PGE + Au according to the former) being 5 to 10 times as high as that for the Merensky Reef. Von Gruenewaldt et al. (1986) point out that the same problem of very high PGE/S and, in addition, high Cu/S and Ni/S ratios occurs with all of the chromitite layers of the middle and upper groups for which they have data.

Gain (1985) has calculated that the present tenors of Ni and Cu in the sulfides of the UG-2 are 22.5 and 9.75 wt percent, respectively. These concentrations correspond to 55 percent pentlandite, 28 percent chalcopyrite, and 17 percent pyrrhotite. A sulfide liquid of this composition would be highly anomalous, and it is unlikely that it could have been in equilibrium with a silicate magma similar in its Ni and Cu contents to that responsible for the sulfides of the Merensky Reef. The only natural ores with pentlandite contents approaching 50 percent are those associated with komatiites, and these ores have very low chalcopyrite contents. A magma such as that responsible for the sulfides of the Merensky Reef, or with Ni and Cu contents resembling the marginal samples analyzed by Harmer and Sharpe (1985) and Davies and Tredoux (1985), would have given rise to sulfides with a maximum Ni content of one-third to one-half of that calculated for the UG-2. (For example, taking Davies and Tredoux's average Ni content of 274 ppm and Rajamani and Naldrett's [1978] partition coefficient for Ni of 275, the resulting sulfide would be expected to contain 7.5 wt percent Ni.) Thus, it would seem that the UG-2 ore has been modified in some manner subsequent to its segregation as a sulfide liquid, and that approximately one-half to two-thirds of the original mass of the sulfide phase present (assuming that all of the original Ni and Cu remain) has been lost. This is illustrated in Table 5.1 where the amount of Fe lost (together with its associated S, which is not shown in the table) to bring the Cu content to that observed in the sulfides of the Merensky Reef (4.65 wt %) has been calculated for all of the middle and upper group chromitites. It is seen that because of its higher sulfide content, the UG-2 has lost much more Fe than the others. The fact that FeS appears to have been lost from sulfides in all of the chromite horizons studied strongly supports a model in which interaction with chromite is responsible for the loss, as suggested by von Gruenewaldt et al. (1986) and Naldrett and Lehmann (1988). The latter have proposed that the Fe has transferred to fill vacancies in chromite and that the S has diffused into the surrounding rocks. As discussed in Section 6.7.5, they show that the available thermodynamic data are consistent with this explanation.

Average PGE tenors for the UG-2 are compared with values for the Merensky (Naldrett, 1981b) and J–M Reefs (Barnes and Naldrett, 1985) in Figure 5.10. It is seen that whereas the UG-2 tenor lies about midway between that of the Merensky and J–M Reefs, reducing the tenor by two-thirds will bring it much closer to the former than the latter.

5.1.2.5 THE PLATREEF

The Platreef occurs on the Potgietersrus limb (northern extension) of the Bushveld Complex where Bushveld igneous rocks cut downward through the Pretoria Series progressively northward along the limb to come to rest on granites of the Archean basement (Wagner, 1929). The Bushveld magma itself appears to have transgressed early cumulates as it crystallized because the lowermost mafic units are only developed at the southern end of the limb and progressively higher units of the Complex come into contact with the floor to the north (Van der Meerwe, 1976). Significant differences exist between rocks of this limb and other sections of the Bushveld and only the major subdivisions can be correlated with certainty.

The Platreef mineralization consists of a zone of sulfide blebs and occasionally more massive

TABLE 5.1. *Sulfides in the Middle and Upper Group Chromitites of the Bushveld Complex*

	I ppm S	II ppm Sulfide	III wt% Cu	IV ppm Sulfide	V Factor	VI ppm Fe
UG-3A	72	183	19.7	775	4.2	359
UG-3	46	117	25.6	664	5.5	320
UG-2	926	2356	9.8	4965	2.1	1578
UG-1	99	251	12.8	688	2.7	265
MG-4A	17	43	60.5	559	13.0	313
MG-3	34	86	24.4	451	5.2	222
MG-2B	62	157	15.3	934	3.3	472
MG2A/B	112	284	10.6	647	2.3	220
MG-2A	63	160	22.5	774	4.8	373
MG-1B	13	33	54.5	387	11.7	357
MG-1A	46	117	20.5	516	4.4	242

I = present S; II = present amount of sulfide; III = Cu in present sulfides; IV = original amount of sulfide if Cu was 4.65%; V = factor of upgrading; VI = Fe lost from original sulfide.

In constructing this table, it was assumed that the S in the analyses of von Gruenewaldt et al. (1986) was present in sulfide that had a metal/S ratio of 1.54, which is the metal/S ratio of pyrrhotite with N_{FeS} = 0.94 in the Fe–S_2 system (a normal magmatic composition). This gives rise to the sulfide contents shown in column II. Assuming that all of the Cu reported in the analyses is in sulfide, the Cu concentration is shown in column III. If one assumes that the original Cu concentration in the sulfides was equal to that in the Merensky Reef sulfides (4.65 wt %), and that the difference between this and column III is due to the loss of FeS, one obtains the original sulfide content shown in column IV. The amount of Fe lost to chromite to make this difference is shown in column VI.

From Naldrett and Lehmann (1988). Data from von Gruenewaldt et al. (1986) and Gain (1985)

stringers sporadically developed within feldspathic pyroxenite and harzburgite. The mineralization is never far from the floor of the intrusion and much of the better ore occurs where dolomites of the Pretoria Series form the floor. Numerous inclusions of highly altered carbonate rocks occur within the ore zone. Some sulfides occur up to 30 m below the basal contact within the dolomites. On the basis of field, textural, and major and trace element studies, Liebenberg (1970), De Waal (1977), Von Gruenewaldt (1979), Gain and Mostert (1982), and Buchanan and Rowe (1984) suggested that interaction of Bushveld magma with the floor rocks (dolomites, banded iron formations, and shales) is responsible for the important development of sulfide in the area. Cawthorn el al. (1985) presented trace element and Sr isotopic data suggesting that fluid or a partial melt derived from nearby granite was more likely to have been the contaminant.

Platinum group minerals include cooperite and sperrylite, always in association with sulfides. As indicated in Table 2.2, estimates of grade range from 7 to 27 ppm total PGE. Pd predominates over Pt, and the ore contains a proportion of Rh to other PGE similar to the Merensky Reef. Recovery grades of 0.36 percent Ni and 0.18 percent Cu are estimated by Von Gruenewaldt (1977).

Mining of the Platreef started in 1926 but ceased a few years later. In 1967, interest in the area was revived and several companies have been active (particularly Johannesburg Consolidated Investments Pty. Ltd.). Economic attractions of the area include a significant tonnage that can be recovered from surface, and Ni and Cu that can account for as much as 50 percent of the value of the ore.

5.1.2.6 DUNITE PIPES

Dunite pipes and associated envelopes of olivine–bronzite–plagioclase pegmatoid occur in both the western and eastern sections of the Bushveld Complex. Mining and geologic studies have been restricted to four pipes in the eastern section (Onverwacht, Mooihoek, Driekop, and Maandagshoek). These pipes have some of the highest grades reported for the Complex. Wagner (1929) cited grades ranging up to 2,050 ppm total PGE. On the 65-ft-level of the Onverwacht pipe, the central 8-m-diameter core of the pipe averaged over 31 ppm Pt, and a 1-m shell surrounding this core graded between 15 and 30 ppm Pt. The average Pt grade for the mine was between 9 and 11 ppm, whereas that for the nearby Driekop pipe was 5 to 6 ppm.

The Onverwacht pipe (Wagner, 1929; Cameron and Desborough, 1964) is typical. A central 20-m-diameter zone of hortonolite dunite (olivine composition, Fo22) tapers downward and is encased within a 100-m-diameter zone of olivine dunite (olivine Fo86–92 at the margin, Fo80–84 near the core). The olivine dunite defines a pipe that is perpendicular to layering in the surrounding bronzitite. Patches of magnetite and ilmenite occur within the pipe; mining revealed the presence of numerous slabs of chromitite at the level at which the pipe transects the LG-6 chromite layer (Steelpoort seam) in the bronzitite. On the basis of textural studies and evidence such as the chromitite, Cameron and Desborough concluded that the pipe resulted from the replacement of bronzitite by hot aqueous solutions that leached SiO_2, Al_2O_3, and Na_2O and introduced FeO, TiO_2, Vanadium (V), and the PGE. Gain (1980) noted that the LG:6 chromitite layer dips markedly in toward a dunite pipe on the farm Maandagshoek, suggesting that simple volume-for-volume replacement is not a sufficient explanation of the emplacement of the pipes. Schiffries (1982) proposed a model in which Cl-rich brines had introduced Pt and other elements and removed Mg. Whereas the Mooihoek pipe is very similar to Onverwacht, the Driekoep pipe differs in that the dunite of the cores is less Fe-rich, containing approximately 70 mole percent forsterite.

5.2 THE STILLWATER COMPLEX

5.2.1 Stratigraphy

Recent studies of the igneous stratigraphy of the Stillwater Complex include the detailed documentation of cross sections by McCallum et al. (1980), Todd et al. (1982), and Raedeke and McCallum (1984), and the mapping of the whole of the Complex by Segerstrom and Carlson (1982). These have been correlated by Zientek et al. (1985) who divided the Complex into the Basal, the Ultramafic, and the Banded series (Fig. 5.1). The Basal series is in turn subdivided into the Basal Norite zone and the Basal Bronzite Cumulate zone, the Ultramafic series into the Peridotite zone and the Bronzitite zone, and the Banded series into the Lower, Middle, and Upper Banded series. Each of the subseries in the Banded series is itself divided into a number of zones (Fig. 5.1) (McCallum et al., 1980).

The Basal series is a laterally continuous but petrologically variable unit consisting of bronzite cumulates (bC) or bronzite plus plagioclase cumulates (bpC). The Peridotite zone of the Ultramafic series consists of an alternating sequence of olivine-, chromite-, and bronzite-bearing cumulate layers. Jackson (1970) recognized between 8 and 21 cyclic units in the Perididote zone, varying in number along strike and changing from ocC (olivine + chromite cumulate) at the base to obC and bC at the top. As a rule, basal ocC contain less than 2 modal percent chromite. Thin-layered intervals (2-cm–1-m thick) containing as much as 80 modal percent chromite occur in some parts of some cyclic units and yet are missing in other parts of the same units. Two of the chromitite intervals are particularly well developed and have been mined.

The Bronzitite zone of the Ultramafic series is a succession of laminated and size-graded bronzite cumulates.

The Banded series is marked by the appearance of cumulus plagioclase, and consists of a sequence of norites, gabbronorites, and anorthosites. Olivine occurs as a cumulus mineral at five places in the stratigraphy; these are referred to as olivine-bearing zones (OB–I to OB–V). The J–M Reef occurs within the lowest olivine-bearing zone.

5.2.2 Mineralization

5.2.2.1 SETTING

The J–M Reef extends for the whole exposed length of the Complex (Fig. 5.11). It occurs within the Lower Banded series approximately 400 m above the base of this series (see Fig. 5.1). The lowermost rocks (the N–I zone) of the series consists of pbC (plagioclase–bronzite cumulate) with 50 to 30 modal percent bronzite. Page et al. (1985) describe these rocks as having modal and locally phase (due to the absence of bronzite) layering. They pass up into a sequence of gabbronorites (GN–I), which are interbanded with anorthositic layers. Based on their work on the west side of the Stillwater Valley, Page et al. (1985) divided the N–I and GN–I zones into seven mappable subzones in which the proportion of cumulus pyroxene decreases upward, although they did not demonstrate any systematic variation in mineral chemistry within individual subzones. Overall the MgNo of bronzite varies between 74 and 83, with a slight upward decrease, whereas the An content of plagioclase is fairly constant at 80 mole percent (Fig. 5.12).

The J–M Reef occurs within a sequence of cumulates, which include olivine-bearing layers, and together constitute olivine-bearing zone 1 (OB–I). This is the lowermost of the five olivine-bearing zones referred to previously.

OB–I is overlain by a sequence of norites (N–II). The data of Barnes and Naldrett (1986) for the Stillwater canyon area indicate that the N–II norites are characterized by bronzites that are less magnesian-rich (MgNos 73–79 in comparison with 77–84) and less Cr-rich (300–1700 in comparison with 1,700–3,400 ppm Cr), and plagioclases that are less anorthitic (An 79–85 in comparison with An 82–88) and richer in Sr (150–170 in comparison with 85–110 ppm Sr), than the GN–I gabbronorites. Although the less magnesian bronzite and less anorthitic plagioclase of the N–II norites could be attributed to fractional crystallization, the higher Sr contents of the plagioclase could not (Sr partitions into plagioclase and should decrease during fractional crystallization). Barnes and Naldrett (1986), therefore, attributed the variations in mineral chemistry to the influx of a second magma. Because there is also a need to explain the formation of rocks in which plagioclase is the sole cumulate phase, or in which it crystallizes first, they followed Todd et al. (1982) and Irvine et al. (1983) in supposing that the new influx was richer in the plagioclase component than the magma already in the chamber.

Todd et al. (1982) have demonstrated the extreme variability of the OB–I zone. This is brought out in Figure 5.13. Near the Frog Pond adit, toward the western end of the Complex, 11 separate olivine-bearing intervals have been observed (Nos. 1 to 10, including O–5A and O–5B). These are interpreted as cyclic units ranging from oC to poC to pC, although pbC and pbaC (plagioclase–bronzite–augite cumulate) are also present. In the vicinity of the West Fork adit and the Anorthosite Ridge just to the west of it, three olivine-bearing intervals occur below the O–5A and O–5B members (Fig. 5.13), but farther east, in the Stillwater canyon, the O–5B member lies directly on the GN–I zone, cutting down through this to lie on the N–I zone. Information on the olivine members of OB–I zone that overlie O–5A and O–5B is not available in the West Fork area. In the Minneapolis adit area, the upper olivine-bearing members are missing, and norites of N–II lie directly on rocks of the mineralized package.

5.2.2.2 J–M REEF

As described previously, where olivine cumulate members occur within the OB–1 zone, these commonly are marked by a sharp base and then grade upward through poC to pC. The J–M Reef is a horizon within the thickest of the olivine-bearing members, O–5B, that is relatively enriched in sulfide.

Details of the mineralized Reef (the 0–5B member) are shown for the Frog Pond, West Fork, and Minneapolis areas in Figure 5.14. The O–5B member at Frog Pond is similar to the one at West Fork. The lower portion consists of about 1 m of cumulus, ameboidal-shaped olivine (20–40 modal percent olivine at Frog Pond, 50–70 percent at West Fork) enclosed in large bronzite oikocrysts, together with interstitial plagioclase. Minor amounts of Cl-rich phlogopite, apatite, and chromite are also present. At Frog Pond this is overlain by poC in which olivine is irregularly distributed in clusters. At West Fork, the basal oC is flanked by 15 to 30 cm of poC and is overlain by pC containing some olivine in places. Because of its alteration, this pC unit consists of three differently colored bands. The pC is coarse grained, with large augite and smaller bronzite oikocrysts, and rare wispy olivine. It is overlain by poC, with small amounts of olivine as discrete grains and as much as 50 modal percent postcumulus augite and bronzite.

The most common sulfides are chalcopyrite,

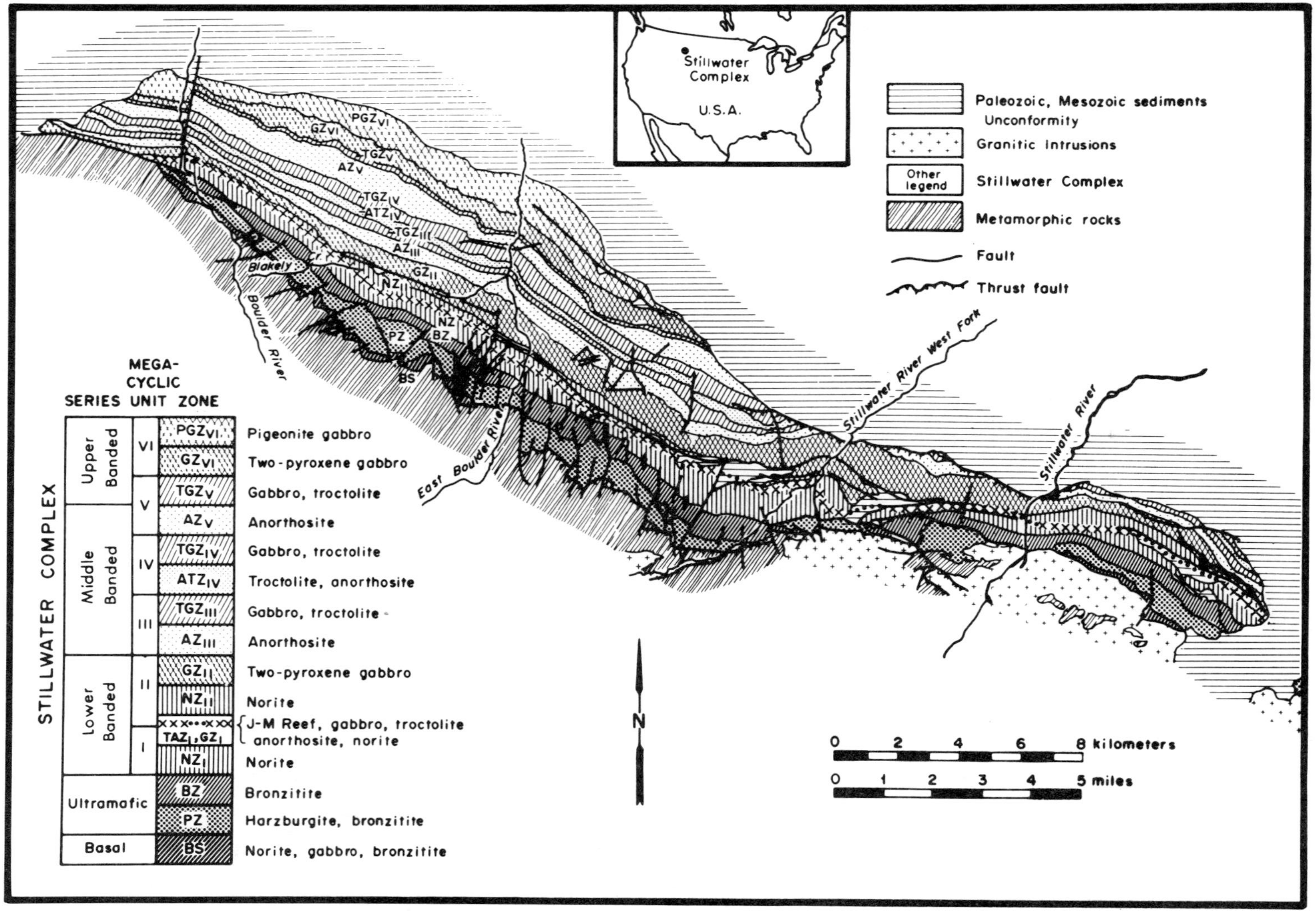

Fig. 5.11. The geology of the Stillwater complex. (From Page et al., 1985.)

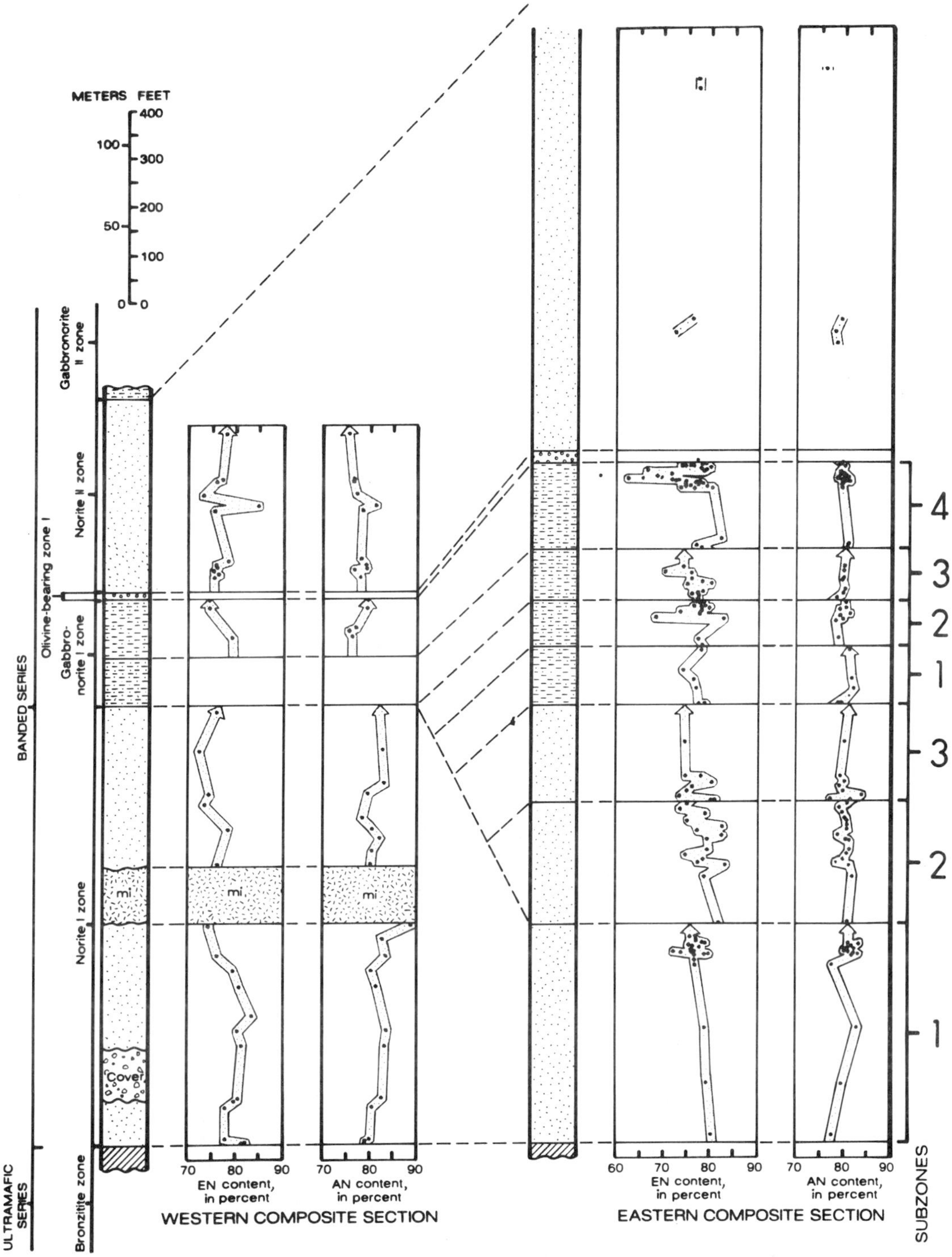

Fig. 5.12. Variation in the En content of bronzite and An content of plagioclase from the Bronzite zone to the GN–II zone of the Lower Banded series of the Stillwater Complex. mi = intrusive rock. (From Page et al., 1985.)

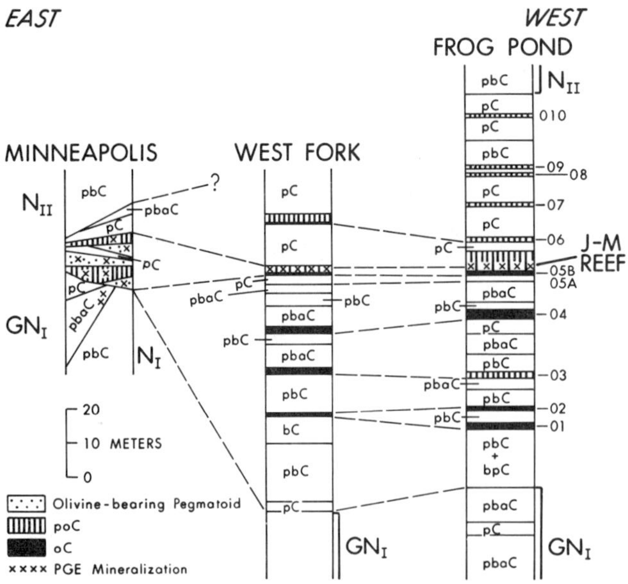

Fig. 5.13. Variation in the stratigraphy of OB–I zone from the Minneapolis adit in the east to the Frog Pond adit in the west. Data from Todd et al. (1982), Leroy (1985), and Turner et al. (1985). (From Naldrett et al., 1987.)

pyrrhotite, pentlandite, and pyrite, which generally occur as blebs or an interstitial network in the lower part of the Reef, and as a fine dissemination toward the top. At both Frog Pond and West Fork, mineralization is concentrated toward the top of the olivine-bearing zone. The highest grades tend to occur toward the top of the poC (at West Fork within 0.6–1.0 m of the top). At West Fork, the upper contact of the mineralization coincides with the contact of the poC with the overlying pC, but at Frog Pond it straddles this contact. The overall grade averages about 19 g/tonne (Pd + Pt) over 1.9 m, with an average Pd/Pt ratio of 3.6. Mann and Lin (1985) report the presence of coarse sulfide segregations in the basal oC layer, but note that many of these contain only traces of Pt and Pd, although discontinuous areas of high grade mineralization have been observed at this horizon in a few places.

The situation is very different in the area of the Minneapolis adit (Turner et al., 1985), which is now the Stillwater Mining Company's mine. The ore-bearing member, which is presumed to be O–5B, lies directly on footwall rocks. In an idealized succession, olivine-bearing pegmatoidal rocks are overlain by poC in which olivine decreases upward as the rock grades into an oikocrystic "mottled anorthosite," and hence into norite of the N–II zone (Fig. 5.15). The idealized succession is rarely seen. A common rock-type is that known as *mixed rock.* This consists of a matrix of medium- to coarse-grained cumulate plagioclase with local intercumulus bronzite and augite forming a net texture, in which there is 20 to 40 modal percent of a coarse ameboidal olivine. Irregular pods and lenses of pegmatitic oC, poC, and pC are present. The proportion of olivine in mixed rock is variable, with bronzite oikocrysts increasing at the expense of olivine toward the west end of the adit.

The poC also contains ameboidal olivine, which commonly shows resorbtion and is jacketed by bronzite.

Raedeke and Vian (1986) describe the sulfides as occurring at four different stratigraphic levels. These are, relative to the base of OB–I: (1) the Footwall zone in Gabbronorite I generally less than 10 m below OB–I, (2) the Basal zone straddling the base of OB–I, (3) the Main zone, within the olivine-bearing sequence and generally less than 3 m from its base, and (4) the Upper zone (Fig. 5.16).

Generally only one mineralized zone is present over any significant strike length, as is illustrated in Figure 5.16, although overlaps are present in places. There are sharp boundaries between higher grade areas, which average 11.2 grade units, and the lower and very low grade areas, which together, average 2.1 grade units.

The Footwall zone mineralization occurs as clotty to finely disseminated sulfides in narrow, 10 to 30 cm, anorthosite and norite layers in the gabbronorite, that, with or without sulfides, form a continuous stratigraphic package toward the western end of the mine.

The base of OB–I is not flat, and a number of depressions, including one major one, have been defined. The major depression was compared by Turner et al. (1985) to potholes in the floor of the Merensky Reef, but it does not have the sharply transgressive features of most of these. Much of the basal mineralization tends to be associated with topographic highs in the base of OB–I, but the Footwall, Main, and Upper zone mineralizations show no readily discernable correlation with features of this contact.

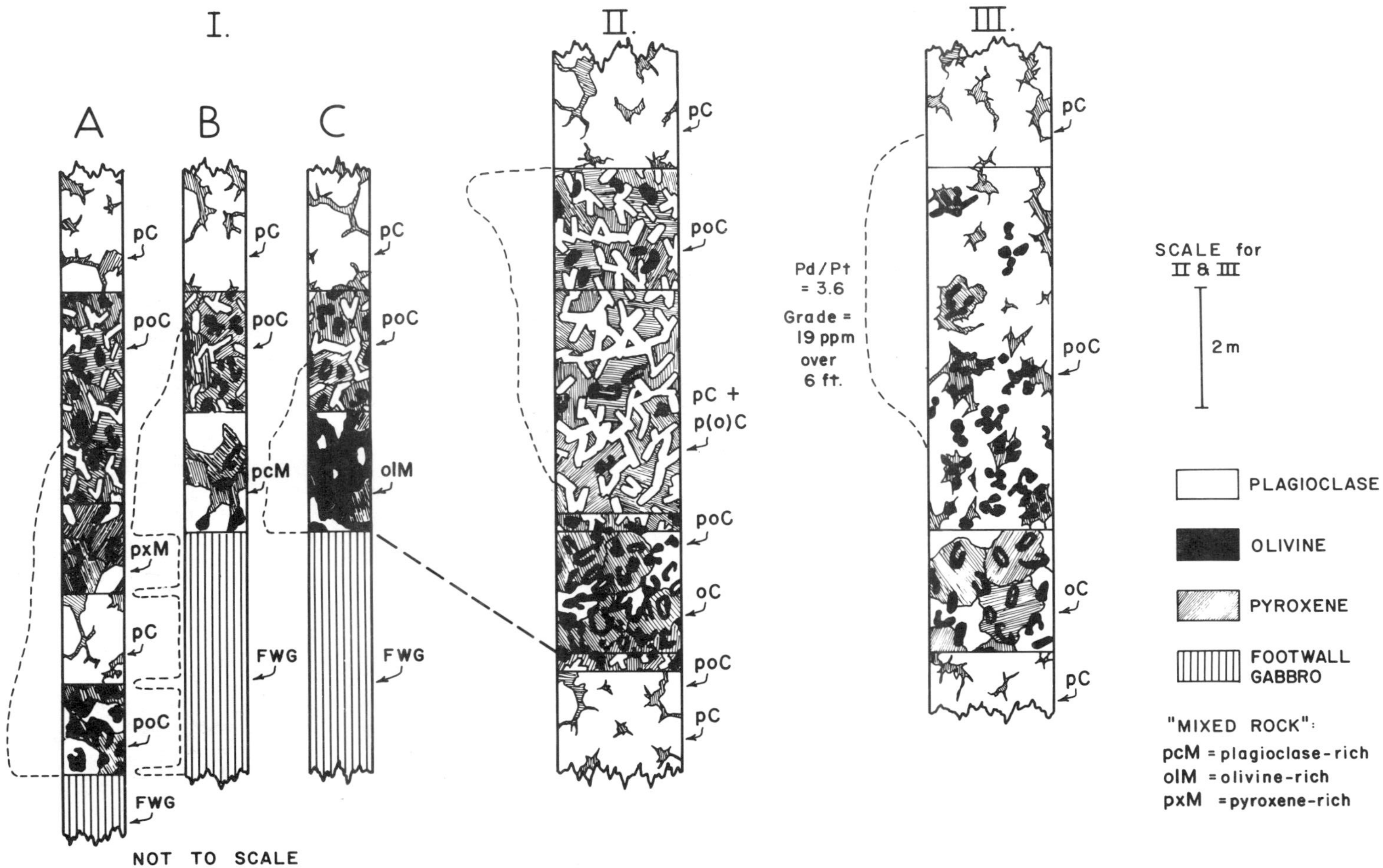

Fig. 5.14. Variation in the J–M Reef (the O–5B member) from east to west. I = Minneapolis adit—A = within pothole, B = pothole margin, C = away from pothole; II = West Fork adit; III = Frog Pond adit. The dashed lines indicate schematically the distribution of PGE within the succession. Data from Turner et al. (1985), Mann et al. (1985), and Leroy (1985). Drawing by B. Murck. (From Naldrett et al., 1987.)

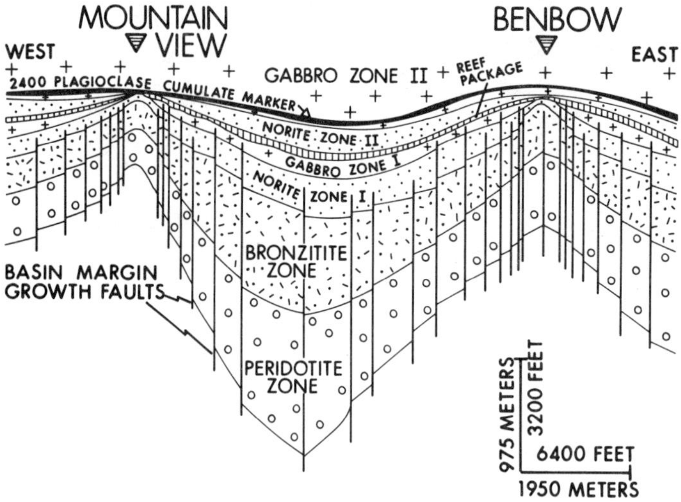

Fig. 5.15. Conceptualized view of basin and ridge development before the Laramide deformation in the eastern part of the Stillwater Complex. (From Turner et al., 1985.)

In general, sulfides constitute much less than 2 modal percent of the rock, and occur as masses up to 6 mm in diameter. In their work on the Minneapolis mineralization, Barnes and Naldrett (1985) described spherical bodies of sulfide within olivine grains, and cuspate masses of sulfide molded about olivine. They interpreted these textures as indicating that the sulfides were present at the magmatic stage. In some cases, cuspate sulfides are entirely enclosed in oikocrystic bronzite,

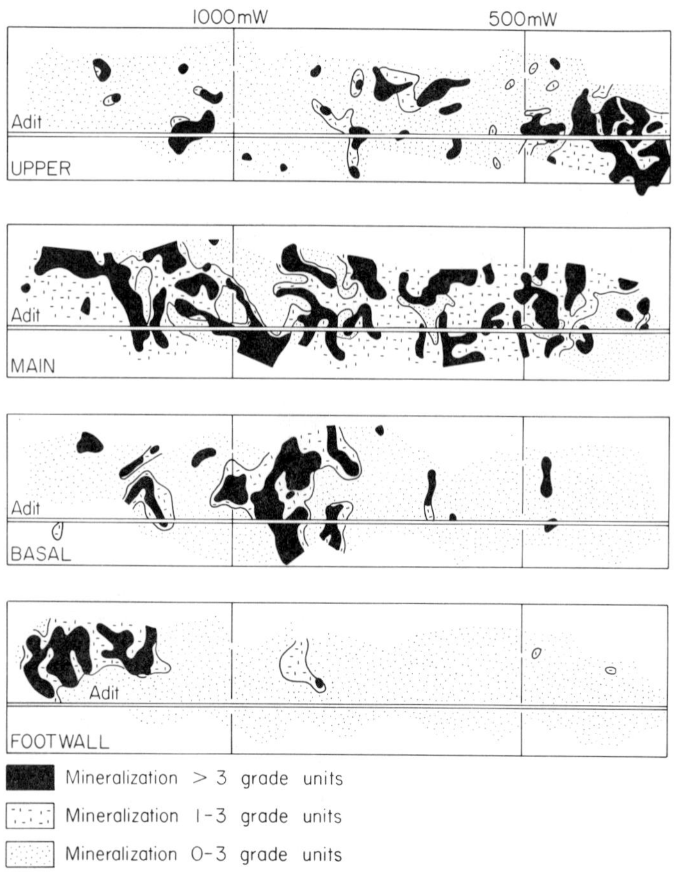

Fig. 5.16. Distribution of mineralization at various levels within the J–M Reef of the Stillwater Complex. (After Raedeke and Vian, 1986.)

which also contains remnants of resorbed olivine, indicating that the resorbtion occurred after the sulfide liquid had solidified (i.e., below 1,100°C). In mixed rock, sulfides occur both in olivine-rich "boulders" and within the matrix.

Barnes and Naldrett analyzed 40 samples of ore representative of sulfides in oC, mixed rock, and pbC, and found a strong correlation between PGE content and percent sulfur, supporting the concept that sulfides have acted as a collector, and emphasizing the homogeneity of the ore with respect to PGE tenor.

Boudreau et al. (1986) have found that apatites from the OB–1 zone, as in the Merensky Reef, have molar Cl/(Cl + F + OH) ratios ranging from 1 to 0.1, greater than for the Upper Zone of the Bushveld, the Skaergaard, and the Kiglapait Complexes.

5.3 VIEWS ON THE ORIGIN OF THE MERENSKY AND J–M REEFS

At the time of the writing of this book, the origin of stratiform PGE-rich ores in layered intrusions is still being hotly debated. This section is not intended as a summary of these debates; it is too early to do this, rather, it presents the current views of this author.

Although there are some differences between the three stratiform PGE deposits described previously, there are many similarities, particularly between the Merensky and J–M Reefs. This section, therefore, focuses on the genesis of these two last named bodies, and the most important similarities are summarized here:

1. Both the Merensky and J–M Reefs occur a few hundred meters above the first appearance of cumulus plagioclase in their respective host intrusions.
2. Mineralization in both Reefs takes the form of interstitital, apparently magmatic, Fe–Ni–Cu sulfides, carrying exceptionally high concentrations of PGE.
3. Both Reefs are characterized by the reappearance of cumulus phases that were dominant in the early crystallization history of both intrusions: olivine and chromite in the Stillwater; bronzite and chromite (and locally olivine) in the Bushveld.
4. Very coarse-grained pegmatoidal gabbroic rocks are common in both reefs.
5. Unusually high concentrations of Cl and Fe in hydrous silicates occur associated with both reefs.

5.3.1 The Problem of High PGE Concentrations

Because both the Merensky and J–M Reefs occur within large igneous complexes, it has been customary for most researchers to attempt to interpret the PGE mineralization in magmatic terms. The sulfides of both complexes, however, contain very much higher concentrations of PGE (particularly Pt and Pd) than the typical "Ni–Cu dominated" magmatic sulfides discussed previously. At the same time, the Ni and Cu contents of the sulfide phase in the Merensky and J–M Reefs are fairly typical for sulfides in equilibrium with a gabbroic liquid (Table 5.2). The recent data of Davies and Tredoux (1985) on the Bushveld and of Zientek et al. (1986) on the Stillwater Complex indicate that the magmas responsible for the two complexes contained similar concentrations of PGE to continental flood basalts (Fig. 5.17). If one is to call on a magmatic model, one has to explain why the partition coefficients from silicate magma into sulfide liquid were so much higher for the

TABLE 5.2. *Ni, Cu, and Precious Metal Concentration Averages in the Merensky and J–M Reefs, Normalized to 100% Sulfides*

	Merensky	J–M	"Typical Sudbury"
Ni%	10.8	7.8	3.6–5.7
Cu%	4.65	5.0	1.2–3.7
Pd (ppm)	130	6,500	0.4–6.0
Pt (ppm)	310	1,900	0.5–3.4

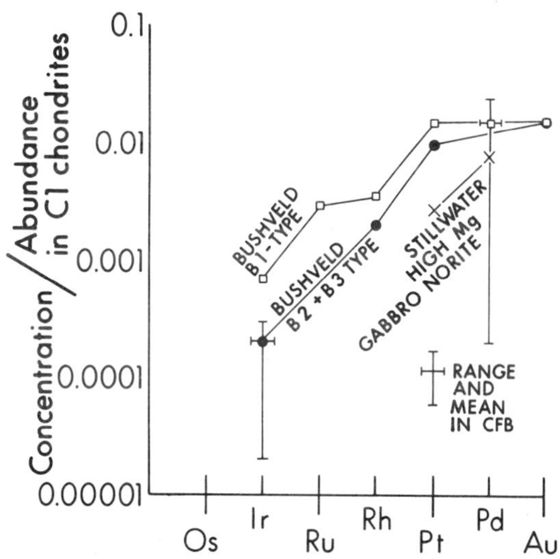

Fig. 5.17. Chondrite-normalized concentrations for PGE in possible Bushveld and Stillwater input magmas. The elongated crosses indicate mean and range for continental flood basalts (from Crocket, 1981). The two Bushveld profiles and the Stillwater profile are drawn for the concentrations that are exceeded by 50% of the samples in each population (data for Bushveld from Davies and Tredoux, 1985; for Stillwater from Zientek et al., 1985). The chondritic abundances used for normalization are those summarized by Naldrett and Duke (1980). (From Naldrett et al., 1987.)

PGE (but not for Cu and Ni) than in other deposits, and also how PGE were scavenged by the sulfides from the very large volumes of mafic magma that they were originally contained within at very low concentrations.

A number of approaches have been taken to these problems. Vermaak (1976a) and von Gruenewaldt (1979) have suggested that the deposits have been enriched in PGE subsequent to the formation of the rocks in which they occur by ascending, fractionated, magmatic fluids that were displaced from the underlying cumulates as these compacted.

Lauder (1970), Stumpfl (1962, 1974), Ballhaus and Stumpfl (1985a), and Boudreau and McCallum (1985) have suggested that the metals have been concentrated by hydrothermal solutions. Ballhaus and Stumpfl point to the fluid inclusions, high Cl and F, and the presence of graphite to support their arguments. Boudreau and McCallum (1985) and Mathez et al. (1985) argue for a hydrothermal origin on the basis of a high halogen content in hydrous minerals of the ore zones, and certain silicate assemblages that are very much out of equilibrium, suggesting metasomatism. Boudreau and McCallum (1985) suggest that the Picket Pin deposit, together with its underlying mineralized pipe, was formed by fluids that became dammed beneath the medium-grained anorthosite that forms the cap of this deposit.

Elliott et al. (1982) and Buntin et al. (1985) have suggested that the Merensky Reef crystallized at very low oxygen fugacities. It is possible that PGE partitioning increases with decreasing fO_2, as Boctor and Yoder (1983) and Doyle and Naldrett (1986) have shown to be the case with Ni.

Campbell et al. (1983) have argued that partition coefficients for the PGE are always very high, perhaps of the order of 10^5. They also point out that the tenor of a given chalcophile metal, which is achieved by a sulfide liquid equilibrating with a silicate magma, is not only a function of the partition coefficient and the concentration of the metal in the magma, but also of the mass of magma with which a given mass of sulfide liquid comes to equilibrium, in other words, the magma-to-sulfide ratio (R) (see Chapters 3 and 6). They argue that any genetic model must, therefore, incorporate a mechanism for bringing a small mass of immiscible sulfide liquid into effective equilibrium with the PGE contained within a very large mass of silicate magma. Naldrett et al. (1986) point out that if the partition coefficients for PGE are of the order of 10^5, and for Ni and Cu are similar to those determined experimentally by Rajamani and Naldrett (1978), raising log R from 3 to 4.5 to 5 will exactly account for the differences in the concentrations of PGE between "normal" and "Merensky-type" magmatic sulfides, but will make no difference to the Cu and Ni concentrations, which is in accord with the observed differences between the two types of magmatic sulfide.

Naldrett et al. (1986) argue that their observation, that the composition of the sulfides in each of the three cyclic units of the Bushveld Complex examined in their study varies greatly between different parts of individual units and also between units (Fig. 5.9), is inconsistent with genetic hypotheses calling on the later concentration of PGE by ascending fluids, whether magmatic or hydrothermal. The amount of PGE that could be trapped in the pores of partially crystalline cumulus rocks is so small, thus the ascending fluids would have had to have derived their PGE from well below the Pseudoreef. On moving upward they would first have encountered sulfides of the Pseudoreef and "tarentaal," then those of the remainder of the Pseudoreef unit, then sulfides within the Merensky Reef, then those of the Merensky unit, then sulfides at the base of the Bastard unit, then those of the remainder of the Bastard. Under these circumstances, the ascending fluids would have progressively relinquished their contained PGE, giving rise to sulfides of highest tenor in the Pseudoreef and tarentaal, and progressively lower tenor upward. This is not what is seen in Figure 5.9. The cyclic variation in Pt/Cu (equivalent to Pt tenor) in the Figure, with the highest values always occurring at the base of a given unit, cannot be accounted for, on the basis of ascending fluids, either magmatic or hydrothermal.

5.3.2 A Possible Model for the Origin of Merensky Reef-Type Deposits

Naldrett et al. (1986) point out that mafic magma chambers develop an internal layering in which horizontal, convecting cells, of differing composition and upwardly decreasing density, communicate with each other through the diffusion of heat and chemical components across the boundary between them (Huppert and Sparks, 1980; Irvine et al., 1983). Sparks et al. (1984) and Barnes and Naldrett (1986) have shown that when a mafic magma differentiates, its density will change in the manner shown in Figure 5.18. Thus a magma that has been crystallizing plagioclase may well have achieved a density slightly greater than its initial density. A fresh pulse of initial magma will, therefore, rise through the fractionated old magma (Fig. 5.19b) to seek its own density level. Campbell et al. (1983) showed that it will rise as a turbulent plume, entraining surrounding old magma into itself. Drawing on the work of Irvine (1975) and Irvine et al. (1983), they pointed out that if both magmas were close to saturation with sulfide, the mixture would be likely to exceed saturation, therefore, immiscible sulfide droplets would form.

These droplets would be swirled around in the

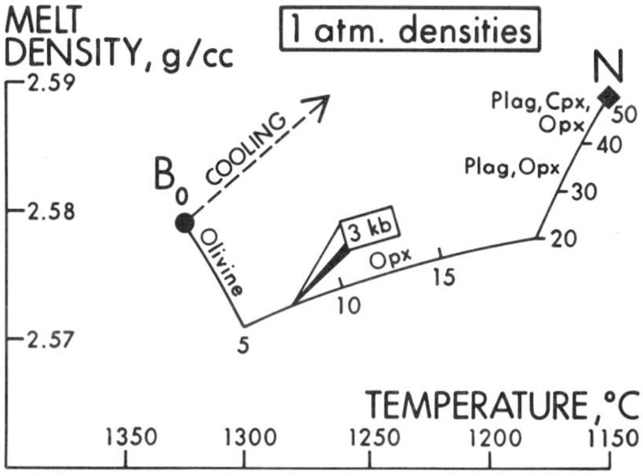

Fig. 5.18. Density variations in fractionating magma. The variations are based on calculated liquids obtained by the computer-simulated fractional crystallization of Cawthorn and Davies's (1983) best representative for the magma responsible for the Lower zone of the Bushveld Complex. The simulation was constrained by the sequence and temperatures of phase appearance and disappearance determined by them on this liquid. (Modified from Barnes and Naldrett, 1986.)

plume, and subsequently in the convecting layer produced by the hot new magma (Fig. 5.19c). The model by Naldrett et al. (1986) is based on the premise that the rapid heat loss at the top of the new convecting layer to the overlying, much colder magma would cause the new layer to convect very rapidly and thus turbulently at first (as the calculations of Huppert and Sparks, 1980, indicate). As heat is lost, crystals of bronzite and/or olivine would form, but as long as convection was turbulent, these, plus any sulfide droplets present, would be kept in suspension (also in accordance with Huppert and Sparks's calculations). Eventually, as the new layer cooled and the rate of heat loss decreased, convection would become laminar, and the suspended crystals, plus sulfides, plus some entrained liquid would sink through the underlying magma as a series of downspouts to spread out over the crystal pile (Fig. 5.19d) as a discrete orthocumulate layer. Huppert et al. (1984) have described experiments with liquids of differing density, temperature, and viscosity. They have shown that when the viscosities are similar, overturn of the kind called on here results in mixing of the two liquids. They describe, however, a set of experiments with liquids of very different viscosity, in which transfer takes place with the lower liquid ending up at the top of the chamber and vice versa with very little mixing. In the case under consideration of descending streams of a mixture of silicate crystals, liquid and minor sulfides, the effective viscosity of the crystal–liquid mixture will be much higher than that of the basaltic melt through which it sinks, providing an analogy with some of the experiments of Huppert et al. (1984) and lending support to the contention

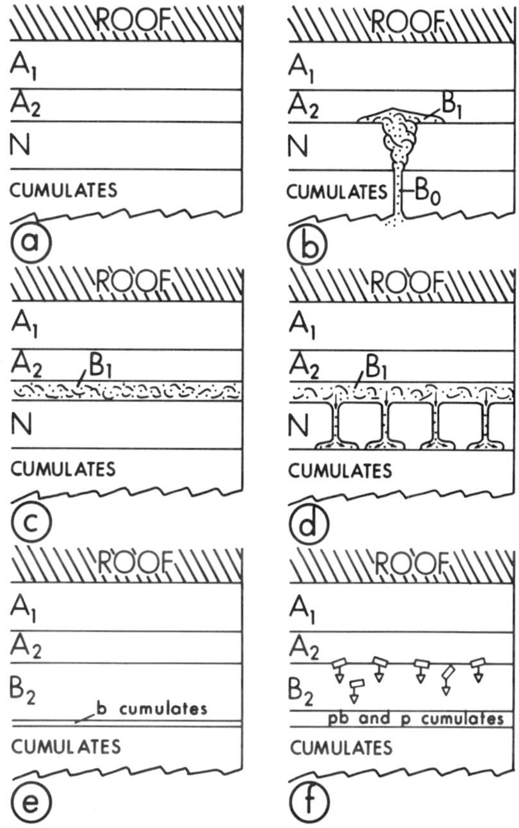

Fig. 5.19. A proposed model for the formation of the Merensky and other cyclic units, showing the formation of the Merensky Reef. ***A.*** Liquid N is crystallizing pbC or pC, and is overlain by two liquid layers (A_1 and A_2, each with plagioclase alone on the liquidus). ***B.*** An injection of fresh, relatively ultramafic magma (B_0) occurs as a turbulent plume, and entrains sufficient N into itself to form liquid B_1, which has bronzite on the liquidus, and at the same time segregates a small amount of sulfide. ***C.*** B_1 spreads out at its own density level and, because it is hot in comparison with the overlying layers of magma, loses heat to them very rapidly and convects vigorously (turbulently); as B_1 cools, it crystallizes bronzite which, together with the sulfide, is kept in suspension by the turbulent convection. ***D.*** As B_1 loses heat, the convection becomes less vigorous and eventually laminar rather than turbulent; at this stage bronzite and sulfides settle to break through the underlying layer (N) as a series of downspouts of crystals + sulfides + entrained liquid, to spread out over the cumulate pile as an orthocumulate (which subsequently crystallizes to form the Merensky Reef). ***E.*** Continued cooling of B_1 results in its density exceeding that of N, whereupon the two liquids finger-mix to form hybrid B_2; this then gives rise to the bronzitite of the Merensky cyclic unit. ***F.*** Cooling of the whole system results in liquid A_2 arriving at its liquidus, at which stage plagioclase separates and sinks into B_2, to join the cumulate pile and give rise to a transitional norite; crystallization of this plagioclase raises the density of A_2, which reaches that of B_2, and the two then finger-mix to form another hybrid from which the bulk of the anorthosite crystallizes. (From Naldrett et al., 1987.)

that the streams will sink to the base of the chamber, rather than mixing on the way down.

Naldrett et al. (1984) proposed that this mixture of bronzite plus sulfide plus liquid crystallized to form the Merensky Reef. Because they have been swirled around within the plume and then in the convecting layer, the sulfides, which in the proposed scenario constitute only a small proportion of the total dissolved sulfide in the magma from which they have segregated, have come into contact with a large amount of this magma (i.e., the system has achieved a high R factor) and thus have developed a high PGE tenor. They have reached equilibrium with the relatively primitive hybrid magma produced by the mixing of the new influx and entrained old magma; thus they have a high Ni/Cu ratio.

Subsequently (Fig. 5.18), cooling of the new layer will raise the density of its magma so that this will sink into the underlying layer of old magma. At this stage the new magma will not be highly charged with crystals because it will have been convecting in a laminar rather than in a turbulent manner, and in consequence the crystals will have been settling out. This new magma will, therefore, have a similar viscosity to the underlying old magma and will mix with it to form a second, less primitive hybrid. This will then crystallize bronzite, followed by bronzite plus plagioclase, to form the bronzitite and norite of the Merensky unit (Fig. 5.19e). Minor amounts of sulfide separating during this stage of crystallization will have a Ni/Cu ratio reflecting the less primitive nature of this hybrid in comparison with that responsible for the Reef itself. They will not, of course, have achieved the high magma/sulfide ratio necessary for high PGE tenors. Similarly, the MgNo of the bronzite will reflect the less primitive compositon of the hybrid. Diffusion and reaction across the boundary between this magma and the underlying Merensky Reef may have caused chromite saturation and thus given rise to the top chromite layer before the newly formed hybrid as a whole had cooled to its liquidus and had started to crystallize bronzite.

Naldrett et al. (1986) regard the Pseudoreef and Bastard units as the consequence of separate influxes of magma that underwent similar life cycles to that discussed previously for the Merensky unit. The Pseudoreef and tarentaal of the Pseudo unit are seen as forming in the same manner as the Merensky Reef, but the proportion of sulfide to silicate was much less, accounting for the lower grade of the mineralization. The Boulder bed, equivalent to the Pseudoreef at Rustenburg, is thought to be a case of downspouting in which the supply of descending material, and the relative viscosities of the descending material and the magma through which it was sinking (Huppert et al., 1984), resulted in the downspouting material breaking up into a series of "drops," rather like water from a dripping tap. Sulfides at the base of the Bastard probably also achieved high magma/sulfide ratios in a plume, but here there is no evidence of downspouting of crystals plus sulfides plus magma, and it is possible that the sulfides settled on their own, becoming modified as they sank through the underlying magma, but retaining enough of their former PGE-rich composition to be distinguishable from those that formed subsequently as the main part of the Bastard unit crystallized.

Referring to some of the major similarities pointed out between the Merensky, and J–M Reefs at the beginning of this section, the model outlined previously explains the occurrence of the reefs a few hundreds of meters above the incoming of cumulus plagioclase, as it is only after plagioclase has been crystallizing for some time that the density of the residual magma will increase above that of the new input, and the sequence of events of the model can occur. The involvement of new primitive magma in the crystallization also explains the reappearance of high temperature cumulus phases. The formation of the Merensky Reef itself through the "downspouting" of suspended crystals means that it will collect as a relatively loosely packed assemblage of grains and interstitial fluid, as an orthocumulate. The magma interstitial to the grains will not have lost low melting point constituents, and will remain partially liquid much longer than the overlying and underlying adcumulate layers. So long as it remains unsaturated in them, any ascending volatiles (H_2O, CO_2, and halogens) will have to dissolve in this layer to pass across it, and the layer will, therefore, serve as a trap for them. This accounts for its recrystallization to pegmatoid, the prevalence of hydrous minerals, and the high concentrations of halogens within them.

5.4 MINERALIZATION IN THE GREAT DYKE OF ZIMBABWE

5.4.1 Geology of the Great Dyke

The Great Dyke of Zimbabwe (2.47 Ga; Hamilton, 1977) is 550 km in length and cuts Archean granites and greenstone belts of the Zimbabwe craton. It consists of two major magma chambers (North and South chambers, Fig. 5.20) (Prendergast, 1988; Wilson and Prendergast, in press) divided into a number of subchambers, each of which has a thick sequence of ultramafic rocks (the Ultramafic sequence) capped by gabbros and norites of the Mafic Sequence.

A transverse section of the Great Dyke is trumpet-shaped with inward dips increasing from the central axis toward the margins but decreasing again near the walls (Wilson and Prendergast, in press). Individual layers thin away from the axis and eventually become incorporated in the Border Group with its layering parallel to the walls (Wilson and Prendergast, 1987). At depth it has a

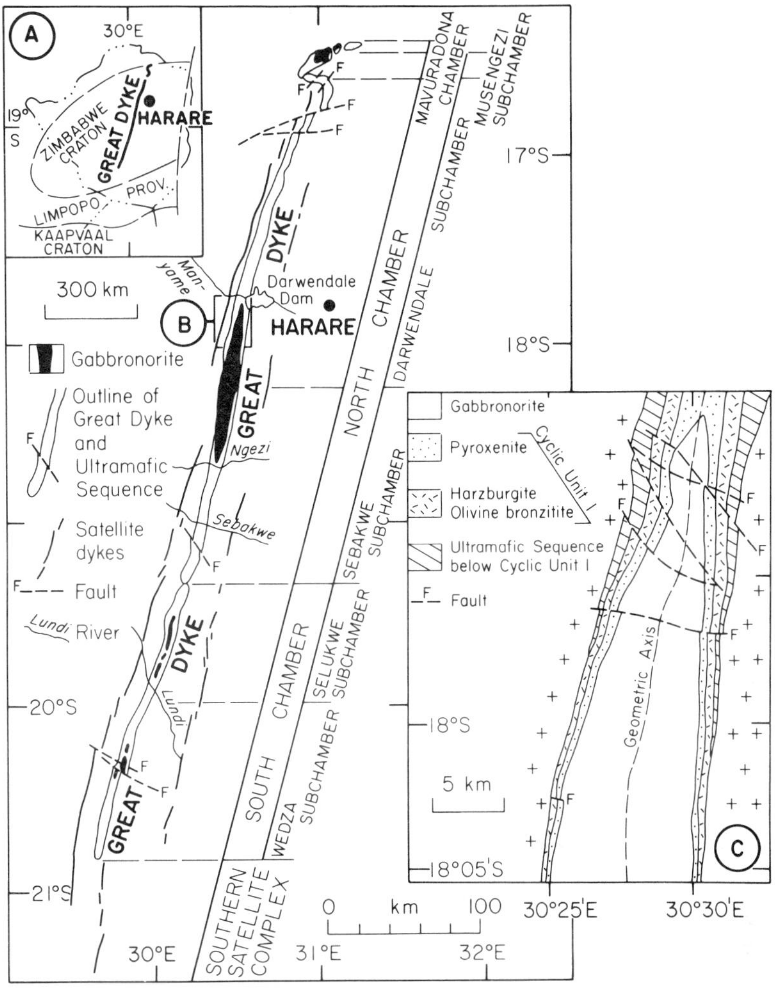

Fig. 5.20. The Great Dyke of Zimbabwe, showing the location of Cyclic Unit No. 1 in the Darwendale Subchamber. (After Wilson et al., in press.)

dikelike feeder (Podmore and Wilson, 1987). In longitudinal section the layers plunge gently toward the center of each subchamber, giving an overall boatlike structure. The Mafic sequence is preserved as remnants in the center of each of the subchambers and has been extensively eroded. In the Darwendale subchamber of the North Chamber, 2,000 m of the Ultramafic sequence is overlain by 1,120 m of the Mafic sequence (Wilson and Wilson, 1981).

Cyclic layering is well developed in the Ultramafic sequence and 14 cyclic units (Jackson, 1970) are recognized in the Darwendale subchamber. The cyclic units in the lower (dunitic) part of the Ultramafic sequence are defined by narrow chromitite layers, whereas those in the upper (bronzitite-rich) part, progress from a basal chromitite, through dunite, harzburgite, and olivine bronzitite, to a well-developed bronzitite layer at the top (Wilson and Prendergast, 1987). The origin of the cyclic units is repeated injections of primary magma followed by fractionation and differentiation (Wilson, 1982).

5.4.2 Sulfide Mineralization

The uppermost cyclic unit (Cyclic Unit 1) of the Darwendale subchamber contains several zones of sulfide mineralization (Worst, 1960; Bichan, 1969; Prendergast, 1988; Wilson et al., in press). This Unit (Fig. 5.21) is 400 m thick at the axis of the Great Dyke in the Darwendale subchamber and consists of a number of subunits defined on the basis of narrow chromitite layers, repeated stratigraphy, and reversals in mineral compositions (Wilson, 1982). Although sulfide mineralization has been reported from the lower subunits (Bichan, 1969), the main metal zones are found in Subunits 1a and 1b and are referred to as Sulfide zones 1(i) 1(ii), 1(iii), 2, and 3 by Wilson et al. (in press) and are also referred to in this way here. Sulfide Zone 1, the Main Sulfide Zone (MSZ) (Bi-

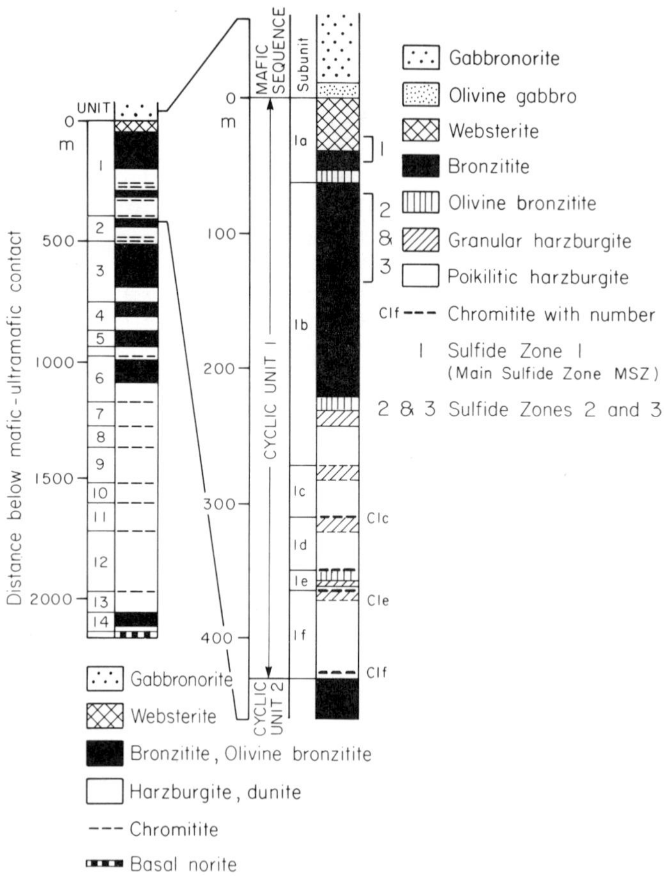

Fig. 5.21. Stratigraphy of the Ultramafic sequence of the Great Dyke in the Darwendale subchamber with special reference to Cyclic Unit No. 1. (After Wilson, 1982.)

chan, 1969), consists of up to 8 percent of disseminated sulfide (pyrrhotite, chalcopyrite, pentlandite, and pyrite), occurs at the top of the main bronzitite layer, and extends into the overlying websterite. Sulfide Zones 2 and 3 are more attenuated than Zone 1 and extend over up to 80 m, but they have much lower concentrations of sulfide (<2 percent by volume). These sulfide zones have been identified in all subchambers of the Great Dyke (Wilson and Prendergast, 1987; Prendergast, 1988), but in detail the lithologies, position, and petrologic characteristics differ from one subchamber to the next and even within the same subchamber. The tabular nature of the sulfide zones and apparent lateral persistence of approximately 400 km makes them a major metal resource for PGE (see Table 5.1).

5.4.2.1 CHEMICAL TRENDS WITHIN MINERALIZATION

Wilson et al. (in press) report analytic results on two drill holes through mineralization in the Darwendale subchamber. Figure 5.22 is based on their data and summarizes vertical variations of chalcophile elements and element ratios in the holes. Wilson et al. pointed out that there are a number of systematic variations in element concentrations and ratios that characterize each of the zones and subzones, which led them to conclude that each shares the same cyclicity. Aspects of this cyclicity are illustrated schematically in Figure 5.23 and are described as follows:

1. The Pt and Pd concentrations in the small amounts of sulfide contained in the rocks immediately underlying each sulfide zone increase as the zone is approached.
2. The Pt and Pd content of the sulfides drops sharply as the percentage of sulfide within the rock increases upward into each sulfide zone. The drop is distinctly sharper for Pd than for Pt.
3. Because of the more rapid drop of Pd than Pt, the Pd/(Pd + Pt) ratio also drops sharply as the sulfide content increases at and just above the base of a zone.
4. The Cu content of the sulfides (not illustrated in Fig. 5.22) remains relatively constant, show-

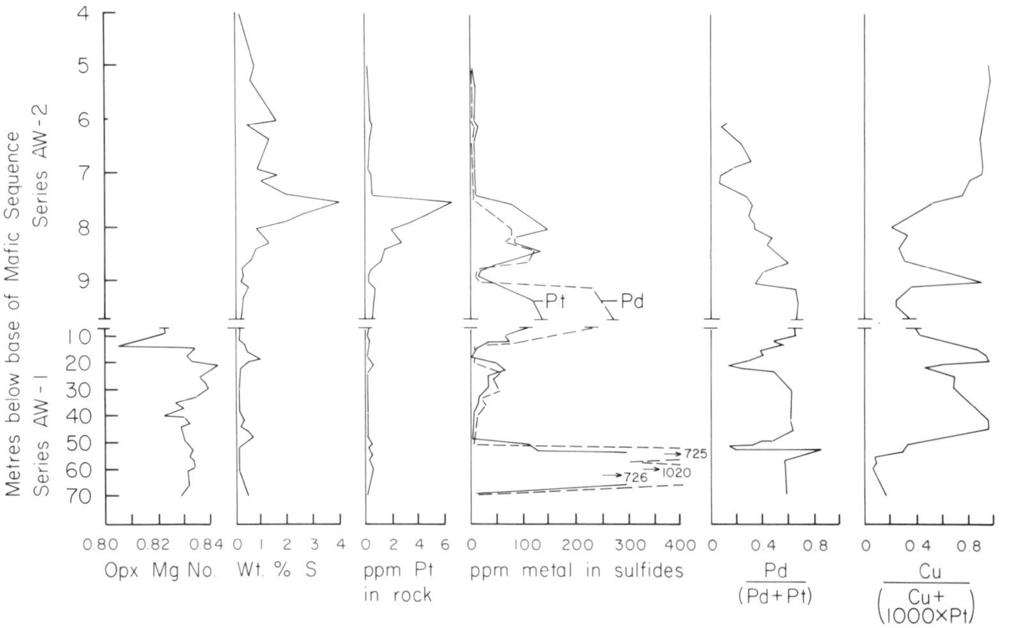

Fig. 5.22. Vertical variations in chalcophile elements and chalcophile element ratios, together with the MgNos of coexisting bronzite, in a transect through sulfide zones 1 (i), (ii), and (iii) (the Main Sulfide Zone), 2 and 3 in the Darwendale subchamber of the Great Dyke, based on data from two drill holes of Wilson et al. (in press).

ing no systematic trends, but because of the drop in Pt the Cu/(Cu + 1,000 × Pt) ratio rises sharply to a value of 0.8 to 0.9 once the sulfide content of the rock increases at the base of a zone.

5. The Pt and Pd content of the sulfides remains low, once this drop has occurred within any given zone, but once the sulfide content of the rock decreases at the top of a zone, Pt and Pd in the sulfides starts increasing. As this occurs, the Pd/(Pd + Pt) ratio rises rapidly to a value of about 0.6 and the Cu/(Cu + 1,000 × Pt) ratio starts declining gently to level out at a value between 0.2 and 0.4.

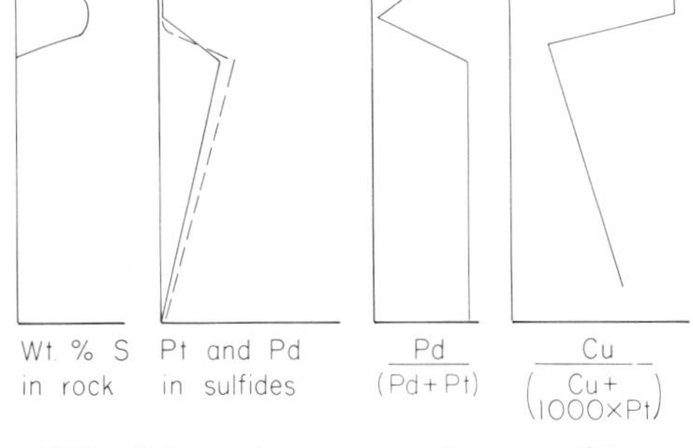

Fig. 5.23. Schematic summary of aspects of the cyclical variation in element concentrations and ratios exhibited by each of the sulfide zones illustrated in Figure 5.22.

5.4.2.2 COMPARISON WITH THE MERENSKY AND J–M REEFS

The PGE mineralization of the P1 pyroxenite of the Great Dyke shows several major differences to the Merensky and J–M Reefs of the Bushveld and Stillwater Complexes.

1. It does not occur at or near the base of clearly defined cyclic units, as do the Merensky and J–M Reefs (Vermaak, 1976; Kruger and Marsh, 1985; Naldrett et al., 1986; Todd et al., 1982).
2. It is not associated with abrupt changes in the composition of minerals, of the order of 10 mole percent Ens in bronzite and 20+ mole percent An in plagioclase as is the case with the Merensky Reef (Kruger and Marsh, 1985; Naldrett et al., 1986).
3. The top of the mineralization is not marked by abrupt decreases in the Pt/Cu ratio, as is the case with the Merensky ore (Naldrett et al., 1986). As outlined previously, cyclic variations in the metal contents of sulfides and metal ratios have been identified in the Great Dyke ore, which have not been described for the Merensky and J–M Reefs.
4. Unlike the Merensky and J–M Reefs, pega-

matoidal texture is not associated with the Great Dyke mineralization.

5.4.2.3 ORIGIN OF CHEMICAL TRENDS

Wilson et al. (in press) note that the systematic and smooth way in which Pd and Pt decrease upward in the sulfides of each zone in the Great Dyke is suggestive of the fractional segregation of sulfide from a column of magma. The fact that Pd becomes depleted more rapidly than Pt is suggestive that its partition coefficient into sulfide is higher than that of Pt. Cu shows essentially no upward depletion in the sulfides; its partition coefficient between sulfide and basaltic magma is known from experiment to be high, about 250 (Rajamani and Maldrett, 1978), which implies that only a very small amount of sulfide can have been removed from the magma. This implies, in turn, that to produce the marked effect that is seen for Pt and Pd, the partition coefficients from silicate magma into liquid sulfide for these metals must be very high indeed, which is consistent with predictions made earlier in this section.

The fractionation hypothesis can be tested by modeling Pt, Pd, and Cu through simple Rayleigh fractionation, using concentrations for Pt and Pd typical for continental flood basalt (10 ppb and 5 ppb, respectively) and partition coefficients of 250 for Cu and $10^{4.5}$ and 10^5, respectively, for Pt and Pd. The results of this calculation are plotted in slightly modified form in Figure 5.24. A reasonable match is obtained between observations and model for most compositional trends.

5.4.2.4 MODEL OF THE MINERALIZATION

Wilson et al. envisaged that a portion of Great Dyke magma became saturated with sulfide (Fig. 5.25) as the result of cooling and crystallization, and that the sulfides formed in this way settled into silicates accumulating on the magma chamber floor to form sulfide zones 3, 2, and 1(iii). These zones have 2.0 percent sulfide or less, which is consistent with their segregating in direct response to magma crystallization. In segregating and settling, they removed the PGE fractionally from their source magma. Wilson et al. calculated from their model that the columns of magma that became sulfide saturated and from which the PGE were fractionated ranged in height from 50 to 500 m.

PGE became exhausted in the magma as a consequence of their fractionation, and before a subsequent PGE-rich sulfide zone could form, this magma required replenishment in PGE. The association of the sulfide replenishment cycles in Figure 5.22 with increases in the MgNo of coexisting bronzite indicate that this was probably achieved by gradual mixing of the depleted magma with slightly more primitive, PGE-rich magma. This pulled the depleted magma off the sulfide saturation curve (Fig. 5.25) toward the sul-

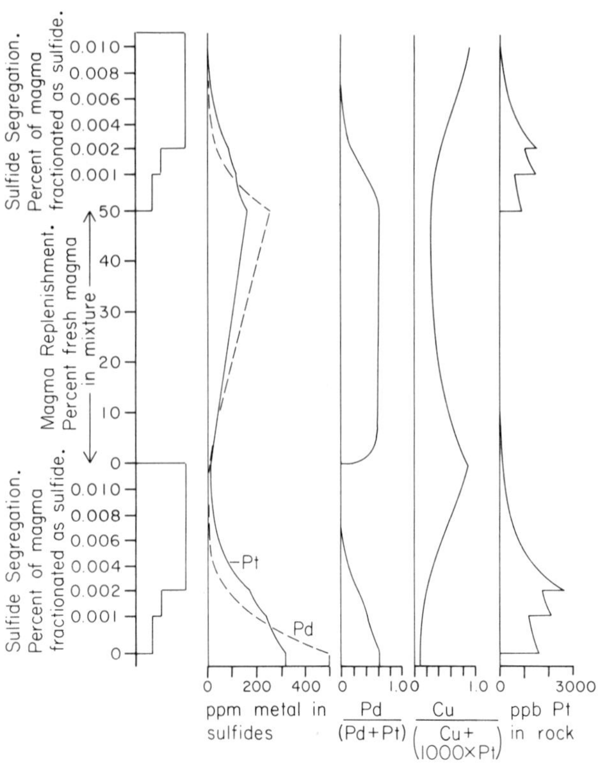

Fig. 5.24. Modeling of the variation of chalcophile element concentrations and ratios in typical sulfide zones as a consequence of fractional segregation of sulfide followed by replenishment of PGE by mixing of the old depleted magma with fresh, PGE-bearing magma.

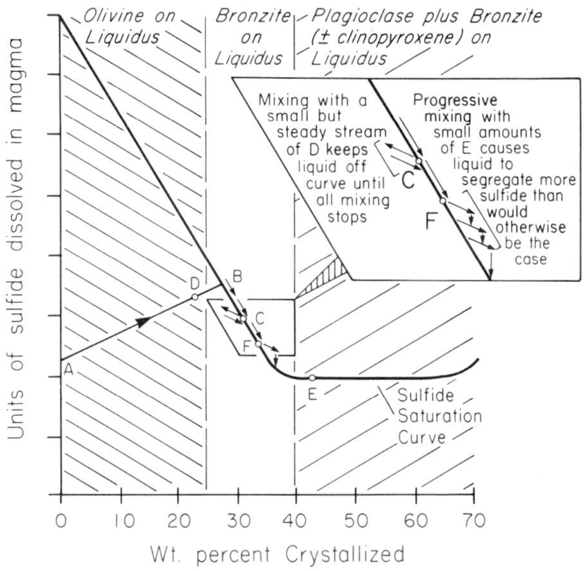

Fig. 5.25. Schematic diagram showing how the solubility of sulfide in the magma of the Great Dyke is likely to vary with fractional crystallization, and mixing with other more and less primitive magmas. This diagram is modified to reflect the more primitive nature of the Great Dyke from that developed for the Bushveld Complex by Naldrett and von Gruenewaldt (1989).

fide unsaturated (left) side, and thus caused a cessation in sulfide segregation while replenishment was going on. It is estimated here that the amounts of fresh magma involved in replenishment ranged from 10 to 15 percent to 40 to 50 percent for different zones. Once replenishment ceased, the magma started to crystallize and thus moved back to the saturation curve, at which stage precipitation of the next sulfide zone started.

Wilson et al. considered that zone 1(i) and perhaps 1(ii) contained too much sulfide to be the result of simple crystallization along the sulfide saturation curve. They proposed that some other more drastic process, capable of inducing sulfides to segregate without the simultaneous crystallization of cotectic proportions of silicates, had to be called upon. Noting that zone 1(i) occurs very close to the contact with the overlying Mafic-layered sequence, they suggested that mafic magma, possibly related to the melting of roofs rocks by the intrusion, gradually mixed with the magma responsible for cyclic unit 1, pulling it off the saturation curve (Fig. 5.24) to the right, into the zone of saturation, thus inducing higher than cotectic proportions of sulfide to segregate.

As noted earlier, the Great Dyke mineralization is very different ot that of the Merensky and J–M Reefs. The model of Wilson et al. cited previously suggests that this could be due to differences in the nature of the magma mixing that gave rise to the different deposits. In the case of the Great Dyke, the new magma appears not to have been very much more primitive than the resident magma, and to have mixed with it in a rather gentle way, leading to transitional effects. As indicated above, with the Merensky and J–M Reefs, rapid inputs of a relatively primitive magma led to the initiation of new cyclic units, the resetting of mineral compositions to more primitive compositions, the batch segregation of PGE-rich sulfides, and the formation of the ore zones as a result of the settling of sulfides intermixed with orthocumulate silicates, which thus gave rise to the pegamatoidal texture of the Reefs.

6. THE COMPOSITION OF MAGMATIC SULFIDE ORES

6.1 INTRODUCTION

Magmatic sulfide ores have, as emphasized previously, a simple mineralogy and composition. Pyrrhotite, pentlandite, chalcopyrite, and lesser amounts of pyrite are the main sulfide minerals; Fe, Ni, Cu, lesser amounts of Co, and very small amounts of the platinum group elements (PGE) and Au are the most important metallic constituents. Magnetite is present in variable amounts; in some cases, this primary mineral reflects the amount of oxygen incorporated in the ore magma, although O may be lost as this magma crystallizes (Naldrett, 1969). In other cases; the ore may be oxidized subsequent to crystallization (see Chapter 3), with the development of secondary magnetite.

6.2 VARIATIONS IN FE, S, AND O CONTENTS

The genesis believed of most Ni sulfide ores places major constraints on the proportion of magnetite and the metal/sulfur ratio of the sulfides that they contain. Where a sulfide liquid first segregates within a silicate magma, it is widely distributed as small droplets in a relatively large volume of the magma. It has been pointed out in Chapter 3 that the composition of the droplets is buffered by the magma, and that the fugacities of O and S and the activities of Fe, Ni, and Cu must be the same in both the droplets and the magma. In the simplified case of an immiscible Fe sulfide–oxide liquid, this phase develops in a silicate magma only where the values of the intensive parameters a_{FeO}, fS_2, and fO_2 in the liquid coincide with those in the magma. Thus, in a magma with a given fO_2 (which will be largely defined, and, to some extent, buffered by the Fe^{+3}/Fe^{+2} ratio) and a given a_{FeO} (which varies between fairly narrow limits for magmas of similar overall composition and FeO content), the fS_2 at which the magma becomes saturated in a sulfide liquid is defined. Reference to Figure 3.6A also shows that specifying the fS_2 and a_{FeO} defines the composition of the sulfide liquid including its S content.

Inasmuch as the Fe^{+3}/Fe^{+2} ratios and FeO contents of most mafic and ultramafic magmas do not vary greatly, the relative proportions of Fe, S, and O in an ore also tend to stay within well-prescribed limits. The presence of Ni and Cu in the associated magma affects the composition of the ore, as these metals are more chalcophile than Fe, but they substitute for Fe on an atom-for-atom basis, and hence the principles just discussed concerning the controls on the relative amounts of metal, S, and O still apply (there are, of course, minor variations due to the different thermodynamic properties of these three metals). Thus, it is rare, for example, for an unaltered ore to contain more than 5 wt percent O (present as magnetite), and the metal/S ratio is not commonly greater or less than that spanned by the field of pyrrhotite at 600°C. Fe and Fe–Ni alloy are not normally consituents of these ores, and troilite, if present, is commonly associated with a more S-rich pyrrhotite phase from which it has exsolved. For the same reason, the amount of pyrite in a magmatic ore should never be more than the amount that can exsolve from pyrrhotite during cooling from liquidus temperature. If appreciable pyrite is present, it is an indication that either the ore is not magmatic or that it has been altered subsequently.

In some cases, a sulfide ore is found in association with an intrusion that has become enriched in carbon by reaction with nearby graphitic slate (Pauly, 1969). Here the mineral assemblage may have an atypically high metal/S ratio but this is the result of an abnormal geologic association leading to an unusually reduced host magma.

6.3 VARIATIONS IN CU AND NI CONTENTS

Compositional data for a large number of ore deposits are listed in Table 6.1. In most cases the averages in this table (those incorporating data from Naldrett et al, 1979; Naldrett and Duke, 1980; Naldrett et al., 1982) are based on samples which were individual and not composite, were relatively large (1–2 kg), and were selected to be as representative as possible for a given deposit. The results of each sample of each deposit have been recalculated as though the sample were composed of 100 percent sulfide and then averaged for the deposit where it was thought that the averaging was appropriate.

Where two or more compositionally distinct ore types are present, information on the relative abundance of the ore types (for example the average Cu/(Cu + Ni) ratio of the ore reserves of the deposit) has been used to weight samples in reaching the average. This calculation has been done in only a very few cases. Data on Cu and Ni are shown in Figure 6.1. The vertical and horizontal bars connected to most data points indicate the standard error of the mean for each element. In essence this figure illustrates all the major element compositional data for the deposits. The S content of most of the ores illustrated is between 38 and 40 wt percent; thus the percentage of Fe plus Cu and Ni accounts for the balance of the major consituents.

It can be seen from Figure 6.1 that ore deposits of a certain association occupy specific fields. Most Archean komatiite-related deposits contain between 10 and 16 wt percent Ni and 0.5 and 1.5 wt percent Cu. Exceptions to this grouping include the Durkin shoot and the Carnilya and Nepean deposits in Western Australia, Epoch in Zimbabwe, and Langmuir No. 1 and McWatters in Ontario. Durkin is described by Ross and Keays (1979) as containing considerable supergene ore, and Nepean has undergone extreme dynamic metamorphism (Barrett et al., 1976). Sulfides at Langmuir No. 1 have been converted to magnetite without destruction of the original net texture of the ore (Green and Naldrett, 1981). Both Epoch and McWatters are in rocks that have been strongly altered to talc and carbonate. In all of these cases, the original Ni/S ratio of the ore appears to have been increased, either by removal of S and Fe or through the addition of Ni, thus accounting for the high level of Ni in sulfide.

Ore deposits related to the Proterozoic komatiites of the Ungava belt are characterized by similar Ni contents (10–16 wt %) but much higher Cu contents (3–4%) than other komatiite-related deposits.

The Sudbury deposits contain between 3 and 6 percent Ni and 1 and 5 percent Cu. The lower limit of 1 wt percent Cu is based on data from the Strathcona deposit and is probably too low as a result of biased sampling. A range of 2 to 5 wt percent for Cu is probably more accurate for Sudbury. The data presented here for Strathcona agree with the studies of Naldrett and Kullerud (1967) and Cowan (1968) who pointed to the strong zoning across the deposit with Cu, Ni, and the Cu/Ni ratio increasing across the deposit from hanging-wall to footwall. This and similar zoning at Levack West parallels zoning in the PGE and is discussed in a later section.

Deposits related to intrusions that are part of flood basalt volcanism are characterized by higher concentrations of Cu than Ni in the sulfides, Cu ranging from 9.7 wt percent at Great Lakes to over 17 at Minnamax in the Duluth Complex, and Ni ranging from 4 wt percent at Minnamax to about 9 in the Noril'sk-Talnakh ores.

Genkin et al. (1980) have emphasized the strong zoning present in the Noril'sk-Talnakh ores; in these deposits the Cu/Ni ratio and the overall Cu content are much higher in massive ores at the base of the associated intrusive bodies and in fractures in the underlying country rocks than in the disseminated ore in the overlying picrite and gabbros. This zonation is very similar to that at Strathcona.

Values for deposits associated with tholeiitic intrusions in greenstone belts (with the exception of the Kanichee deposit) range from 4.5 to 9 wt percent Ni and 1.3 to 2.5 wt percent Cu.

Layers of weakly disseminated sulfide, rich in PGE, that occur in layered intrusions tend to be richer than normal than those forming more massive deposits associated with similar rocks. This is particularly true of the Upper Group-2, although there is a special reason for this as discussed in Section 6.7.5.

6.4 COBALT

In the Archean komatiite-related deposits, Co values are mostly in the range of 0.22 to 0.34 wt percent. The Proterozoic komatiites from Ungava have values of 0.21 and 0.23. The Sudbury ores are characterized by values between 0.15 and 0.20. Co data for deposits related to tholeiitic gabbros in greenstone belts is limited, but values are somewhat over 0.2 wt percent. Very few Co data are available for flood basalt-related deposits.

6.5 PLATINUM GROUP ELEMENTS

6.5.1 Introduction

Data for the PGE are presented in Table 6.1, recalculated to concentrations expected in 100 percent sulfides as discussed earlier. They are also presented diagramatically in Figures 6.2 to 6.9, together with Cu and, in most cases, Ni and Co. Here, to smooth out interelement differences and facilitate comparison of the profiles for different deposits, the analyses of Table 6.1 have been normalized with respect to abundances in type 1 carbonaceous chondrites, which are also given in Table 6.1.

6.5.2 Archean Komatiite-Related Deposits

Data are illustrated in Figure 6.2 for nine deposits of this type from the Kambalda camp of Western Australia (data from Cowden et al., 1986, as recalculated by Naldrett and Barnes, 1986). All deposits exhibit similar, relatively flat profiles for the PGE with near or somewhat above chondritic abundances and relatively low Pt/(Pt + Pd) and (Pt + Pd)/(Ru + Ir + Os) ratios.

Cowden et al. (1986) have shown that massive ores, which occur at the bases of ore-bearing sequences of the Kambalda camp, tend to be signif-

TABLE 6.1. *Analytical Data on Magmatic Sulfide Ores Normalized to 100% Sulfide*

	Average % Sulfide	Ni	Cu	Co	As	Sb	Se	Pt	Pd
Archean, flow-related komatiites									
Langmuir No. 2	61	14.6	0.67	0.27	45		18	890	1,600
		(1.31	(0.12)	(0.020				(138)	(300)
Langmuir No. 1	34	30.2	0.95	0.59			100	2,660	6,350
		(5.89)	(0.25)	(0.009)			(29)	(45)	(2,194)
McWatters	60	21.0	0.52	0.32				1,134	2,391
		(2.56)	(0.07)	(0.065)				(537)	756)
Marbridge		6.5	0.38					360	920
Epoch									
mass		21.3	0.35	0.27				624	1,310
		(1.61)	(0.064)	(0.014)				(82)	(154)
diss		30.0	nr	nr				4,202	9,111
		(1.52)						(768)	(1,792)
Shangani									
mass		13.7	0.39	0.16				837	2,762
		(0.55)	(0.13)	(0.017)				(229)	(354)
diss		23.7	nr	nr				2,511	5,260
		(2.70)						(356)	(765)
Trojan									
mass		10.7	0.52	0.22				572	2,184
		(1.45)	(0.14)	(0.025)				(144)	(2,793)
diss		16.3	nr	0.43				1,628	2,783
		(0.88)		(0.02 6)				(76)	(170)
Lunnon		12.8	1.00					1,533	1,875
Hunt		11.1	0.86					1,515	1,375
Juan		13.8	0.92					1,822	1,863
Fisher		13.5	0.85					1,438	1,773
Long		14.4	1.04					1,095	1,740
McMahon		12.2	0.94					818	1,607
Ken		16.1	1.33					1,497	2,303
Durkin		18.2	1.36					2,024	3,067
Jan		16.2	1.21					1,489	3,256
Carnilya		21.4	1.29					1,717	2,015
Nepean		18.6	1.06					1,172	2,082
Mt. Edwards		12.3	0.93					1,032	1,397
Mt. Windara							16		
Perseverence							11		
Archean, komatiitic dunite lens									
Betheno	2.4	35	0.6				20		3,098
Proterozoic komatiites									
Donaldson West	40	15.5	3.7	0.23	136	16	99	4,130	15,530
		(0.31)	(0.26)	(0.007)				(682)	(3,886)
Katiniq	43	10.5	3.11	0.21				3,320	6,710
		(0.36)	(0.83)	(0.005)				(434)	(462)
Manitoba komatiites									
Pipe	83	3.96	0.16	0.156	139		28	54	122
		(0.13)	(0.016)	(0.003)				(7)	(14)
Manibridge		10.00	0.75					950	2,600
Bucko Lake	6	20.6	2.37					2,441	4,968
Sudbury									
Levack West								2,490	2,867
Main ore	57	5.73	3.72	0.16	<3.5	<0.25	92	1,150	1,250
		(0.31)	(1.95)	(0.008)				(108)	(225)
Cu-rich	93	5.52	27.6	0.027	<3.1	<0.2	194	11,430	13,670
Stringers		(0.95)	(2.59)	(0.009)					
Strathcona	55	3.63	1.23	0.15				420	370
		(0.13)	(0.18)	(0.012)				(69)	(60)
Cu-zone		1.72	24.67	0.05				4,910	6,970
Little Stobie 1	24	3.83	4.41	0.19	160	<2.4	94	1,930	2,120
		(0.08)	(0.65)	(0.003)				(415)	(344)
Little Stobie 2	40	4.00	3.60	0.17	93	<1.3	10	2,130	3,170
		(0.18)	(0.65)	(0.004)				(871)	(829)
Falconbridge	71	5.35	1.52	0.22				550	380
		(0.14)	(0.39)	(0.014)				(109)	(95)

Rh	Ru	Ir	Os	Au	$\delta^{34}S$	Zn	Pb	n	Ref.
230	760	240	414	181	0.3	30	8	31	4
(32)	(123)	(31)	(73)	(59)					
460	1,240	420	1,020	460	−1.9	14	6	7	10
(109)	(294)	(91)	(220)	(138)					
726	1,466	562	532	156				5	14
(118)	(272)	(117)	(103)	(69)					
80				180					3
1,436	4,938	598	668	155				14	14
(136)	(414)	(49)	(66)	(21)					
1,160	3,138	461	511	695				5	14
(210)	(568)	(47)	(50)	(109)					
673	1,024	98	93	56				15	14
(69)	(181)	(13)	(15)	(24)					
683	1,476	318	428	312				5	14
(88)	(189)	(108)	(164)	(41)					
363	1,227	159	177	56				10	14
(77)	(224)	(37)	(36)	(14)					
372	1,109	235	240	176				18	14
(28)	(119)	(29)	(25)	(22)					
248	632	189	422	4,112				*	*
223	803	189	341	4,654				*	*
315	1,028	239	530	581				*	*
275	779	190	424	4,760				*	*
310	972	236	408	515				*	*
291	916	229	422	366				*	*
337	1,028	248	461	456				*	*
503	1,557	347	837	678				*	*
391	1,038	227	499	467				*	*
475	1,462	436	789	407				*	*
347	1,192	392	709	277				*	*
314	926	228	395	420				*	*
					−1.9				8,10
					−2.2				7,8
			775	130					1
800	1,980	320	380	460	3.5	42	11	23	4
(103)	(350)	(53)	(55)	(94)					
640	1,364	230	350	260				23	4
(61)	(206)	(30)	(45)	(64)					
35	217	55	107	57	4.1	12	9	25	4
(8)	(22)	(3)	(10)	(11)					
260				240					14
547	1,548	433	704	776				6	23
168	69	47		1,198					13
186	60	47	22	150	2.58	227	31	21	4
(32)	(14)	(11)	(4)	(43)					
<10	<3	0.3	<2	8,211	3.13	3,176	258	4	2,13
	(0.12)								
30	21	12	8	54		233	26	37	5
(4.7)	(4.4)	(2.4)	(1.6)	(11)					
4	3	0.2	2	4,880		2,750		21	5,13
120	120	62	29	860	1.08	100	15	23	4
(12)	(16)	(6)	(3)	(23)					
300	250	110	46	860	0.46	96	14	11	4
(28)	(40)	(14)	(7)	(314)					
290	230	140	40	170		120	14	23	5
(28)	(31)	(14)	(6)	(60)					

TABLE 6.1. *Analytical Data on Magmatic Sulfide Ores Normalized to 100% Sulfide* (*Continued*)

	Average % Sulfide	Ni	Cu	Co	As	Sb	Se	Pt	Pd
Other gabbro-related deposits									
Montcalm	63	4.31	1.38	0.24	<3		36	58	17
		(0.21)	(0.22)	(0.014)				(7)	(3)
Espedalen	16	6.65	2.01					330	250
		(0.43)	(0.48)					(84)	(36)
Pechenga		4.6	2.45					600	500
Kanichee	54	6.63	8.09	0.27				3,870	13,670
		(0.594)	(2.28)	(0.021)				(691)	(2,493)
Kenbridge	40	8.92	2.21	0.21				772	243
		(0.37)	(0.53)	(0.017)				(194)	(43)
Flood Basalt									
Minnimax	19	3.91	17.2	0.37				2,640	8,840
		(0.32)	(1.60)	(0.082)				(950)	(4,534)
Great Lakes	8.5	5.23	9.76	0.20				5,580	22,450
		(0.45)	(0.87)	(0.004)				(792)	(2,622)
Insizwa (exc. Cu-rich ore)	39	8.32	7.56	0.22				4,495	2,393
		(0.83)	(1.06)	(0.028)				(824)	(361)
Cu-rich ore	60	7.92	17.41	0.149				4,452	3,810
		(0.84)	(6.28)	(0.044)				(1,387)	(1,252)
Talnakh									
Average		7.60	10.90					13,700	36,000
Komsomolsky	86	5.50	8.72	0.15				4,460	19,330
Noril'sk									
Picritic ore		9.0	10.8					18,000	47,000
Taxitic ore		7.6	10.9					15,200	42,100
Large-layered Complexes									
Bushveld									
Merensky Reef	10.9		4.65					279,000	120,000
UG-2 (E. Bushveld)	22.5		9.75	0.55				1,235,000	1,112,000
Platreef (Drenthe)									
Upper		5.56	5.57	0.19				9748	16,718
		(0.54)	(0.56)	(0.012)				(1604)	(1936)
Lower		8.70	6.05	0.32				37,132	43,878
		(1.39)	(0.69)	(0.082)				(8035)	(7615)
Stillwater									
J–M Reef	13.2	(0.86)	6.49					1,560,000	5,410,000
Picket Pin			(0.55)					15	21
Chondrites									
Average Cl. Chondrite	10.8		0.0108	0.051				1,021	545

Ni, Cu, Co in wt percent, As, Sb, Se, Zn, and Pb in ppm, PGE and Au in ppb
nr = not reliable; mass = massive ore; diss = disseminated ore; n = number of samples included in average
Figures in parenthesis below a value indicate standard error of mean.
Ni, Cu, S determined by XRF by X-Ray Assay Laboratories Ltd.; Co, Zn, Pb determined by atomic absorption; As, Se determined by INAA; PGE and Au determined by fire assay and INAA; δ^{34} S determined by Geochron Laboratories Inc.

REFERENCES

1. Groves and Keays, 1979
2. Hoffman et al., 1979
3. Naldrett and Cabri, 1976
4. Naldrett and Duke, 1980
5. Naldrett et al., 1982
6. Seccombe et al., 1981
7. D. T. Groves, unpublished data
8. Donnelly et al., 1977
9. Barrett et al., 1976
10. Seccombe et al., 1977
11. Gorbunov, 1968

Rh	Ru	Ir	Os	Au	$\delta^{34}S$	Zn	Pb	n	Ref.
2.4	3.7	0.7	<1.5	82	−1.81	44	17	21	4
(0.29)	(0.56)	(0.09)		(12)					
36	48	26	27	280				6	4
(17)	(16)	(9)	(19)	(71)					
				200					11
550	450	550	390	2,370				14	14
(82)	(74)	(101)	(74)	(793)					
30	19	15	15	562				9	14
(4.7)	(5.3)	(1.4)	(3.6)	(264)					
120	120	48	<53	1,170				22	4
(24)	(26)	(10)	(13)	(292)					
600	310	74	120	2,590				29	9
(66)	(43)	(9)	(22)	(285)					
263	498	297	190	1,344				15	14
(30)	(120)	(55)	(44)	(267)					
463	738	239	152	1,526				6	22
(160)	(482)	(103)	(65)	(863)					
2,240	1,500	1,500	950	1,600					12
1,210	324	104	67	600				2	14
1,700				1,600					3
1,400				1,700					3
19,000	35,000	7,800	5,200	22,000					17
198,000	371,000	99,000	37,000	34,000					18
327	806	117	98	3408					This Study
(62)	(117)	(21)	(15)	(652)					
2534	3226	490	342	7425					This Study
(544)	(721)	(103)	(62)	(1727)					
48,000	16,000	6,000	7,000	200,000					19
									20
200	690	540	514	152					21

12. Smirnov, 1977
13. Coats & Snajdr, 1984
14. Naldrett, 1981a
15. von Gruenewaldt et al., 1986
16. Gain, 1985
17. Ni and Cu from Ref. 3, PGE and Au from Ref. 15
18. Ni, Cu, and Co from Ref. 16, PGE and Au from Ref. 15
19. Barnes and Naldrett, 1985
20. Page, et al., 1985
21. Ni, Cu, and Co from Palme et al., 1981; PGE and Au from Ref. 4
22. Lightfoot et al., 1984
23. Good (1985)

*These data on W. Australian deposits were determined from composite samples of monthly production periods, samples for two periods being taken and averaged in most cases. The data are those published by Cowden et al. (1986) but have been recalculated here to conform with the method used for all other samples in this table (as described by Naldrett and Duke, 1980).

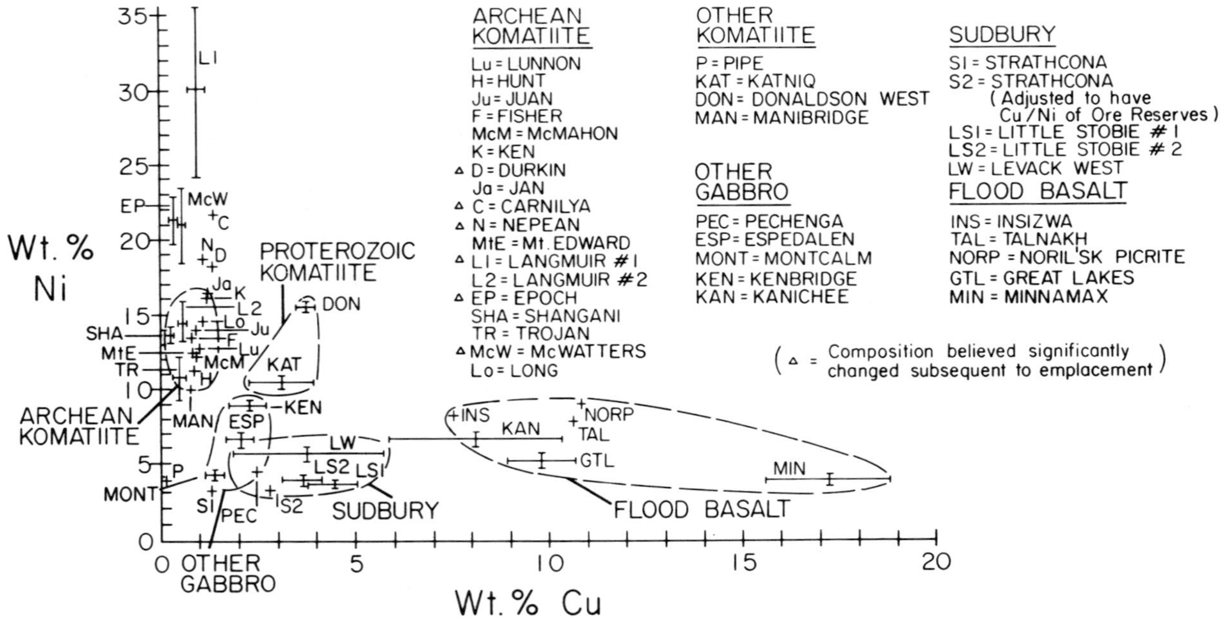

Fig. 6.1. The Ni and Cu contents of sulfides from a variety of Ni–Cu sulfide ores. The different types of ores tend to fall within well-defined areas as shown, with the exception that altered, Ni-enriched Archean komatiite-related deposits are not included within this field and the Kanichee deposit is not included with the field defining deposits related to gabbros in greenstone belts. The bars represent the mean plus and minus the standard error of the mean where this is known. (Modified after Naldrett, 1981a.)

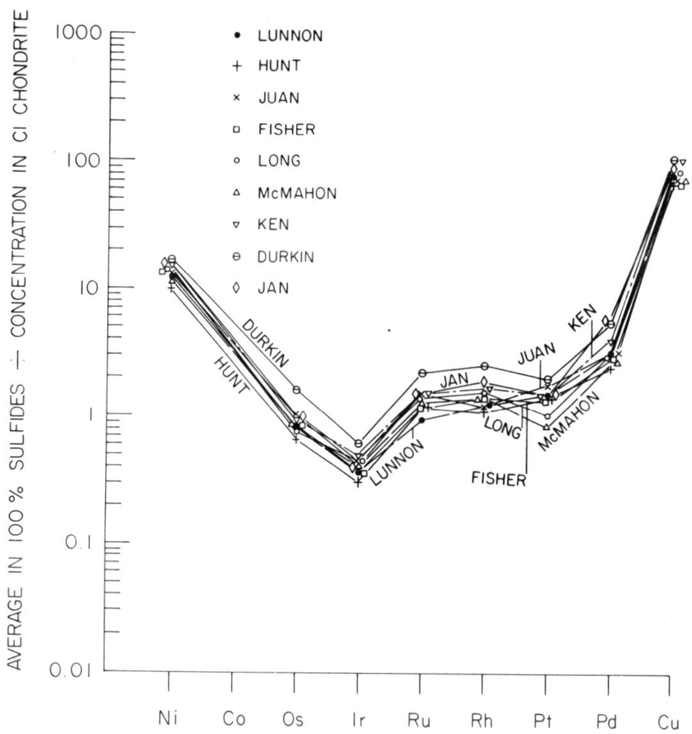

Fig. 6.2. Analytic data for Co, Ni, Cu, and the PGE for nine Archean komatiite-related deposits from the Kambalda mining camp, Western Australia. Data are presented recalculated to 100 percent sulfides and then normalized by the chondritic abundances for each element. The chondritic abundances used for the PGE are those compiled by Naldrett and Duke (1980), those for Co, Ni, and Cu are from Palme et al (1981); they are listed in Table 1. Analytical data are after Cowden et al. (1986).

icantly poorer in Pt, Pd, and Au than the immediately overlying net-textured disseminated ore. In some cases, very high concentrations of Pt and Pd occur at the extreme base of the ore-bearing sequences. Os, Ir, Ru, and Rh tend to be more uniformly distributed in the ore profile. Keays and Davidson (1976), Ross and Keays (1979), and Keays et al. (1981) report that Cu-rich stringers in basalt underlying the massive ore in certain Kambalda deposits are also enriched in Pd. In addition, they report a systematic increase in Pd and a decrease in Ir upward in disseminated ore overlying the massive ore.

6.5.3 Other Komatiite-Related Deposits

Data for two deposits related to the Proterozoic komatiitic rocks of Ungava and the Bucko Lake and Manibridge deposits of the Manitoba Ni Belt are illustrated in Figure 6.3. The deposits exhibit similar PGE contents. The profiles differ from the Archean field primarily in their higher levels of Pd and higher (Pt + Pd)/(Ru + Ir + Os) ratios. The key to this difference is thought to lie in the lower MgO content (<18 wt %) of these magmas in comparison with the Archean magmas (24–27 wt %), as discussed in Section 6.7.1.

The Pipe deposit from the Manitoba Ni belt is also illustrated in Figure 6.3. The deposit exhibits the flat profile characteristic of the Archean deposits, but contains three to ten times lower concentrations of PGE than the latter. Reasons for this are discussed in Section 6.7.3.1.1.

6.5.4 Sudbury

Data for the Little Stobie 1 and 2 and Levack West deposits of the Sudbury district are illustrated in Figure 6.4. All of the deposits are characterized by Pt and Pd values greater than chondritic abundances and by Ru, Ir, and Os abundances very much lower than chondritic, leading to high values of the (Pt + Pd)/(Ru + Ir + Os) ratio. For most elements, Levack West has the lowest and Little Stobie 2 the highest concentrations.

Data for the Strathcona and Falconbridge deposits are compared with those for the three preceding deposits in Figure 6.5. Strathcona shows a similar steep profile, but the absolute concentrations are only one-third to one-tenth of the levels at Levack West. These data are unusual in view of the geographic and geologic similarities of the two deposits.

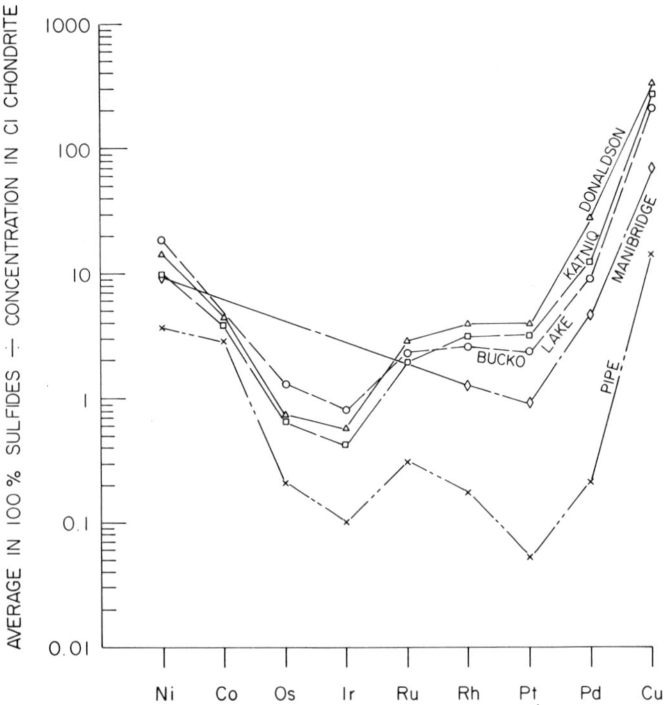

Fig. 6.3. Chondrite-normalized plot for Proterozoic komatiite-related deposits. Data from Naldrett and Duke (1980), Naldrett (1981b), Good (1985). Donaldson and Katiniq are from the Ungava belt, Manibridge, Pipe and Bucko Lake from the Manitoba belt. See legend to Figure 6.2 for further explanation.

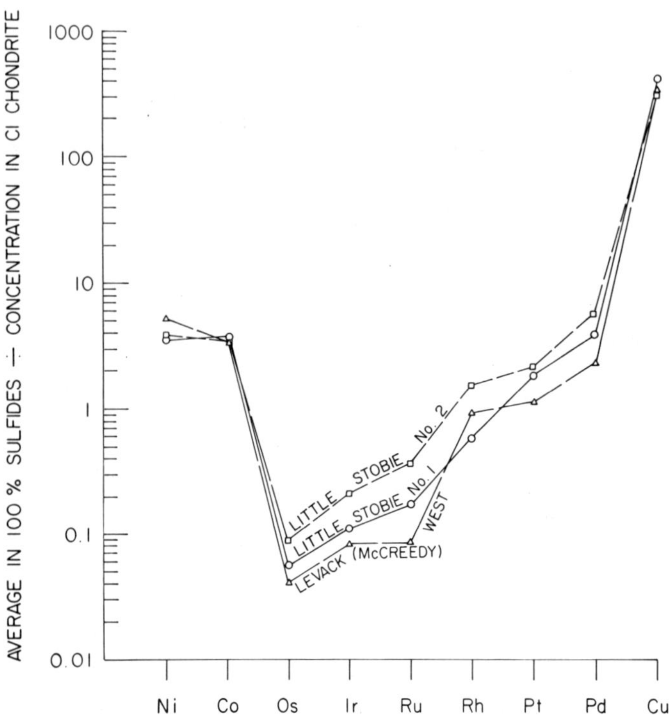

Fig. 6.4. Chondrite-normalized plot for the Little Stobie No. 1 and No. 2 and Levack (McCreedy) West deposits at Sudbury, Ontario. Data from Hoffman et al. (1979). See legend to Figure 6.2 for further explanation.

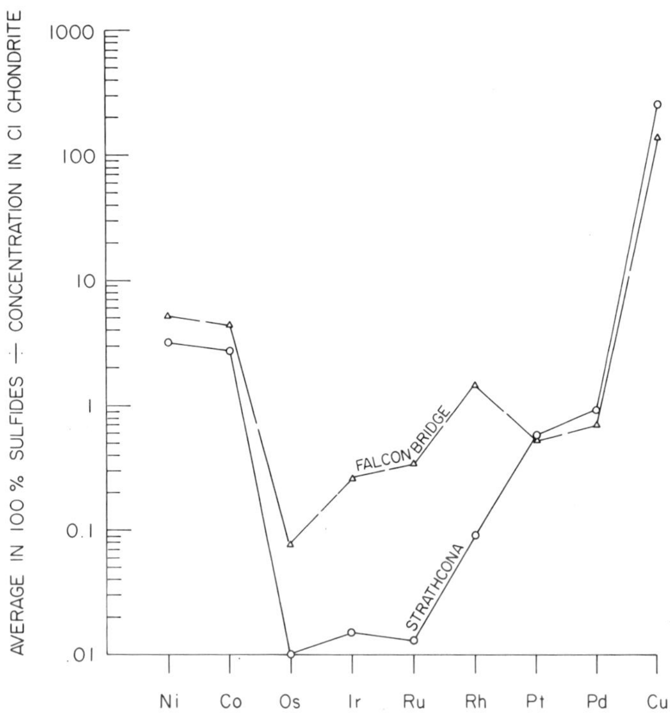

Fig. 6.5. Chondrite-normalized plot for the Strathcona and Falconbridge deposits at Sudury, Ontario. Data from Naldrett et al. (1982). See legend to Figure 6.2 for further explanation.

The Falconbridge deposit is also unusual in that, although the Os, Ir, Ru, and Rh levels are very similar to the richest reported for Sudbury to date, the concentrations of Pt and Pd are very much lower, approaching those of Strathcona.

The PGE data for Strathcona also show a pronounced zoning within the deposit (Fig. 6.6A), a feature that parallels that previously described for Ni and Cu. Levack West shows a very similar zoning (Fig. 6.6B). In both deposits, ore within or closest to rocks of the Sudbury Igneous Complex is characterized by shallow profiles corresponding to low (Pt + Pd)/(Ru + Ir + Os) ratios, whereas that farthest into the footwall shows the steepest profiles and highest ratios.

6.5.5 Deposits Associated with Flood Basalt Magma

Data for deposits with this association are compared in Figure 6.7. The deposits include: the Early Triassic Noril'sk-Talnakh deposits of Siberia (data from Smirnov, 1977, calculated as described by Naldrett and Cabri, 1976); the Minnamax deposit in the 1.1 Ga-old Duluth Complex in Minnesota; the Great Lakes Ni deposit also associated with a 1.1 Ga-old intrusion west of Thunder Bay, Ontario; and the Waterfall Gorge deposit of the Insizwa Complex, Transkei. All four deposits are characterized by significantly higher values of Pt, Pd, and Au than those typical of Sudbury. Insizwa differs from other deposits in having a much higher Pt/(Pt + Pd) ratio of 0.65 and lower (Pt + Pd)/(Ru + Ir + Os) ratio of 6.99. There are indications in the data that these ratios are a function of the degree of differentiation of the host magma, Insizwa being the least differentiated, as discussed in Section 6.7.1.

6.5.6 Other Deposits Associated with Gabbros

All of these deposits show the steep slope characteristic of gabbroic magmas (Fig. 6.8), although the variation in PGE levels is very large, ranging from a high in the Kanichee deposit, which has similar concentrations to those deposits associated with flood basalts, to a low in the Montcalm ore in which the PGE levels are almost as low as in some mafic and ultramafic rocks.

6.5.7 PGE-rich Stratiform Deposits Associated with Layered Intrusions

These deposits may be characterized by extremely high concentrations of PGE. In general those associated with chromitite layers tend to have higher concentrations of PGE than those lacking this association, as illustrated by the contrast between the Upper Group-2 data and that of the Merensky Reef in Figure 6.9. Naldrett and Leh-

A

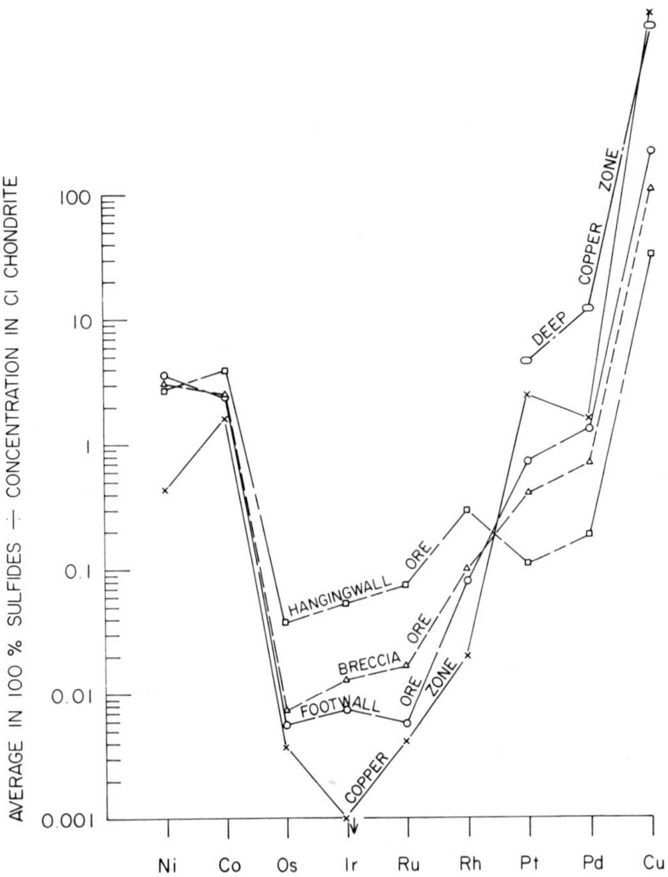
AVERAGE IN 100 % SULFIDES ÷ CONCENTRATION IN CI CHONDRITE
100
10
1
0.1
0.01
0.001
Ni
Co
Os
Ir
Ru
Rh
Pt
Pd
Cu
HANGINGWALL ORE
BRECCIA ORE
FOOTWALL ORE
COPPER ZONE
DEEP COPPER ZONE

B

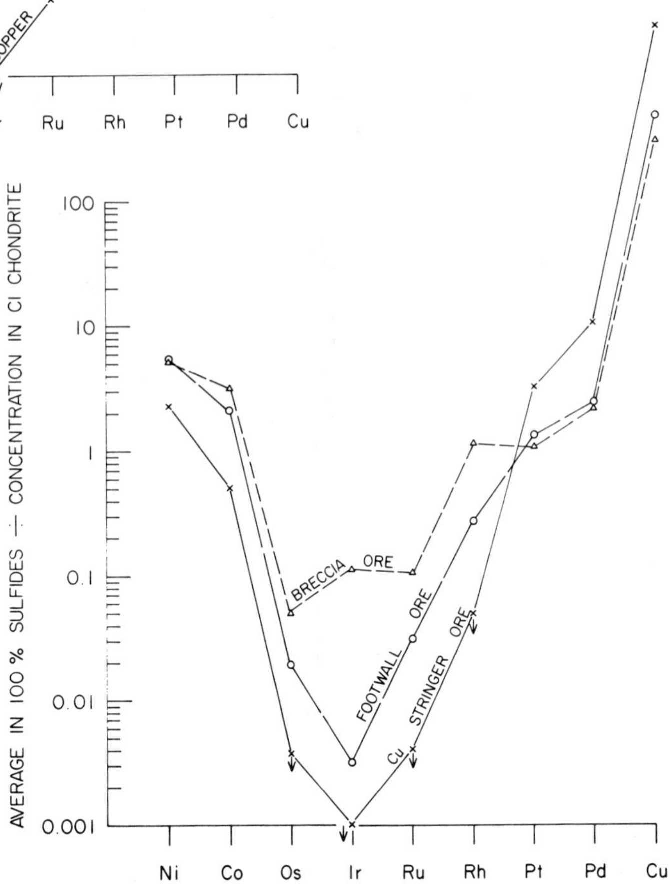
AVERAGE IN 100 % SULFIDES ÷ CONCENTRATION IN CI CHONDRITE
100
10
1
0.1
0.01
0.001
Ni
Co
Os
Ir
Ru
Rh
Pt
Pd
Cu
BRECCIA ORE
FOOTWALL ORE
Cu STRINGER ORE

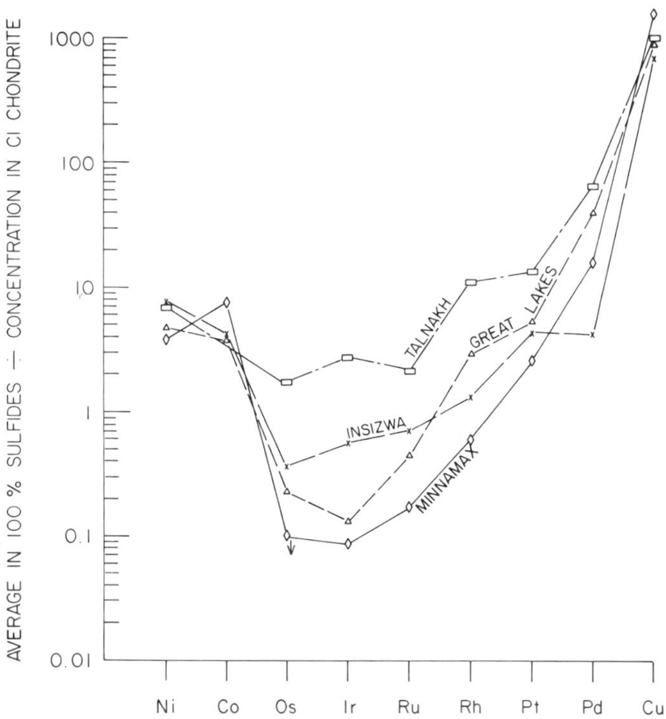

Fig. 6.7. Chondrite-normalized plot for four flood basalt-related deposits. Data from Naldrett (1981a). See legend to Figure 6.2 for further explanation.

mann (1988) have pointed out that the enrichment characteristic of the PGE does not apply to Ni and Cu in the case of the Merensky and J–M Reefs, but does apply to some extent in the case of chromite-hosted deposits, as can be seen in Figure 6.9.

Chromite-hosted deposits also tend to have lower (Pt + Pd)/(Ru + Ir + Os) ratios than those hosted by silicates; this is indicated by the contrast in slopes in Figure 6.9.

6.6 SUMMARY OF COMPOSITIONAL CHARACTERISTICS

The following points regarding composition appear to have general application to Ni–Cu sulfide deposits.

1. When Cu and Ni are included with Fe, the atomic percent metal to atomic percent S is relatively constant and corresponds to that found in most natural pyrrhotites. The proportion of O is also relatively low (<5 wt %). Both of these characteristics are attributable to the buffering effect of silicate magmas on the sulfide liquids segregating from them.
2. Deposits associated with ultramafic rocks generally have higher proportions of Ni and lower proportions of Cu than those associated with mafic rocks. A correlation between the Ni/Cu ratio and magma type has been known for a long time (e.g., Wilson and Anderson, 1959). When one looks at the actual metal contents of the ores, however, rather than the metal ratios, certain anomalous features become apparent. For example, despite its komatiite association, the Pipe deposit contains less than 4 wt percent Ni in its sulfides, approximately one-third of the value typical for deposits of this association. Deposits in altered rocks commonly have higher than normal levels of Ni and PGE, which is an apparent consequence of the alteration.
3. Some deposits show a marked internal variation in the proportions of certain constituents. Cu, Pt, Pd, Au, and to a lesser extent, Ni, increase progressively downward into basal segregations of massive sulfide and then into stringers in subjacent footwall in the komatiite-hosted deposits at Kambalda and the Sudbury deposits; Fe, Co, Rh, Ru, Ir, and Os show the reverse trend. These observations are also

Fig. 6.6. *A.* Chondrite-normalized plot of the hanging-wall ore, footwall breccia ore, footwall ore, Cu zone ore, and deep Cu zone ore at the Strathcona deposit. The arrows pointing down signify that the values plotted are maximum values. *B.* Similar plot of the footwall breccia and footwall ores and the Cu-rich stringers associated with the footwall ore at the Levack (McCreedy) West deposit. Data after Naldrett et al. (1982). See legend to Figure 6.2 for further explanation.

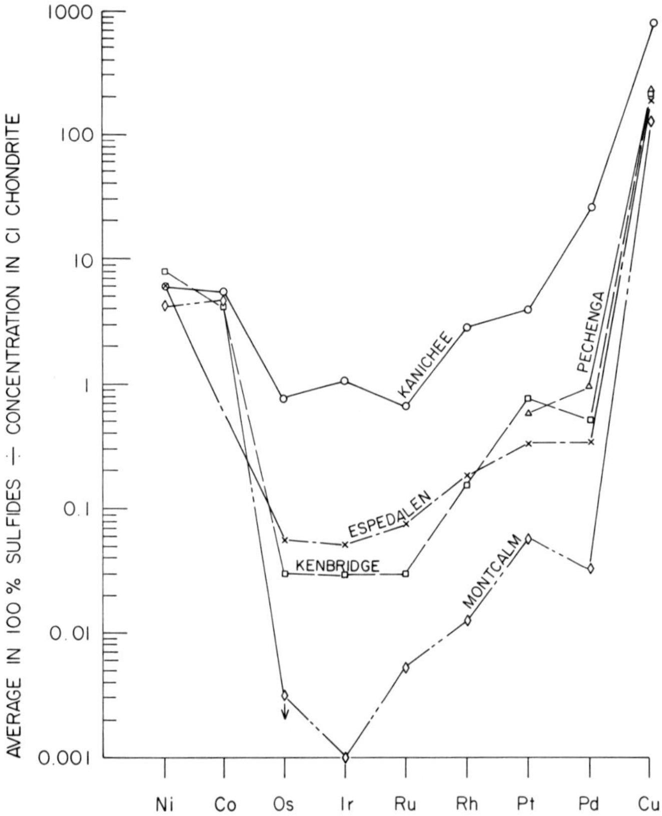

Fig. 6.8. Chondrite-normalized plot of a series of Precambrian gabbro-related deposits. Data from Naldrett and Duke (1980). See legend to Figure 6.2 for further explanation.

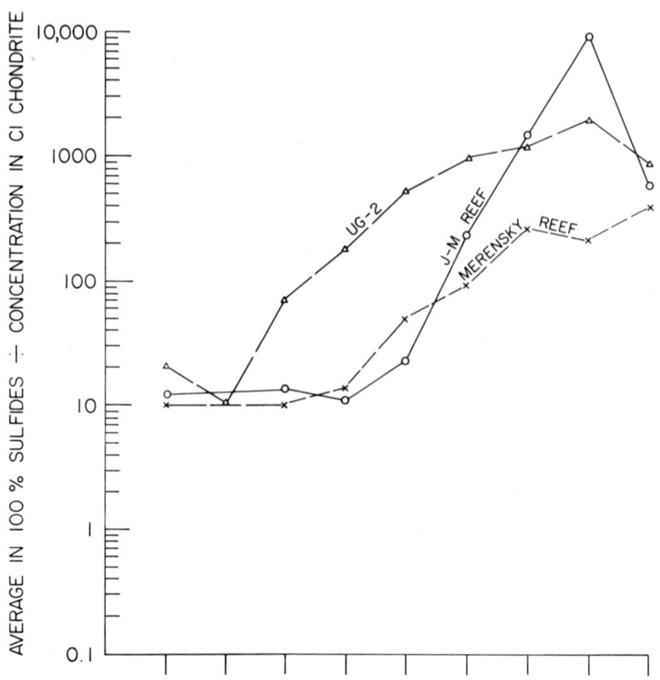

Fig. 6.9. Chondrite-normalized plot for the J–M Reef of the Stillwater Complex and UG-2 and Merensky Reef of the Bushveld Complexes. Data from Naldrett (1981b). See legend to Figure 6.2 for further explanation.

true of the Waterfall Gorge mineralization of the Insizwa Complex (Lightfoot et al., 1984) and apply to the contrast between disseminated and basal massive ore at the Noril'sk and Talnakh deposits of the U.S.S.R. (Genkin et al., 1981).

4. A negative covariation of the ratios Pt/(Pt + Pd) and Cu/(Cu + Ni) exists in Ni–Cu sulfide deposits associated with tholeiitic magmas (Fig. 6.10). Deposits associated with komatiites are an exception to this rule.
5. The chondrite-normalized PGE patterns of magmatic sulfides are systematically related to the composition of the igneous host-rocks. In particular, the (Pt + Pd)/(Ru + Ir + Os) ratios are lowest in sulfides associated with rocks that have crystallized from the most ultramafic magmas. The sulfides associated with peridotitic komatiites, which have crystallized from magmas containing more than about 20 wt percent MgO, have (Pt + Pd)/(Ru + Ir + Os) ratios of 3.0 or less. Ratios of 12 or more are characteristic of magmatic sulfide deposits related to gabbroic rocks that have crystallized from magmas containing less than about 12 wt percent MgO.
6. Despite the similarity of the PGE pattern for deposits of a particular magmatic association, the absolute concentrations vary greatly in a way that shows no obvious relationship to magmetic affiliation. This variability is particularly pronounced in the compositional differences between stratiform PGE-rich ores from layered complexes and almost all other magmatic sulfides, as illustrated in Figure 6.11. It is also seen in this Figure that Co, Ni, and Cu show very much less variation than the PGE. It will be recalled from Chapter 5 that the PGE content of the magmas involved in the Bushveld and Stillwater Complexes is no higher than for other types of continental basaltic magmatism.
7. Sulfides hosted within certain chromitites tend to be very rich in PGE, and also considerably richer in Ni and Cu than are those of most magmatic sulfide ores.

6.7 DISCUSSION OF REASONS FOR COMPOSITIONAL VARIATIONS

6.7.1 Variation in Ni, Cu, Co, and PGE in Magmas During Fractional Crystallization and Partial Melting

The fact that Ni and Co are compatible in mafic silicates, and that Cu is relatively incompatible, has been fundamental to much modern geologic interpretation of igneous processes (for a recent

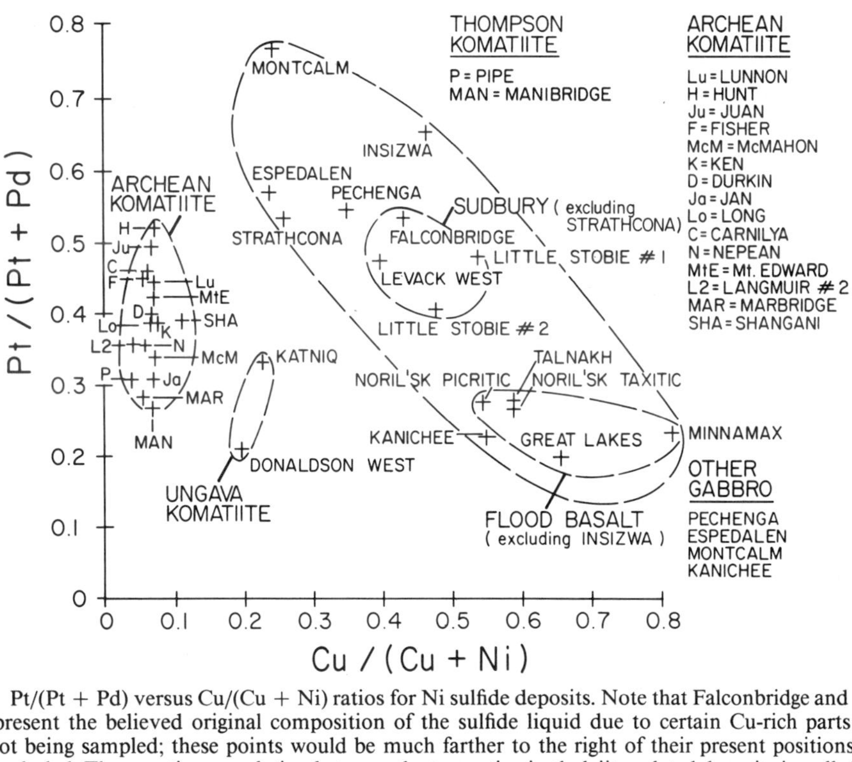

Fig. 6.10. Pt/(Pt + Pd) versus Cu/(Cu + Ni) ratios for Ni sulfide deposits. Note that Falconbridge and Strathcona do not represent the believed original composition of the sulfide liquid due to certain Cu-rich parts of the ore deposits not being sampled; these points would be much farther to the right of their present positions if this ore had been included. The negative correlation between the two ratios in tholeiite-related deposits is well shown, as is the exception of the komatiite-related deposits. (From Naldrett, 1981b.)

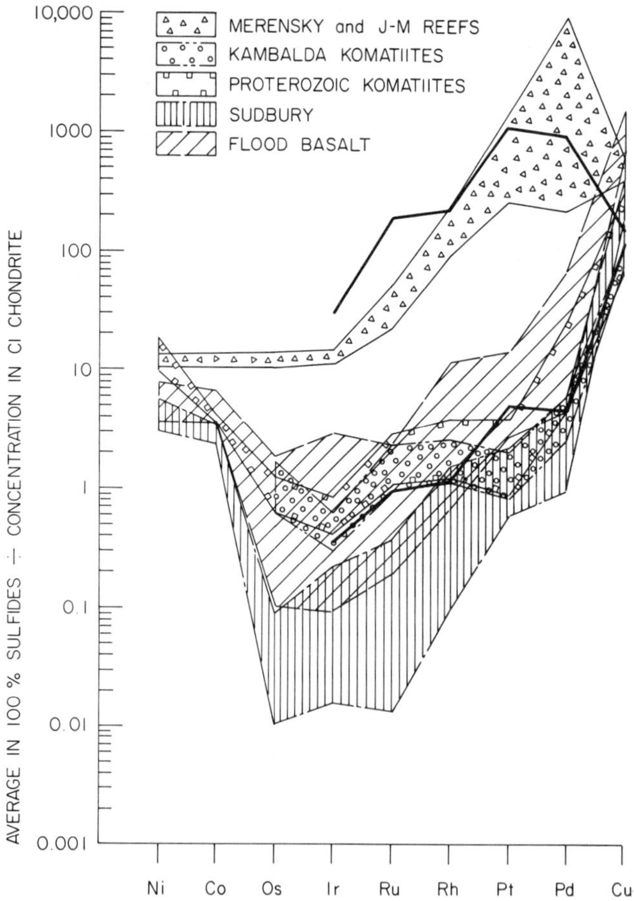

Fig. 6.11. Comparison of data presented on Archean and Proterozoic komatiite, flood basalt, and the Sudbury deposits with that of the Merensky and J–M Reefs. The heavy dark lines superimposed on the Figure show the PGE profiles to be expected as a result of the equilibration of sulfides with a magma that is the equivalent of the Bushveld chill composition (Davies and Tredoux, 1985) at magma/sulfide ratios of 5.3 (upper line) and 2.5 (lower line).

review, see Basaltic Volcanism Study project, 1981). There is also little argument that the relative concentrations of PGE in silicate magmas vary with magma type, in particular with its degree of fractional crystallization (see reviews of Crocket, 1979; Barnes et al., 1985; Hamlyn and Keays, 1985). There is, however, considerable debate as to whether the PGE are compatible in early forming silicates or oxides (Keays and Davison, 1976; Crocket, 1979; Naldrett et al., 1979) or existed as very fine metallic particles or sulfides that were suspended in the magma, served as nuclei for silicates or chromite, and were thus removed from the system (Barnes, 1983; Stockman and Hlava, 1984; Barnes et al., 1985).

In a study of the distribution of PGE in komatiites, Brugmann et al. (1987) showed that their data are explicable if Cu, Au, and Pd are incompatible in olivine and Ni, Ru, Ir, and Os are compatible, the last three with partition coefficients of about 1.8. Naldrett and Barnes (1986) have shown that the Cu/Ni, Pd/Ir, and Pd/Rh ratios of a number of sulfide deposits that are associated with five variably fractionated flood basalt intrusions can be explained if Cu and Pd are incompatible in the crystallizing phases, and if Ni, Ir, and Rh are compatible with respective partition coefficients between basaltic magma and olivine of 8, 10, and 5; and between basaltic magma and clinopyroxene of 2, 10, and 5 (Figs. 6.12 and 6.13). They argue that their deduction for Ir is not necessarily in contradiction with that of Brugmann et al. (1987), as the latter's value was determined for an Mg-rich komatiitic melt; they suggest that the coefficient for Ir varies from about 2 to 8 as the MgO content of the magma decreases from about 28 to 8, similar to the way in which the coefficient for Ni between magma and olivine changes with MgO content (Arndt, 1977).

They used their coefficients to model the partial

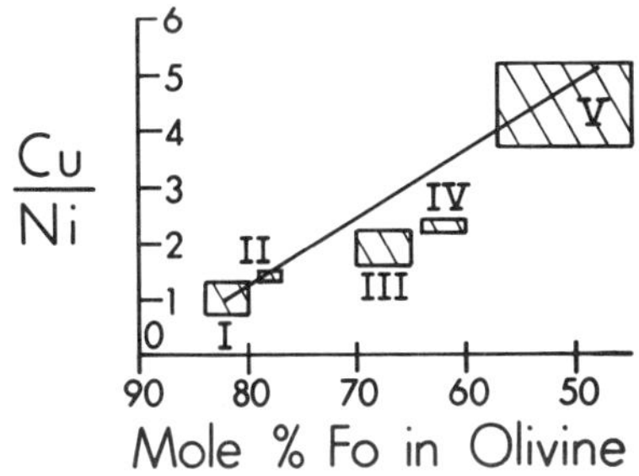

Fig. 6.12. Variation in Cu/Ni ratios of sulfide ore with forsterite content of olivine close to, but not immersed in, the sulfides, for a series of flood basalt-related deposits. The sizes of the boxes represent the extremes in Cu/Ni ratio resulting from using plus and minus the standard errors of the means of both elements to produce the maximum variation (vertical axis), coupled with the range in composition observed within the olivines (horizontal axis). The model curve is drawn assuming fractional crystallization of olivine, clinopyroxene, and plagioclase in the proportions 28:28:44. The curve is essentially independent of R in the case of Cu/Ni. I = Insizwa (data from Naldrett, 1981a); II = Talnakh (data from Naldrett, 1981a); III = Great Lakes (data from Naldrett, 1981a); IV = Waterhen (data from Mainwaring and Naldrett, 1977); V = Minimax (data from Naldrett 1981a). (From Naldrett and Barnes, 1986.)

melting of spinel–herzolite mantle. Figures 6.14 and 6.15 illustrate the variations that they obtained for Ir and Pd, assuming typical mantle concentrations of 4.1 and 4.4 ppb, respectively (Basaltic Volcanism Study Project, 1981). The results are in reasonable agreement with the measured concentrations in a wide variety of mafic and ultramafic magma types except for flood basalt magmas, where their predicted concentrations tend to be rather higher than those observed. The increase in Ir and decrease in Pd with partial melting results in an increase in the Pd/Ir ratio in less magnesian magmas. Because Pt tends to follow Pd, and Ru and Os tend to follow Ir, the variation in (Pt + Pd)/(Ru + Ir + Os) from Archean to Proterozoic komatiite-related and then tholeiite-related deposits can thus be explained.

Naldrett and Barnes (1986) and Hamlyn and Keays (1985, 1986) pointed out that if a small amount of sulfide remains behind in the mantle as the partial melt is removed, this will greatly increase the bulk partition coefficient of the PGE into the mantle residue, reducing the concentration in the resulting melt, as shown in Figures 6.14 and 6.15. Many flood basalts appear to have been derived from mantle in which some sulfide remained behind, and the same may be true of the

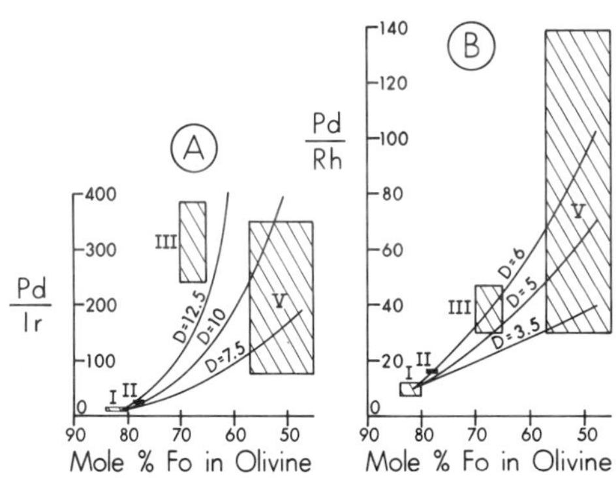

Fig. 6.13. Variation of Pd/Ir (**A**) and Pd/Rh (**B**) in sulfides with Fo content of olivine (see legend to Fig. 6.12 for further explanation). Model curves (solid lines) are drawn assuming the incompatibility of Pd and bulk partition coefficients for Ir and Rh between magma and mafic silicates as shown. All of the PGE have been assumed to have had the same sulfide-magma partition coefficients, and the curves are thus independant of R. (From Naldrett and Barnes, 1986.)

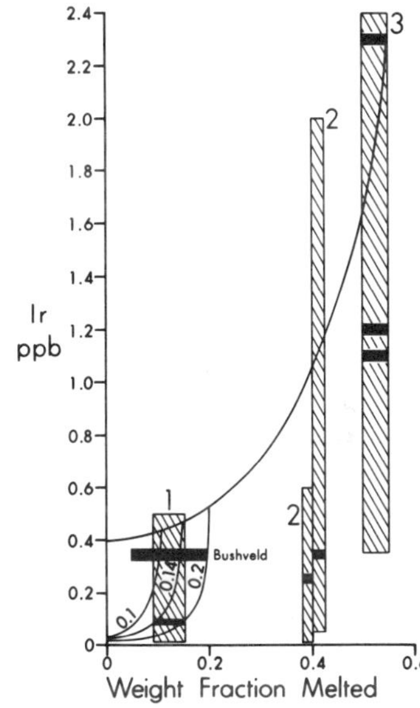

Fig. 6.14. Calculated Ir content of mafic and ultramafic magmas as a function of the degree of melting, compared with the Ir content of natural rocks. The upper curve is for the melting of a sulfide-free mantle. The curves labeled 0.1, 0.14, and 0.2 are for mantle containing this amount in wt percent of sulfide, respectively. Because Ir has a very high partition coefficient, it is held back by the sulfide in the mantle until this has been eliminated by dissolving in the magma. The curves are drawn on the assumption that a mafic magma can dissolve 1.0 wt percent sulfide. The boxes illustrate the range in composition of natural rocks plotted with the numbering as follows: (1) = flood basalts (Crocket, 1981); (2) = pyroxentic komatiite from "Fred's" flow, Munro Township (Crocket, 1981); (3) = spinifex-textured material from Munro Township, Mt. Clifford and Kambalda (Crocket, 1981). An area appropriate to the Bushveld Complex is also indicated in the figure. (Modified after Naldrett and Barnes, 1986.)

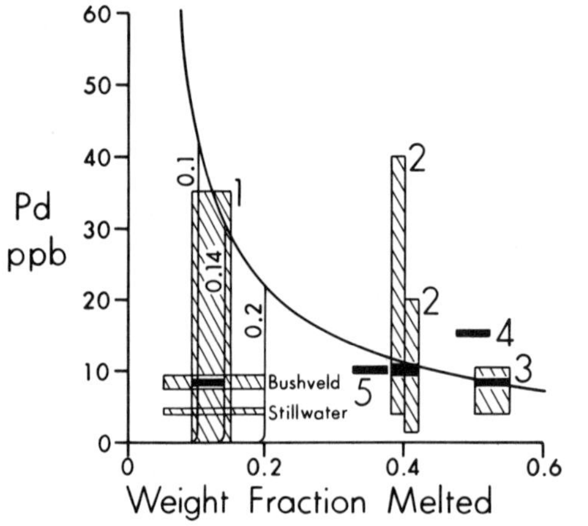

Fig. 6.15. Calculated Pd content of mafic and ultramafic rocks as a function of the degree of mantle melting. (4) = Komatiites from Gorgona Island (Hamlyn and Keays, 1985); (5) = high magnesian basalts from Disko Island (Hamlyn and Keays, 1985). See legend to Figure 6.14 for further explanation.

Bushveld and Stillwater Complexes, if the initial magmas were as low in PGE as the data of Davies and Tredoux (1985) and Zientek et al. (1985) suggest (see later discussion). Naldrett and Barnes (1986) stress that although the removal of sulfide during magma ascent can also greatly reduce its PGE content, the resulting Ir/Pd ratio will be very different to that resulting from sulfide remaining in the mantle, because during ascent the only partition coefficients involved are those into sulfide, whereas when considering PGE retention by residual mantle sulfide one also has the partitioning of Ir into mantle silicates to contend with.

Davies and Tredoux (1985) and Zientek et al. (1985) present data on the PGE contents of possible magmas for the Bushveld and Stillwater intrusions, respectively. Their data show a wide spread, but 50 percent of the samples of Davies and Tredoux contained at least 8 ppb Pd and 0.35 ppb Ir, and at least 50 percent of the samples of Zientek and Mei of Stillwater gabbro norite (thought by Helz, 1985, to be the best approximation to the magma composition) contained at least 4.3 ppb Pd. As seen from Figures 6.14 and 6.15, it is not difficult to account for concentrations of this order through partial mantle melting. There is no evidence that either magma underwent enrichment in PGE before emplacement.

With these concepts in mind, reasons for the variation in the composition of deposits associated with different types of rocks will be discussed. Before doing this, however, it is necessary to discuss the compositional variation that may be imposed on a deposit internally, and can, if the sampling is not complete, give rise to false estimates of its bulk composition.

6.7.2 Internal Zoning of Deposits

The variations in Cu, Ni, and the PGE within the Strathcona and Levack West deposits at Sudbury have already been referred to (Figs. 6.6A and 6.6B). The comparison between the different ore types of individual deposits is also brought out in Table 6.2, in which absolute concentrations of metals (recalculated to 100 percent sulfides) are shown for the different ore zones in the deposits. There is a progressive increase in Ni, Cu, Pd, Au, and Zn and decrease in Co, Rh, Ru, Ir, and Os from hanging-wall through granite breccia to footwall ore at Strathcona. Parallel, although less dramatic, changes in the same metals are present between granite breccia and footwall ore at Levack (McCreedy)West, with the exception of Au, which is lower in the footwall ore. The Cu-rich stringer ore at Levack West appears to be the natural continuation of this process, with all metals following their previous trend (except Ni, which decreases and Ru and Os for which precise values were unobtainable). The Cu zone at Strathcona is not analogous geologically to the Cu-rich stringers at Levack West, for it occurs very much farther into the footwall. Naldrett et al. (1982) pointed out that it did not appear to be analogous chemically either, for the concentrations of PGE and Au are very much lower than in deep-zone ore, in contrast to the dramatic increases in these metals in the Cu-rich stringers at Levack West. The resampling, however, that accompanied the mining of this zone has shown it to contain considerably higher levels of Pt, Pd, and Au than originally thought (Coats and Snajdr (1984) report 4,910 ppb Pt, 6,970 ppb Pd, and 4,880 ppb Au for an average of 21 composite samples of drill hole material). These data answer the question of the "missing" PGE that was raised by Naldrett et al. (1982).

Naldrett et al. (1982) have reviewed earlier papers on the subject of this zoning and suggest three main processes as likely to influence the distribution of metals across a deposit such as that at Strathcona.

1. The different ore types can be interpreted as having been emplaced successively from a single body of magma with the lowermost ore type settling into position first. The ore types would then consist of successive segregations of sulfides from the magma, and variations in composition would depend on the partitioning of the metals into these successive segregations. The more chalcophile metals would partition more strongly into the initial segregation, become depleted in the host magma and, therefore, be less concentrated in subsequent segregations of sulfide.

2. The ore types can be viewed as resulting from the fractional crystallization of a sulfide melt. Monosulfide solid solution (Mss) would be

TABLE 6.2. *Comparison of Average Metal Concentrations in Different Ore Types at Strathcona and Levack West*

Ore Type	Hanging-wall Ore	Granite Breccia Ore		Footwall Ore		Cu-rich Stringers	Cu Zone
Deposit	Strathcona	Levack West	Strathcona	Levack West	Strathcona	Levack West	Strathcona
Wt %							
Ni	3.1	5.6	3.4	5.8	4.0	5.52	1.72
Cu	0.37	3.3	1.2	5.3	2.3	27.6	24.67
Co	0.21	0.17	0.14	0.11	0.13	0.027	0.05
ppb							
Pt	114	1,098	413	1,334	750	11,434	4,910
Pd	105	1,223	380	1,350	701	13,670	6,970
Rh	60	228	20	53	16	<10	4
Ru	52	73	12	22	4	<3	3
Ir	29	62	7	17	4	0.3	0.2
Os	20	26	4	10	3	<2	2
Au	19	166	78	97	112	8,211	4,880
ppm							
Zn	122	153	160	529	227	3,716	2,750[a]
Number of Samples	10	16	18	5	8	4	1

[a]Estimate only, based on reported presence of 0.5 percent sphalerite containing 54.9 wt percent Zn; Abel et al., 1979.

the first phase to crystallize from most natural sulfide melts, and the lower melting point constituents as well as those constituents less compatible in the Mss structure than the melt, would become enriched in the latter. Hawley (1965) first suggested this model to explain Fe–Cu–Ni zoning at the Frood deposit, and it was subsequently favored by Keays and Crocket (1970) and Chyi and Crocket (1976) to account for the zoning at Strathcona.

3. The metals in the ores may redistribute themselves by diffusion in response to a thermal gradient. Naldrett and Kullerud (1967) originally suggested this mechanism to explain zoning in Ni and Cu at Strathcona, and Hoffman et al. (1979) favored it to explain the zoning in PGE at Levack West.

The first process can account for the progressively lower Ni, Cu, Pt, and Pd contents in successively higher ores at Strathcona, and for the trend of decreasing Pt/Ni and Pd/Ni ratios, provided that one assumes that the partition coefficients between sulfide and silicate melts for Pt and Pd are much higher than for Ni. It is difficult to explain the variation in the Cu/(Cu + Ni) ratio in this way as the partition coefficients for these two metals between a sulfide melt and a silicate magma of basaltic composition are approximately equal (Rajamani and Naldrett, 1978).

A far more serious objection is that, although Co is not as chalcophile as Ni and Cu; and Rh, Ru, Ir, and Os may not be as chalcophile as Pt and Pd; all of these metals are nevertheless highly chalcophile and would partition strongly into any sulfide liquid. All of them should, therefore, show depletion in successive sulfide segregates rather than the enrichment that is observed at both Strathcona and Levack West (Table 6.2). The process of successive segregation is, therefore, ruled out as an explanation for the observed zoning.

As mentioned, diffusion down a thermal gradient can account for the enrichment in Ni, Cu, Pt, and Pd away from the heat source (the Sudbury Igneous Complex) into the footwall of many Sudbury deposits. It is difficult to envisage, however, how Rh, Ru, Ir, Os, and Co could diffuse back against the thermal gradient to explain the enrichment of these metals in the hanging-wall ore. It is possible that Fe is also mobilized and would diffuse down the thermal gradient, perhaps less rapidly than Ni and Cu as Naldrett and Kullerud (1967) have suggested. If Rh, Ru, Ir, Os, and Co were less mobile than Fe, they could become enriched in this way. The degree of enrichment in these metals, however, is such as to require more than 50 percent of the original sulfides to diffuse out of the hanging-wall ore zone in response to the thermal gradient, leaving all of their original Rh, Ru, Ir, Os, and Co behind. The texture of the hanging-wall ore at Strathcona indicates that no large-scale mobilization of sulfides has occurred out of this ore, and this process is also ruled out as an explanation.

Naldrett et al. (1982) conclude, as Keays and Crocket (1970) and Chyi and Crocket (1976) had earlier, that fractional crystallization of a sulfide melt is the most likely process responsible for the metal distributions. The composition of the ore is such that the Mss would be the first phase to crystallize. Liquidus relations in the Fe–Ni–S and Cu–Fe–S (see Chapter 3) systems indicate that, although Ni and Cu would substitute in the solid–solution, both would be more concentrated in the liquid. Fractional crystallization of the initial sulfide melt would, therefore, give rise to a liquid significantly enriched in these metals.

This interpretation carries implications also with respect to the preference of other metals for the sulfide liquid versus the Mss structure. Pt, Pd, Au, and Zn (as well as Ni and Cu) should prefer the sulfide melt and Co, Rh, Ru, Ir, and Os the Mss structure. The work of Skinner et al. (1976) on the Fe–Pd–S and Fe–Pt–S systems is consistent with the requirement that Pt and Pd become enriched in the residual liquid. The study by Mackovicky et al. (1986) demonstrated that Pt will be concentrated in the residual liquid in the Cu–Fe–Pt–S system, which can concentrate 6 to 15 wt percent Pt at 900°C. At this temperature the liquid contains up to six times as much Pt as the coexisting pyrrhotite. Mackovicky et al. point out that their work supports the hypothesis that Pt and Pd become concentrated in the Cu-rich residual liquid resulting from the crystallization of Mss from a sulfide magma.

The study of Skinner et al. (1976) of the Pt–Pd–S system implies, but does not prove, that during crystallization of an Fe-rich sulfide melt, the residual sulfide liquid would become steadily enriched in Pd over Pt. The lack of any observed variation in the Pt/(Pt + Pd) ratio from hanging-wall to footwall between many of the ore types listed in Table 6.2 suggests that this postulated enrichment is not very pronounced during much of the crystallization history of a sulfide melt, although the markedly lower Pt/(Pt + Pd) ratio of the Cu-rich stringers at Levack West suggests that it occurs in the final stages.

In their work on PGE and the Fe–Ni–S system, Distler et al. (1977) have studied the distribution of the light elements (Pd, Rh, and Ru) between Mss and a phase, pentlandite type I, that they regarded as the result of the peritectic reaction between the solid–solution and an Ni-rich residual liquid lying on the S-deficient side of the solid–solution. In view of Kullerud's (1963) study of pentlandite, it would seem more likely that their S-deficient liquid reacted first to produce a high temperature $Ni_{(3\pm x)}S_2$ phase, which then reacted at an unquenchable rate with Mss to produce pentlandite. Regardless of the correctness of the interpretation of Distler et al. on the origin of the type I pentlandite, their data provide evidence that Pd not only favors the liquid but that Rh and Ru favour the solid-solution structure, Ru more strongly than Rh. In view of the chemical similarities between Pt and Pd on one hand and Ir, Os, and Ru on the other, the heavy PGE are likely to behave similarly, with Pt favoring the liquid and Ir and Os the solid–solution structure.

The increase in the Cu/(Cu + Ni) ratio toward the footwall shown by the different ore zones at Strathcona and Levack West requires that Cu favor the liquid over the Mss to a greater degree than Ni. This is consistent with the data of Craig and Kullerud (1969) on the Fe-Ni-Cu-S system. The depletion in the absolute Ni content of the Levack West Cu-rich stringers with respect to the footwall ore requires that, as the residual liquid becomes richer in Cu, there is a reversal in the tendency for Ni to favor the liquid and that it begins to favor the Mss structure. This is again consistent with Craig and Kullerud's study (1969); they reported that at 850°C "the sulfide liquid contains much more Cu and less Ni than the coexisting Mss."

Liquidus data on the Fe–Co–S system are unavailable. The work of Kuznetsov et al. (1966) indicated the existence of a $CoS_{(1+x)}$ phase with an NiAs-type structure. Experience with the Fe–Ni–S system would suggest that this phase is likely to form a solid–solution with FeS analogous to the solid–solution between FeS and NiS. The slightly higher maximum melting point of the $CoS_{(1+x)}$ phase (1,195°C) with respect to the $Fe_{(1-x)}S$ phase (1,188°C) indicates that if this is so, the Fe–Co solid–solution would pull away from the Co–S join first upon cooling, leading to an enrichment of Co in the early forming solid–solution with respect to the liquid. Although this reasoning is speculative, particularly when applied to the much more complex natural system, one can say that available data are again consistent with the interpretation of fractionation of a sulfide liquid.

Studies of the FeS–ZnS liquidus (Kullerud, 1953) indicated that, where only small amounts of ZnS are involved, ZnS prefers the sulfide liquid to substituting into the FeS structure.

In summary, all of the existing phase equilibria data indicate that much of the zoning at Strathcona and Levack West resulted from progressive fractionation of a sulfide liquid, with the remaining liquid moving steadily away from the Sudbury Igneous Complex toward the footwall. The precise mechanism by which this fractionation has occurred poses a problem. The solid–solution would be denser than the sulfide liquid and one might suppose that it would sink and the liquid float. On the other hand, the ores have not crystallized from a pool of sulfide liquid, but instead are interstitial to silicates within the matrix of brecciated footwall gneiss and in fractures in the underlying gneiss. The present author envisages the process as one of cooling and crystallization coupled with the filter pressing of the fractionating liquid progressively down into the brecciated footwall and away from the more coherent rocks of the Complex in the hanging-wall.

Fractional crystallization of a sulfide magma may also explain the stringers enriched in Cu, Pt, Pd, and Au at Kambalda, the Cu-rich veins near the base of the Insizwa Complex, and some of the Cu-enrichment toward the footwall of the Talnakh deposits. The fact that these examples do not show the same progressive zoning that is exhibited at Sudbury may be because progressive zoning requires special conditions in the way of brecciation and fracturing of the footwall, if the fractionating sulfide liquid is to be progressively filter pressed away from early crystallizing Mss.

6.7.3 Relationship Between the Composition of Ore and That of the Host Magma

6.7.3.1 KOMATIITE-RELATED ORES

A number of different factors can account for the compositional variations in komatiite-related ores, assuming that they have all been derived from magmas with a very similar initial composition. These variations are discussed in turn.

6.7.3.1.1 Effect of Variations in Magma/Sulfide (R)

The current literature indicates that a Ni content of 1,500 ppm, a Cu content of 50 ppm, and a Co content of 75 ppm are normal for an Archean komatiitic liquid containing 27 wt percent MgO. Data summarized by Crocket (1981) suggest that concentrations of the order of 5 ppb Pt and 7.5 ppb Pd are also normal.

Figure 6.16A illustrates how the Ni and Cu contents of the sulfide segregating from this magma will change as the value of the magma/sulfide (see Chapter 3), shown here as $\log_{10}$ R, varies between 10 and 1. Figure 6.16B illustrates the associated changes in Ni and in Cu in the magma itself as a function of R. It can be seen by comparison with Figure 6.1 that the Ni and Cu contents of many komatiite-related ores lie on or close to the line in Figure 6.16A representing the effect of variation in R on sulfides in equilibrium with the model magma. This relationship suggests that these ores may be the result of reaction of variable quantities of sulfide with silicate magmas of relatively similar compositions. The proposition is tested for other elements in Table 6.3 where the calculated Ni, Cu, Co, Pt, and Pd contents of sulfide liquids resulting from equilibration with the same silicate magma at different R values are listed. It is seen, provided that the assumed partition coefficients for the PGE of about 10^5 are appropriate, that the composition of the Lunnon shoot is con-

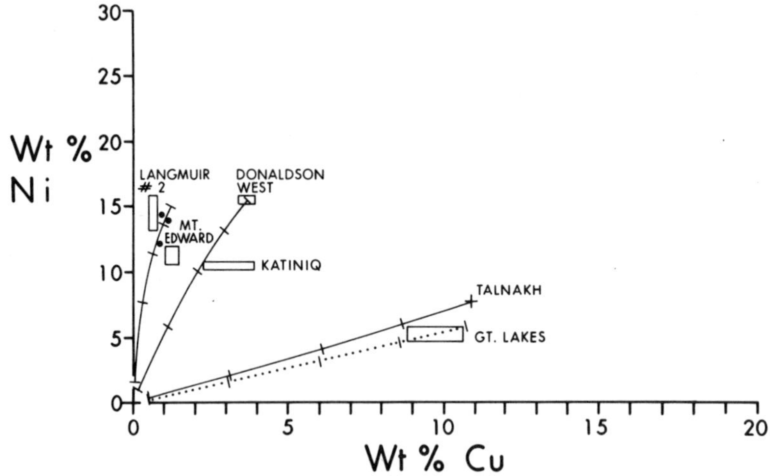

Fig. 6.16. ***A.*** Modeling of how the Ni and Cu contents of ores related to certain specific magma types will change in response to variation in R. The curved lines refer to variations in R, the tick at highest Ni and Cu being equivalent to log R >5 and subsequent ticks indicating log R = 3, 2.5, 2, and 1 in progression. The dotted line indicates the same variation for the Great Lakes deposit, assuming that the Pt and Pd values for sulfides in equilibrium with the host magma at a value of log R = 10 were the same as at Talnakh. The ore then calculates to have equilibrated at log R = 3.34.

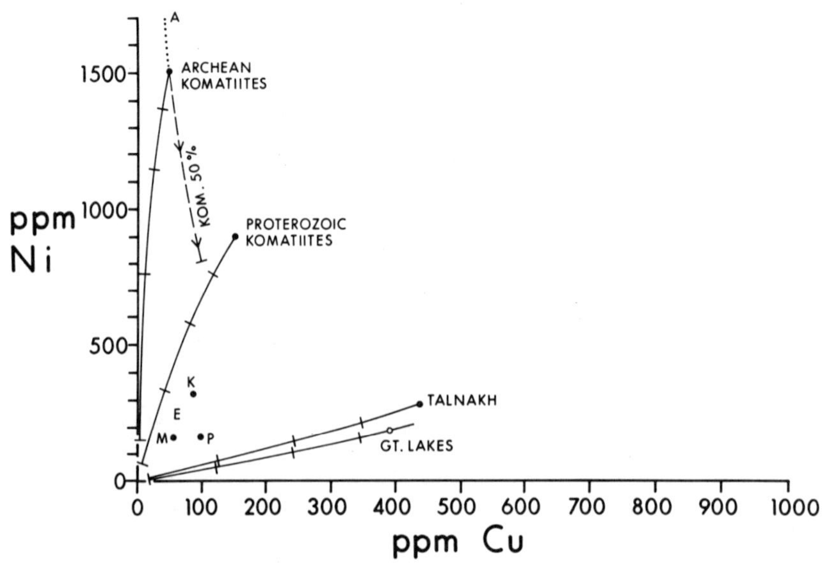

Fig. 6.16. ***B.*** Compositions of silicate liquids in equilibrium with sulfide ores illustrated in Figure 6.16A. The dashed line indicates the effect of the fractional crystallization of 50 percent olivine from the composition marked "Archean komatiites." Point A and the dotted line indicates the fractionation trend of the liquid if that marked "Archean komatiites" were itself the result of the fractionation of 20 percent olivine. (Modified after Naldrett, 1981a.)

sistent with equilibration at a value of log R between 3.0 and 2.5, most probably 2.5 (R = 316). Assuming that all of the sulfide involved in the deposit was originally in solution in the host silicate magma, this ratio amounts to 0.32 percent sulfide, or about one-third of the total sulfide believed to be soluble in the magma at low pressures.

The Pipe orebody is associated with an ultramafic rock considered to be komatiitic in origin (Peredery et al., 1982). This ore can be modeled quite closely at a value of log R = 1.53 (R = 34).

TABLE 6.3 *Compositional Variation of Sulfide and Silicate Melts with Variation in R*

		Silicate Liquid					Sulfide Melt					Typical Deposits for Comparison						
	Log R	Ni	Cu	Co	Pt	Pd	Ni	Cu	Co	Pt	Pd	Ni	Cu	Co	Pt	Pd		
		ppm			ppb		wt %			ppb		wt %			ppb			
Archean komatiite	10	1500	50	75	5.0	7.5	15.0	1.25	0.30	500	750							
$D_i^{Sulfide/Silicate}$	6	1500	50	75	4.55	6.82	15.0	1.25	0.30	455	682							
Ni = 100	5	1500	50	75	2.50	3.75	15.0	1.25	0.30	250	375							
Cu = 250	4	1490	49	75	0.45	0.68	14.9	1.22	0.30	45	68							
Co = 40	3	1370	40	73	0.05	0.074	13.7	1.00	0.29	5.0	7.4							
Pt = 10^5	2.5	1140	28	68	0.016	0.024	11.4	0.70	0.27	1.6	2.4	12.2	1.04	0.31	1.45	1.62	Lunnon	Naldrett (1981a)
Pd = 10^5	2	760	14	55	0.005	0.0076	7.6	0.36	0.22	0.50	0.76							
	1.53	390	6	35	0.0017	0.0026	3.9	0.15	0.14	0.17	0.26	4.0	0.16	0.16	0.054	0.12	Pipe	Naldrett (1981a)
	1	150	2	18	0.0005	0.0008	1.5	0.05	0.07	0.05	0.08							
Proterozoic komatiite	10	890	180	40	5.0[a]	15.0[a]	17.8	4.50	0.23	500	1500							
$D_i^{Sulfide/Silicate}$	6	890	180	40	4.6	13.7	17.8	4.50	0.23	455	1365							
Ni = 200	5	890	180	40	2.5	7.5	17.8	4.49	0.23	250	750							
Cu = 250	4	875	175	40	0.45	1.35	17.5	4.39	0.23	45	135							
Co = 570	3	740	144	38	0.05	0.15	14.8	3.60	0.22	5.0	15.0	15.5	3.7	0.23	4.13	15.5	Donaldson W.	Naldrett (1981a)
Pt = 10^5	2.6	595	110	33	0.02	0.06	11.9	2.77	0.20	2.0	6.0	10.5	3.4	0.21	3.32	6.71	Katiniq	Naldrett (1981a)
Pd = 10^5	2	300	52	26	0.005	0.015	6.0	1.30	0.15	0.50	1.5							
	1	47	8	6	0.0006	0.007	0.93	0.19	0.038	0.055	0.17							

[a]Figures based on Crocket (1981)

Naldrett et al. (1979) pointed out that low R values such as this require more sulfide than is believed to be soluble in komatiitic magmas, and concluded that assimilation of country-rock S had occurred.

6.7.3.1.2 Effect of Fractionation of the Host Magma

The dashed line in Figure 6.16B is drawn on the basis of the calculations of Duke and Naldrett (1978) and Duke (1979); it illustrates the effect of fractional crystallization of olivine on the Ni and Cu contents of the peridotitic komatiite composition modeled previously. Suppose that this particular peridotitic komatiite was itself the consequence of the fractional crystallization of approximately 20 percent olivine from an even more magnesian komatiite (the dotted line in Fig. 6.16B). If this were so, the original komatiite would have contained about 1,700 ppm Ni, 42 ppm Cu, 4.2 ppb Pt, and 6.3 ppm Pd. Equilibration of this magma with a sulfide melt at an R value of 250 (log R = 2.4) would have resulted in the sulfide melt containing 12.2 wt percent Ni, 0.52 wt percent Cu, 1,060 ppb Pt, and 1,580 ppb Pd, which is within or very close to the limits of error on the composition of the Langmuir No. 2 deposit. This particular case is presented as an example; similar combinations of fractional crystallization and equilibration at different R values can be used to account for most deposits related to Archean komatiites that are believed to represent primary magmatic compositions.

6.7.3.1.3 Effect of Oxygen Fugacity

In the preceding discussion, it has been assumed that the partition coefficients between sulfide and silicate magma do not change. In Chapter 3, experimentally determined variations with fO_2 in $K_D^{\text{Sul. melt/Olivine}}$ for Fe–Ni exchange (Boctor, 1981, 1982) have been discussed, together with the contention by Doyle and Naldrett (1987) that the O content of a sulfide melt controls the value of $K_D^{\text{Sul. melt/Olivine}}$, increasing 0_2 causing a diminution in the value.

Cowden et al. (1987) criticized the explanation of Naldrett (1981) of the ore compositions at Kambalda in terms of R and magma fractionation, pointing out that neither is able to account for all of the variation observed. They proposed a model in which increasing oxidation leads to increased partitioning of Ni and decreased partitioning of Fe from a silicate magma into the sulfide liquid. This trend is the opposite of that observed by Boctor (1981, 1982) between olivine and sulfide, and extended by Doyle and Naldrett (1987) to silicate melts and sulfide.

Significantly, in the view of this author, Cowden et al. (1987) fail to consider the combined effects of fractionation and variable R factor, which can explain most of their compositional variation, as has been shown previously. It must be admitted, however, that it is difficult to explain all of their new compositional data, in particular deposits with Ni greater than 18 wt percent, without either calling upon some variation in D, or supposing that the deposits richest in Ni have had their Ni contents upgraded in a secondary manner. Cowden et al. (1987) do not consider the latter possibility likely.

It would seem, therefore, that $K_D^{\text{Sul. melt/Olivine}}$ and $K_D^{\text{Sul. melt/Sil. magma}}$ are likely to vary with increasing fO_2, but that there is no consensus as to whether they increase or decrease. The partition coefficients used in Sections 6.7.3.1.1 and 6.7.3.1.2 can explain much of the compositional variation, and appear to provide a reasonable first approximation to what occurs in nature.

6.7.3.2 THE SUDBURY ORES

Naldrett et al. (1982) have discussed explanations for compositional variations among certain Sudbury deposits and have shown that starting with the same source magma, it is possible to explain many of these differences using the principles of fractional and batch equilibration of sulfides. As has already been discussed, assimilation of siliceous material by an ultramafic or mafic magma results in the segregation of sulfide dissolved in the magma as an immiscible sulfide liquid; varying degrees of assimilation result in varying proportions of sulfide being forced out of solution. This concept was called upon by Naldrett et al. to produce a stepwise, progressive modification of the source magma. A revised version of their scheme, modified to account for the new perception that the values of $D_{PGE}^{\text{Sul. melt/Sil. magma}}$ are of the order of 10^5 (Campbell and Barnes, 1984), is shown in Figure 6.17.

It is proposed that the initial magma had olivine on the liquidus and was almost but not quite saturated in sulfide, containing 0.15 wt percent S, 240 ppm Ni, 169 ppm Cu, 21.5 ppm Co, and 2.25 ppb Pd (labeled "source magma" in Fig. 6.17). A portion of this magma, intruding beneath the North Range, precipitated 0.22 percent of its total load of 0.45 percent dissolved sulfide (equivalent to equilibrating at a value of log R = 5). Modeling indicates that at this stage, the magma would have had the composition labeled "magma 1" in Figure 6.17. Assimilation followed by precipitation of 18 percent of the remaining dissolved sulfide (log R = 3.1) would have given rise to the sulfide liquid labeled "Levack West" in Figure 6.17 and compared with the observed composition at Levack West in Table 6.4. Assimilation followed instead by precipitation of 68 percent of the dissolved sulfide would have given rise to the composition labeled "Strathcona" in Figure 6.17 and compared with the observed Strathcona composition in Table 6.4.

A second portion of the source magma intruding toward the South Range underwent fractional crystallization with the precipitation of 7 percent

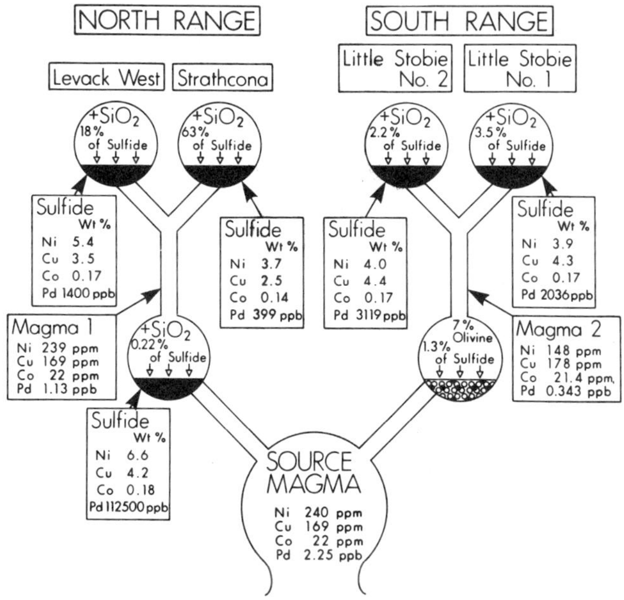

Fig. 6.17. Flow sheet illustrating the effect of fractionally crystallizing olivine (in the first event contributing to the South Range ore) and of abruptly segregating a given percentage of the sulfide dissolved in the magma (in all other cases) on the composition of a possible source magma for Sudbury and on the sulfides segregating from it. The combination of these sequences of events can give rise to ores with compositions closely resembling those estimated for Levack West, Strathcona, and Little Stobie 1 and 2, with the exception of the Cu content. The abrupt precipitation of sulfide is envisaged as the consequence of mixing of magma with silica-rich material.

olivine and 1.3 percent dissolved sulfide. Assuming a value of $D_{Ni}^{Olvine/Sil.\ magma}$ for Ni of 9, this fractionation would have reduced the Ni content to 149 ppm, increased Cu to 178 ppm, decreased Pd to 0.334 ppb, and reduced Co only slightly. Assimilation followed by precipitation of 3.5 or 2.2 percent of the total dissolved sulfide (equivalent to values of log R of 3.8 and 4.0, respectively) would give rise to the compositions given in Figure 6.17 and Table 6.4 under Little Stobie 1 and 2, respectively. It can be seen from Table 6.4 that these compositions are close to the observed compositions.

It is emphasized that this scenario is only one of an infinite number that could account for the compositional variability in the Sudbury and other ores. It does, however, illustrate that the observed variability is fully explicable on the basis of the magmatic model.

6.7.3.3 DEPOSITS EXTREMELY ENRICHED IN PGE

In deposits extremely enriched in PGE, such as the Merensky Reef and UG-2 in the Bushveld Complex and the J–M Reef in the Stillwater Complex, the enrichment is reflected in the very high

TABLE 6.4. *Average Compositions of Sulfide Liquids Responsible for Certain Ores*

	Ni	Cu	Co	Pt	Pd	Rh	Ru	Ir	Os	Au
	wt %		ppb							
Strathcona	3.6 (3.7)	1.2(2.8)[a] (2.5)	0.15 (0.14)	420	372 (389)	30	21	12	8	54
Levack West	5.7 (5.4)	3.7 (3.5)	0.16 (0.17)	1,154	1,253 (400)	186	60	47	22	150
Little Stobie 1	3.8 (3.9)	4.4 (4.3)	0.19 (0.17)	1,930	2,120 (2,036)	119	123	61	29	862
Little Stobie 2	4.0 (4.0)	3.6 (4.4)	0.17 (0.17)	2,130	3,170 (3,119)	303	247	113	46	860

Figures in parentheses are compositions modeled as discussed in text.

[a]This is the Cu content if Cu is calculated by applying the Cu/(Cu + Ni) ratio of the ore reserves to the average Ni content for the deposit, thus allowing for the believed unrepresentative sampling (Naldrett et al, 1982).

PGE tenor of the sulfides making up the deposits. Base metals, although rich in these sulfides, are not enriched to nearly the same extent over other magmatic sulfides (see Fig. 6.11). As described in Chapter 5, some geologists have interpreted the PGE enrichment as the consequence of hydrothermal fluids (e.g., Stumpfl, 1974; Boudreau, 1986; Johan and Watkinson, 1985). Others regard them as magmatic; Campbell et al. (1983) have suggested that they are a special case of magmatic segregation in which the sulfides have achieved a very high magma/sulfide ratio and have outlined a mechanism (described in Chapter 5) whereby this could have occurred. This point is illustrated in Figure 6.11, where the two solid lines illustrate the equilibration of Davies and Tredoux's (1985) analyses of Bushveld B1-type chilled samples (the concentration exceeded in 50 percent of the samples analyzed), the lower bar with $R = 10^{2.5}$ and the upper bar with $R = 10^{5.3}$. The same magma can account well for widely differing sulfide compositions simply as a result of variations in the magma/sulfide ratio.

Naldrett et al. (1987) have considered the quantitative aspects of this suggestion for the Merensky Reef and the UG-2 of the Bushveld Complex. Their treatment of Pt in the Merensky Reef is shown in Figure 6.18A. The values used for the concentration of Pt in the magma (X_i) are those of Davies and Tredoux (1985). Two examples are modeled: one in which X_i is the concentration that is exceeded by 50 percent (15 ppb Pt) and in the other by 10 percent (24 ppb Pt) of the samples analyzed; the partition coefficient has been assumed to be 10^5. The value of log R required for the modeled sulfides to match the observed value is 4.2 for the higher concentration and 4.4 for the lower. The height of the vertical column of magma with which the Merensky sulfides must have equilibrated has been calculated from these R values and is 300 m for the lower R (equivalent to 24 ppb Pt in the original magma) and 475 for the higher R (equivalent to 15 ppb Pt in the original magma).

Figure 6.18B indicates that the present Pt tenor in the sulfides of the UG-2 at Maandagshoek (Gain, 1985) cannot be attained with the lower of the two initial magma concentrations modeled (15 ppb Pt). It can be attained for the higher of the two (24 ppb Pt) but only at a high log R value of 5.8, which in turn implies equilibrating the sulfides with a 1,000-m-high column of magma. If, as discussed later, the Pt tenor of the sulfides has been doubled subsequent to their equilibration with their source magma by FeS-loss, the original lower tenor of 1,000 ppm can be explained with $X_i = 15$ ppb at log $R = 5.3$, which, in this case, implies equilibration with a 630-m-high column of magma, or with $X_i = 24$ ppb at log $R = 4.9$, implying a 250-m column of magma.

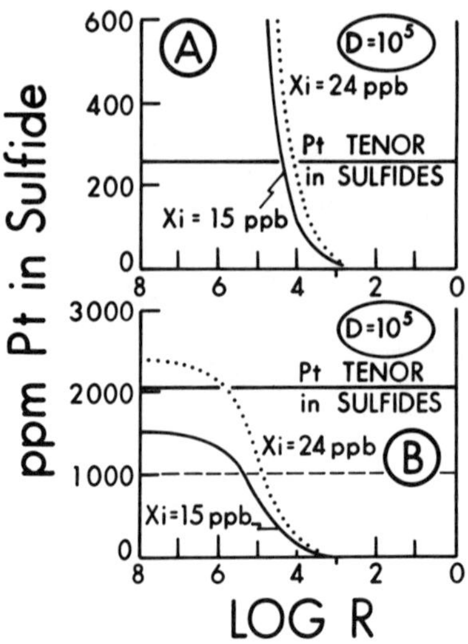

Fig. 6.18. Variation of the concentration of Pt in the sulfides of the Merensky Reef (A) and UG-2 (B), with log R (R = magma/sulfide mass ratio) assuming certain values of the partition coefficient, D, and the observed abundance of Pt in the initial magmas. The horizontal dashed line in (B) indicates the original Pt tenor of the sulfides if, as discussed in the text, their original mass was two times greater than their present mass, the excess consisting of FeS that has been lost due to reaction with chromite (see text).

6.7.4 Recognition of the Controls That Have Influenced the Composition of an Ore

A number of factors have been shown to influence the composition of a sulfide ore. The most obvious factor is the composition of the host magma, both with respect to its content of metals such as Ni, Cu, Co, and the PGE, and with respect to its major elements and their influence on the values of the sulfide melt–silicate magma partition coefficients. The composition is a function of the partial melting that gave rise to the magma at its source, and subsequent modifications due to the fractional crystallization of silicate minerals and the fractional segregation of a sulfide liquid. Three other factors have been shown to be important: (1) the fractional crystallization of Mss ($Fe_{(1-x)}S$–$Ni_{(1-x)}S$) from a sulfide liquid leading to a modification in the composition of the residual liquid, (2) the magma/sulfide ratio operative during the segregation of a sulfide liquid from a silicate magma, and (3) the removal of successive batches of sulfide liquid from a given batch of magma, leading to a progressive depletion in the concentration of the chalcophile elements in the magma.

Variations in composition resulting from fractional crystallization of Mss will be dependant on the relative partitioning of different elements between liquid sulfide and Mss. Thus progressive fractionation of Mss should result in an increase in the concentrations of Cu, Ni, Pt, and Pd, and an increase in the Cu/(Cu + Ni), Cu/Co, and (Pt + Pd)/(Ru + Ir + Os) ratios in the liquid. In the preceding discussion of the variations between different ore types at both Strathcona and Levack West (Section 6.7.2), it is shown that the Cu/(Cu + Ni) and (Pt + Pd)/(Ru + Ir + Os) ratios appear to be particularly reliable indices of the compositional changes resulting from Mss fractionation. These two parameters have been used to construct Figure 6.19, in which the lines connecting the different ore types from a single deposit indicate the kind of compositional change to be expected as a result of Mss fractionation.

Compositional variations resulting from variation in the magma/sulfide ratio and the removal of successive batches of sulfide from a given batch of magma depend on the partition coefficients between sulfide melts and silicate liquids of the elements involved. It is shown that these coefficients increase in the order Fe, Co, Cu approximately equal to Ni, and then, very much higher, the PGE. Thus the successive removal of a sulfide liquid from a silicate melt would be expected to deplete all of these elements in the melt; however, relative to Cu and Ni, this effect should be least marked for Fe, and less marked for Co, and very much more marked for the PGE. In the sulfides themselves, variations in composition due to variations in Fe content in the host magma would not be detectable. In comparison with an early sulfide segregate from a given silicate melt, however, a late segregate should have lower overall concentrations of these elements, similar Cu/(Cu + Ni) and (Pt + Pd)/(Ru + Ir + Os) ratios, and lower Cu/Co and PGE/(Cu + Ni) ratios. Equilibration at successively lower values of R would produce similar compositional variation.

Because the Cu/(Cu + Ni) and (Pt + Pd)/(Ru + Ir + Os) ratios should not be overly sensitive to the successive segregation of sulfide or variations in R value, the diagram constructed using these two parameters (Fig. 6.19) should provide information about some of the other processes affecting the composition of sulfide ores, unobscured by the segregation of differing proportions of sulfide. Variations resulting from the fractionation of Mss have already been discussed. Deposits segregating from initially similar magmas should plot in the same place (as do Strathcona and Levack West). Little Stobie deposits No. 1 and No. 2 also plot together, although they have higher Cu/(Cu + Ni) ratios than Levack West and Strathcona. This observation is interpreted in section 6.7.3.2 as the result of the fractionation of olivine from the host magma, thereby lowering the Ni content without affecting the Cu or PGE contents greatly. The data of Cochrane (1984) on the Copper Cliff offset deposits at Sudbury indicate that the 810, 830, and 865 deposits plot fairly close together, although with a somewhat lower (Pt + Pd)/(Ru + Ir + Os) ratio than the others. Naldrett (1984) suggests that the area occupied by all of the above deposits represents the "norm" for Sudbury.

The Falconbridge deposit at Sudbury is clearly an exception to the postulated norm, having much lower values of both ratios, and resembling the product of the earliest fractionation (the

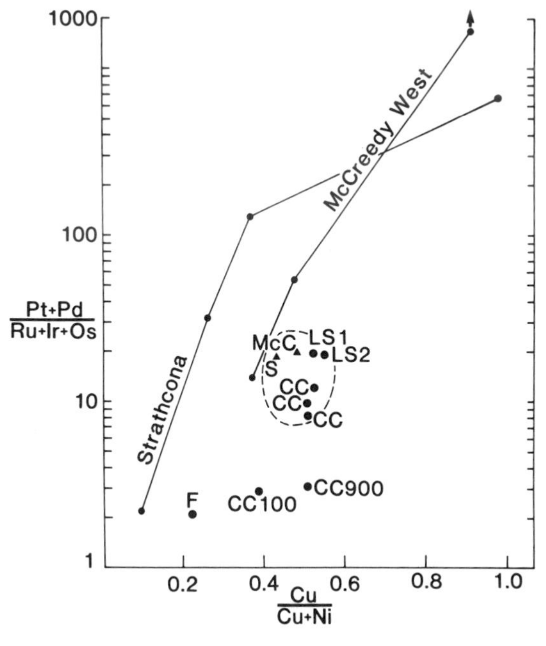

Fig. 6.19. Plot of (Pt + Pd)/(Ru + Ir + Os) versus, Cu/(Cu + Ni) for several Sudbury ore deposits. The points connected by straight lines indicate compositional changes that are believed to result from fractional crystallization of Mss from a sulfide melt. Average compositions of many deposits are rather similar and fall within the oval. Some deposits are distinctly atypical and plot well below the circle. In the case of Falconbridge, it is believed that a significant proportion of fractionated ore is missing from the deposit. F = Falconbridge #5 shaft; CC100, CC900, = Copper Cliff mine; McC = Levack (McCreedy) West; S = Strathcona; LS1, LS2 = Little Stobie Numbers 1 and 2. (From Naldrett, 1984.)

Hangingwall ore) at Strathcona. Naldrett et al. (1982) pointed out this feature, arguing that the present Falconbridge deposit represented the early crystallization products of a much larger ore body, of which the late crystallizing portion, rich in Pt and Pd, had been faulted away, and either had been removed by erosion, or remained to be discovered. The Copper Cliff offset deposits 100 and 900 also stand out as anomalous. It is possible that they also represent early crystallizing fractions and that ore richer in Pt and Pd than that sampled either remains to be sampled, or originally occupied a now-eroded portion of each of these shoots.

6.7.5 Sulfides Hosted in Chromitite

The high PGE concentrations of sulfides associated with chromitites (Section 6.5.7), such as the UG-2 chromitite of the Bushveld, can be explained in part as the result of their having achieved high magma/sulfide ratios, as described for the Merensky Reef. As noted in Section 5.1.2.4, however, they are also characterized by high Ni and Cu contents. The high Ni may be due to some of this occurring in solid–solution in the chromite, but the high concentration of Cu cannot be explained in this way.

Gain (1985) suggested that the sulfides of the UG-2 had been enriched in PGE, subsequent to their deposition, as a result of the streaming through the chromitite layer of a volatile component rich in O_2. This oxidized the sulfides and removed SO_2, thereby greatly reducing the mass of sulfide. Von Gruenewaldt et al. (1986) suggested that Fe from the sulfides had replaced Mg in the chromite, the Mg entering the intercumulus liquid, thus leading to an enrichment in Ni, Cu, and PGE.

Naldrett and Lehmann (1988) have pointed out that chromite that has crystallized from basaltic magma contains vacancies that are due to the presence of excess Fe_2O_3 and Al_2O_3 in its structure. As chromite cools, its ability to tolerate vacancies decreases and these will decrease in number, either as a result of the incorporation of divalent cations into its structure from other sources or, once the solvus is reached, through the exsolution of the excess sesquioxide. Noting that a potential source of a divalent cation is the Fe of nearby sulfide, Naldrett and Lehmann proposed the following reaction:

$$\tfrac{4}{3}Fe_2O_3^{(spinel)} + \tfrac{1}{3}FeS = Fe_3O_4^{(spinel)} + \tfrac{1}{6}S_2 \qquad (6.1)$$

Assuming that ΔG can be approximated by $(\Delta H - T\Delta S)$ over a temperature range, the activities of Fe_2O_3 ($[Fe_2O_3]_2$) and Fe_3O_4 ($[Fe_3O_4]_2$) after the sulfide–chromite assemblage has cooled through a temperature interval (from T_1 to T_2) can be related to the activities of FeS ($[FeS]_{1 and 2}$) and the fugacities of S_2 ($(fS_2)_{1 and 2}$) at the start and finish of the cooling, respectively, and the activities of Fe_2O_3 and Fe_3O_4 at the start ($[Fe_2O_3]_1$ and $[Fe_3O_4]_1$), by the expression:

$$\left(\frac{[Fe_2O_3]_2}{[Fe_2O_3]_1}\right)^{4/3} \times \frac{[Fe_3O_4]_1}{[Fe_3O_4]_2} = \exp\left(\frac{\Delta H}{R}\left(\frac{1}{T_2} - \frac{1}{T_1}\right)\right) \times \left(\frac{(fS_2)_2}{(fS_2)_1}\right)^{1/6} \times \left(\frac{[FeS]_1}{[FeS]_2}\right)^{1/3} \qquad (6.2)$$

Naldrett and Lehmann assumed that sulfides of the Merensky composition were the precursor to those in the UG-2 (a reasonable assumption, as these have formed from the Bushveld magma at a horizon close to that of the UG-2 and are not associated with large amounts of chromite). Their calculation of the quantities of Fe that must have been transferred from sulfide to chromite through this reaction to account for the enrichment in Cu in the sulfides of the Bushveld Middle and Upper Group chromitites are shown in Table 5.1. It is seen that because of its much higher overall sulfide content (both present and initial sulfide content), much more Fe transfer is required in the case of the UG-2, than is necessary to explain the Cu-enrichment of the other chromitites. Naldrett and Lehmann, therefore, chose to investigate the UG-2 as an extreme case. According to their model the amount of Fe transferred from sulfide to chromite is 1,578 ppm, which amounts to a mole fraction of 1.8×10^{-3} of a rock composed of 100 percent chromite of UG-2 composition.

They solved equation (6.2) using the spinel model of Lehmann and Roux (1986) to relate the change of activity of Fe_2O_3 with temperature with the change in vacancy content of the spinel. The extent to which Fe can be removed from the sulfide depends on the S fugacity and the mass ratio of sulfide to spinel. If it assumed that the S fugacity follows the normal path of sulfide cooling at constant composition, reaction (6.2) requires an initial vacancy content in the spinel of 3×10^{-3}. Naldrett and Lehmann also used the Lehmann–Roux spinel model to show that this concentration of vacancies was realistic in a basalt of normal composition and oxidation state.

7. USE OF GENETIC CONCEPTS IN EXPLORATION

Ideas concerning the genesis of an ore deposit have always been an important aspect of the planning of an exploration program. The magmatic model for the formation of Ni–Cu deposits carries with it concepts that have many implications for exploration—some simple, qualitative, and in common use and others not so simple, more quantitative, and not yet in common use. Examples of the application of these concepts form the subject of much of this chapter. Platinum group elements (PGE) deposits have different characteristics, and bring with them different criteria for recognizing likely ore-bearing environments. To avoid confusion, Ni–Cu sulfides are discussed first, followed by a discussion of PGE deposits.

7.1 NI–CU ORES

7.1.1 Qualitative Application of Concepts

The presence of so many magmatic ore bodies at the base of their host intrusions suggests that gravitational settling of an immiscible sulfide liquid is the most important mechanism in Ni–Cu sulfide concentration. Most of the exceptions to the rule of basal control occur in areas that have undergone extreme deformation (e.g., Thompson, Manitoba; Shebandowan, Ontario) where plastic flow of sulfide has occurred, perhaps coupled with dissolution, transportation, and reprecipitation via aqueous fluids. For sulfides to be concentrated effectively as a result of gravitational settling, some process must cause a significant amount of sulfide liquid to become immiscible at a particular stage in the crystallization of a silicate magma—much more than would segregate at any one time as a result of the normal crystallization of the magma. In this way, the excess sulfide can then settle to form a concentration that is relatively undiluted by silicate grains. This process of enhanced immiscibility must be related to controls on the solubility of S in silicate melts.

Experimental studies suggest that the solubility of S in a mafic melt can be reduced by:

1. *Oxidation.* MacLean (1969) has shown that the oxidation of Fe^{2+}, with which S associates in the silicate magma, to Fe^{3+}, with which it does not associate, lowers the amount of S that can dissolve in an FeO silicate liquid.
2. *Silicification.* In Chapter 4, it is pointed out that the addition of silica to a mafic magma depresses the S solubility, and (according to Irvine, 1975) it is suggested that the assimilation of a large amount of country rock, fractured and heated by meteorite impact, may account for the abundant Ni sulfide ores at Sudbury.

The ores at Noril'k-Talnakh and Duluth, Minnesota have been cited (Chapter 4) as examples where the addition of S from an external source has contributed to the total S content of the host intrusions and enhanced the formation of an immiscible sulfide liquid.

Naldrett (1973) and Naldrett and Cabri (1976) have suggested that the association of Ni sulfides with peridotitic komatiites is due to the derivation of these magmas from a level in the Archean mantle that was particularly enriched in sulfide. Lesher et al. (1984) and Huppert and Sparks (1985a) suggest that some of the S has been derived by assimilation as the lavas traversed across sulfide-bearing country rocks. Whatever the ultimate source of the S, geologic evidence indicates that the lavas were saturated in sulfide at the time that they deposited their sulfides, so that these sulfides (and therefore, the contained metals) in any given ore deposit were not derived from the volume of rock immediately overlying the deposit, but from a much larger volume of magma.

In summary, magma anomalously rich in sulfide may be produced directly from the mantle, but they are frequently the result of the addition of further S as magma enters the crust. Unusual amounts of immiscible sulfide have also been produced by the silicification of magmas that are not, overall, unusually rich in sulfide, and, in theory, by oxidation of magma, although actual examples of immisciblity induced by oxidation have not been documented.

7.1.1.1 SIMPLE CONCEPTS IN CURRENT USE

Concepts and axioms in active use by all exploration geologists include:

1. The almost universal association of Ni–Cu deposits with mafic and ultramafic rocks, which can commonly be located by airborne magnetic surveys.

2. The relationship of ore to the basal intrusive contact. This initial empirical observation is one of the principle arguments in favor of a magmatic model that incorporates the gravitational settling

of dense sulfide melts. The model has been established so well that it has become customary to establish the stratigraphic base of an igneous body and concentrate most exploration along this zone.

3. The relationship of deposits to certain petrotectonic settings. Archean greenstone belts are particularly important as they contain periodotitic komatiites. Although, in the past, rifting may not have served as a guide to much exploration outside of the U.S.S.R., it is now recognized that rifting is commonly accompanied by magmatic activity, and at Noril'sk-Talnakh and Duluth (see Chapter 4) rifting has allowed mafic magma to come into contact with S-rich upper crustal sediments. In some cases, the S-rich nature of the sediments themselves is the consequence of rifting (e.g., the formation of evaporites during early stages in the development of a rift zone).

4. The relationship of mineralized bodies to stratigraphy. This point, which is extremely important in certain areas, is an empirical observation rather than a deduction from general principles. Komatiites occur in the lower parts of mafic to felsic volcanic cycles that characterize many greenstone belts. Specific horizons within a given komatiite succession form particularly important targets for reasons that are not fully understood. These horizons are sometimes rich in Fe sulfide-bearing sediments (e.g., Kambalda; Ross and Hopkins, 1975; Gresham and Loftus-Hills, 1981), suggesting that S assimilation may have played a role, although direct evidence of this hypothesis is lacking. Alternatively, the horizons may represent ultramafic rocks derived from the tapping of a particularly S-rich portion of the mantle (Naldett, 1973).

The stratigraphic approach has also proved important in exploration for komatiite-related deposits in Zimbabwe and Canada, and in the search for tholeiite-related deposits at Pechenga, U.S.S.R.

5. Recognition of favorable and unfavorable mafic and ultramafic environments. Naldrett (1973) pointed out that, because they lacked significant associated Ni–Cu sulfide bodies, certain types of mafic and ultramafic bodies appear to be unfavorable hosts. These consist of (1) Alpine ultramafic bodies, including both cumulate and metamorphic-textured portions of ophiolite complexes, (2) Alaskan-type ultramafic bodies, (3) those mafic and ultramafic rocks associated with alkalic ring complexes. The belief that the metamorphic-textured harzburgites, lherzolites, and dunites of ophiolite complexes represent portions of the mantle led, on the basis of genetic ideas, to the conclusion that significant concentrations of magmatic sulfides are unlikely to have formed within these rocks. The rarity of deposits within Alaskan-type rocks, or in the cumulate portions of ophiolite complexes remains, however, enigmatic from a genetic point of view.

7.1.1.2 POTENTIAL AIDS TO EXPLORATION—RECOGNITION OF SULFIDE-SATURATED MAGMAS

When faced with the task of evaluating a mafic or ultramafic body as an exploration target, one of the first questions that exploration geologists must attempt to answer is whether the magma, or a portion of the magma, responsible for the body was ever saturated in S. The question is raised because:

1. The demonstration that many important deposits result from S assimilation should lead geologists to investigate whether the magma could have come into contact with sulfide or sulfate deposits during its ascent from its source.
2. The experimental demonstration that oxidation or silicification may result in the "salting-out" of immiscible sulfides should lead them to investigate whether there is evidence that either of these processes was active throughout, or within a particular portion of the body that they are considering.

Any of the above aspects might lead to the conclusion that a mafic or ultramafic body had a reasonable chance of sulfide saturation in its magma, but will not answer with any certainty the question of whether it actually became saturated. A number of approaches have been made recently to answer this question. At first sight it would seem logical that the best method of recognizing a sulfide-saturated magma would be to determine the S contents of rocks thought to be representative of the initial magma of the igneous body, and then to compare these with the experimentally based estimates of the saturation level expected in magmas of similar composition. This approach was attempted by the author on a series of spinifex-textured lavas from the Abitibi greenstone belt, but was unsuccessful because of the extreme mobility of the sulfides during metamorphism, even under the zeolite–facies metamorphism that occurred in the Munro Township area, Ontario. Sulfur may also be lost as a result of oxidation in the case of subaerial lavas (Naldrett and Goodwin, 1977). It is concluded that the best approach is to monitor the composition of the silicates of the lava or intrusion, as discussed in the following section.

7.1.2 Quantitative Modeling of Sulfide Segregation

The quantitative understanding of how metals partition between sulfide and silicate melts has reached a stage where modeling their behavior can assist in exploration by identifying favorable districts, and by distinguishing favorable from less favorable bodies of mafic or ultramafic rock and, in special circumstances, favorable from less favorable zones within a given body of rock (Naldrett et al., 1984). It can also provide an estimate of the Ni tenor of the ore that might be discovered within a given intrusion.

The basic premise in attempting to determine whether a particular body of rock has crystallized from a magma from which molten sulfide had previously segregated or not is straightforward; *because chalcophile elements partition strongly into sulfide in preference to silicate melts and minerals, magmas from which sulfide has segregated are depleted in these elements.* The magnitude of depletion depends on the degree to which the particular element partitions into the sulfide as well as on the amount and manner in which the sulfide was segregated.

In principle, the elements with the highest values are the most sensitive indicators of sulfide segregation. In practice, it is found that Ni is the most useful indicator, not only because of its high partition coefficient, but also because it is normally present in readily detectable quantities (unlike Pt) and is relatively immobile during alteration (unlike Cu). Accordingly the examples that are discussed in this section deal with Ni.

7.1.2.1 TYPES OF SULFIDE SEGREGATION

Sulfide may react with and segregate from a magma in a number of ways, of which batch equilibration and fractional segregation can be regarded as two end-members.

7.1.2.1.1 Batch Equilibration

In batch equilibration, a significant proportion of the internal dissolved sulfide becomes immiscible or there is a significant addition of external S that induces this immiscibility. The critical feature of this process is that immiscible sulfides form and equilibrate with the silicate melt in a single stage. The process is illustrated schematically in Figure 7.1. Sulfide becomes immiscible throughout the magma (Fig. 7.1a), forming droplets that equilibrate with the silicate liquid, settle and accumulate (Fig. 7.1b), after which the magma begins to crystallize olivine as shown in Figure 7.1c. Sulfide saturation may occur in only a small portion of the magma, in which case the process is called "localized batch equilibration."

Obviously, the degree of depletion of Ni (or other chalcophile element) in a given mass of magma that has equilibrated with a batch of sulfide is related to the relative masses of the magma and sulfide (i.e., to the value of the magma/sulfide ratio or R as defined in Chapter 3). If the proportion of sulfide is very small, it removes very little Ni from the silicate magma and the Ni content of this magma is much the same as it was before liquation and equilibration with the sulfide. On the other hand, if the proportion of sulfide is large, a great deal of the available Ni is removed, lowering its concentration in the silicate magma appreciably.

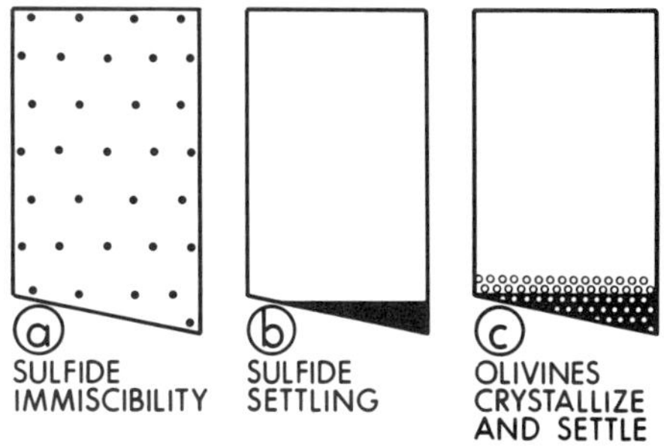

Fig. 7.1. Stages in the formation of a sulfide concentration by batch crystallization. ***a.*** Sulfide immiscibility occurs due to some sudden change in intensive parameters such as the activity of SiO_2 or other felsic component (due to hybridization with country rocks or magma mixing), increase in fS_2 due to sulfide or S assimilation increase in fO_2 or some other cause. ***b.*** Sulfides settle and in this way are removed from interaction with the magma. ***c.*** Silicate minerals either start or resume crystallization. (After Naldrett et al., 1984.)

The concentration of Ni in a typical basaltic magma that initially contained 300 ppm but has undergone batch equilibration with sulfide is shown in Figure 7.2. It can be seen that the Ni depletion is almost negligible where the R value is high, but increases rapidly once the value drops below about 1,000. As a first approximation, an R value of 1,000 or less is required for batch equilibration to produce *conspicuous* Ni depletion in basaltic magma.

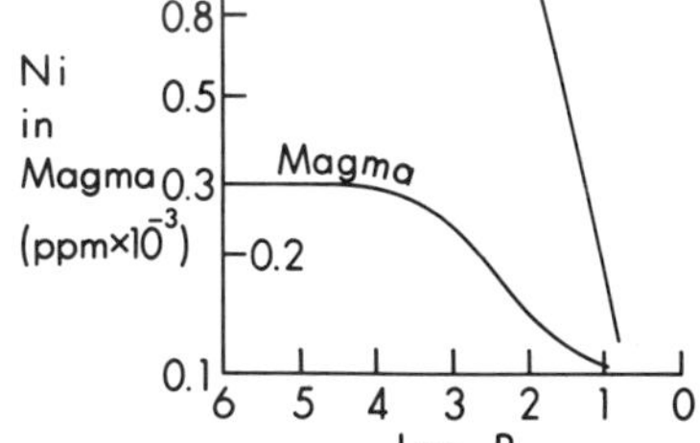

Fig. 7.2. Variation in the Ni content of primitive basaltic magma and olivine in equilibrium with the magma resulting from batch equilibration with sulfide as a function of the magma-sulfide ratio (R). Values of $D_{Ni}^{Sul.melt/Sil.\ magma} = 275$, $D_{Ni}^{Olivine/Sil.\ magna} = 7.23$.

The Ni depletion is reflected in the olivine that crystallizes subsequent to batch equilibration and the Ni content of the first olivine to crystallize is also shown in Figure 7.2.

7.1.2.1.2 Fractional Segregation

In fractional segregation, very small amounts of sulfide become continuously immiscible, equilibrate with the silicate magma, and then settle or are removed in some other way so that further effective communication with the magma is precluded. The typical situation in which fractional segregation would occur would be during the fractional crystallization of a sulfide-saturated silicate magma as depicted schematically in Figure 7.3b. Because the silicate melt contains the maximum possible amount of sulfide in solution, crystallization of even a small amount of olivine (or other S-free minerals) increases the sulfide concentration in the remaining liquid, and thereby, in most cases, forces some of the dissolved sulfide out of solution.

The variation in concentration of elements in a magma undergoing simultaneous fractional crystallization of olivine and fractional segregation of sulfide can be calculated using the Rayleigh Fractionation law. Obviously, a critical factor is the relative proportion of sulfide to olivine being removed. In most cases this proportion appears to be in the range of 1:50 to 1:400 (Naldrett et al., 1984).

The compositional trends of silicate magmas derived by the fractionation of a komatiitic starting liquid that initially contained 1,750 ppm Ni and 32 percent MgO are plotted in Figure 7.4 for sulfide-unsaturated and a series of sulfide-saturated conditions. Although Mg is not incorporated into sulfide, it partitions strongly into olivine, and fractional crystallization of olivine depletes the magma in Mg as well as Ni. For this reason the MgO concentration is commonly used as an index of fractionation for mafic and ultramafic rocks. It can be seen from Figure 7.4 that there is a major difference in the covariance of Ni and MgO between the sulfide-saturated and the sulfide-unsaturated cases.

7.1.2.2 EXAMPLES OF THE APPLICATION OF THEORETICAL MODELS IN NATURE

7.1.2.2.1 Komatiites

Nickel–sulfide deposits occur associated with flows and intrusions of the komatiite magma suite in Western Australia, Zimbabwe, and Canada. Not all areas of komatiitic volcanism appear to have associated deposits and it is important to develop criteria for distinguishing sequences of komatiites with high Ni potential from those with low potential.

Accumulations of komatiitic lavas are characterized by cyclical variations in composition such that the flows in each cycle become less magnesian upward. It has been shown (Naldrett and Smith, 1981; Arndt and Nesbit, 1982) that the intracycle variation is due to low-pressure fractional crystallization of olivine as illustrated in Figure 7.3a. As olivine crystallizes and accumulates at the base of the shallow crustal magma chamber, the magma becomes less magnesian and is tapped off from time-to-time to extrude on the surface as

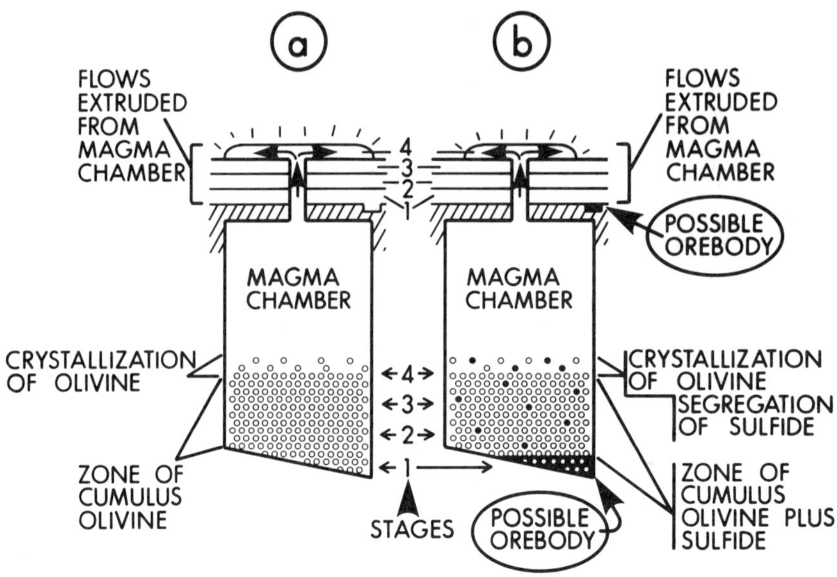

Fig. 7.3. Fractional crystallization of olivine and fractional segregation of sulfide from a cooling komatiitic magma. ***a.*** Sulfide-unsaturated case. ***b.*** Sulfide-saturated case. Note that the magma is assumed to erupt periodically to give rise to a series of komatiitic flows of successively lower MgO content. (After Naldrett, et al., 1984.)

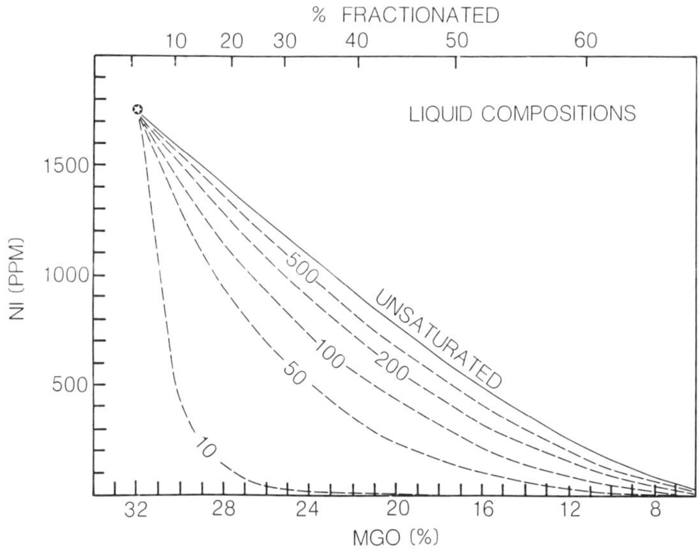

Fig. 7.4. Computer simulation of the fractionation of a komatiitic magma containing 32 wt % MgO as a consequence of the separation of olivine (unsaturated) or a mixture of different proportions of olivine + sulfide. The numbers indicate the weight ratio of olivine/sulfide separating. (Modified after Duke and Naldrett, 1978.)

lava flows. Phenocryst-free, spinifex-textured samples in the volcanic succession provide a guide to the composition of the magma in the chamber as it fractionates. A komatiite magma that gives rise to magmatic sulfide deposits must be sulfide saturated as in Figure 7.3b.

7.1.2.2.1.1 Barren Komatiites. The compositions of spinifex-textured komatiites and basaltic komatiites from apparently unmineralized komatiite sequences in greenstone belts in Canada, Australia, Zimbabwe, and Finland are shown in Figure 7.5, together with the fractional segregation model curves of Figure 7.4. The analyses define a reasonably coherent trend that agrees well with the unsaturated trend calculated for the model parent komatiite composition. The agreement between the trends is important because it

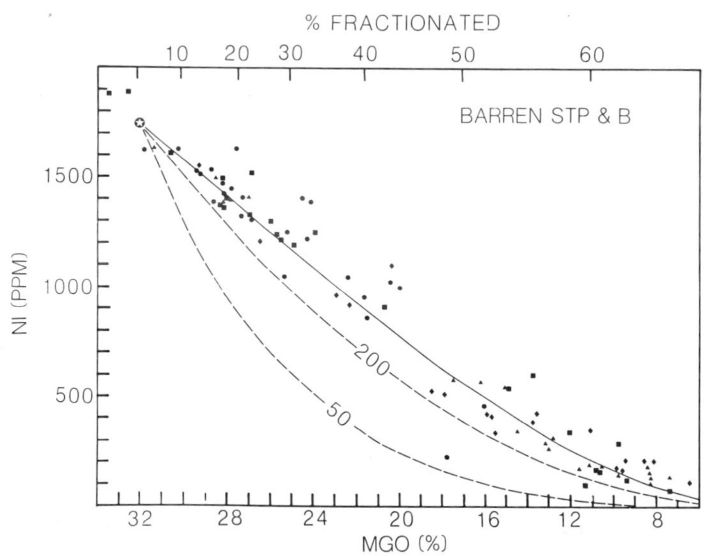

Fig. 7.5. The compositions of "barren" spinifex-textured peridotites and komatiitic basalts from Yakabindie, Western Australia (triangles); Belingwe, Zimbabwe (circles); the Abitibi belt, Canada (squares); and Barberton, South Africa (diamonds). The data are in reasonable agreement with the sulfide-unsaturated trend from Figure 7.4. (After Naldrett, 1984.)

implies that the computer-based model curves may be used as a standard against which to compare "liquid equivalent" komatiites in general.

7.1.2.2.1.2 Kambalda, Western Australia. The deposits of the Kambalda camp, described in Chapter 4, are typical of Ni–sulfide ores of the komatiite association. The numerous small but high-grade deposits occur mainly at the base of a sequence of komatiite flows from which the sulfides are believed to have settled. The sequence averages 500 m in thickness and becomes progressively less magnesian upward. Lesher et al. (1981) analyzed a series of spinifex-textured samples from the Kambalda sequence and compared these to the model fractionation curves of Duke (1979). Their data for Kambalda are compared in Figure 7.6 with the model curves of Figure 7.4. The Ni depletion is quite apparent by comparison, for example, with the barren komatiites of Figure 7.5.

7.1.2.2.1.3 Scotia, Western Australia. The Scotia deposit is located about 120 km north–northwest of Kambalda and has been described by Stolz and Nesbitt (1981) as a typical example of the komatiite ore association. Their analyses of spinifex-textured rocks are plotted in Figure 7.7. The resulting pattern is very similar to the Kambalda data: The rocks are significantly depleted in Ni in comparison with the barren komatiites and to the model sulfide-unsaturated trend.

7.1.2.2.1.4, Dumont Intrusion, Quebec. The Dumont Ni deposit, located 62 km northeast of Rouyn, Quebec, occurs within a differentiated sill of komatiitic affinity (see Chapter 4). The sill, described by Duke (1985, 1986), is at least 7 km long and up to about 1 km wide, and comprises a lower ultramafic zone 450 m thick and an upper mafic zone 250 km thick. Disseminated magmatic sulfides grading 0.45 percent Ni occur over a 3,000-m strike length within a dunite subzone of the ultramafic zone.

The composition of olivines from this dunite and the overlying peridotite are plotted in Figure 7.8. Also shown are model trends calculated for the olivines, assuming a parent magma containing 27.5 wt percent MgO. The olivines from the mineralized portion of the dunite (shown by dark circles) display an extreme range of Ni and yet a rather restricted range of forsterite contents. Some of the olivines are drastically depleted in Ni, whereas a few are moderately enriched. What is interesting from the standpoint of possible exploration guidelines is the fact that the olivine in samples of the unmineralized dunite and from the overlying peridotite (shown by open circles) are distinctly depleted in Ni in comparison with the sulfide-unsaturated model curve. Thus, the segregation of even a rather modest quantity of magmatic sulfide is recorded in the overlying, sulfide-free rocks. In this case, a sample equivalent in composition to the liquid part of the magma involved is not required. A guide to how Ni was varying with respect to MgO in the magma is preserved in the composition of the olivines formed from the magma.

7.1.2.2.2 Gabbros of Northern Maine

Thompson and Naldrett (1984) have studied the segregation of sulfide from the Moxie and Katah-

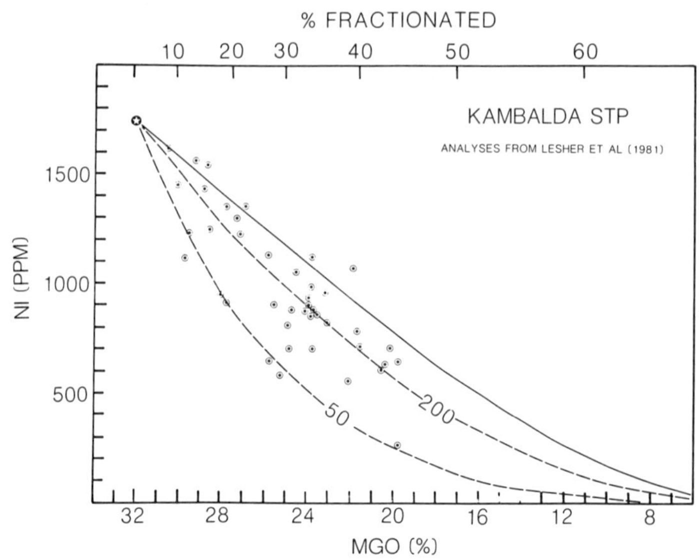

Fig. 7.6. The composition of spinifex-textured peridotites from Kambalda, Western Australia (from Lesher et al., 1981) are plotted together with model trends from Figure 7.4. The rocks are significantly depleted in Ni in comparison with the sulfide-unsaturated trend, and fractionation under sulfide-saturated conditions with an average olivine/sulfide ratio of 200:1 is suggested. (After Naldrett et al., 1984.)

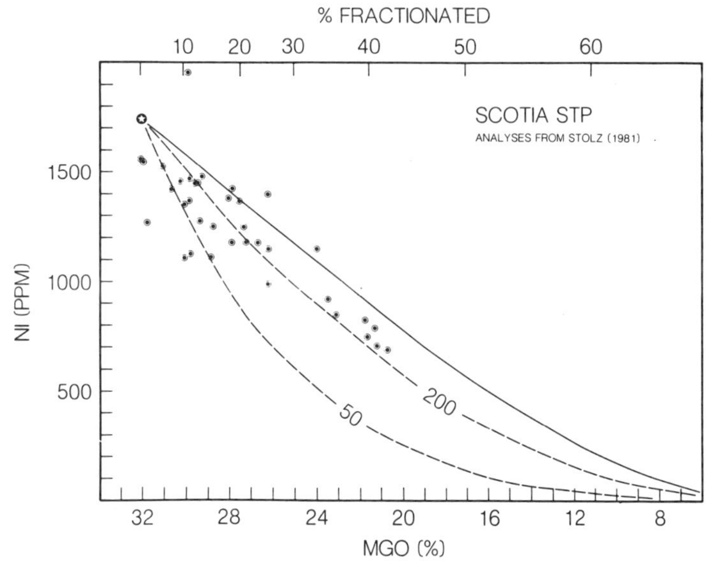

Fig. 7.7. The composition of spinifex-textured periodotites from the vicinity of the Scotia mine, Western Australia (data from Stolz, 1981). The rocks are significantly depleted in Ni in comparison with the sulfide-unsaturated trend. (After Naldrett et al., 1984.)

din plutons of northern Maine. These are Devonian-aged bodies of peridotite and olivine gabbro that were intruded into Siluro–Devonian sedimentary rocks during the Acadian orogeny (Fig. 7.9). They contain uneconomic deposits of Ni sulfide that, nevertheless, are interesting examples of the principles discussed in this section.

7.1.2.2.2.1 The Moxie Pluton. This body is a 2- to 8-km wide and 70-km long, irregular sill-like mass that appears to have been intruded from west to east during the Acadian orogeny. Disseminated sulfides are ubiquitous in the rocks of the western part of the intrusion and three small sulfide showings have been discovered in the western

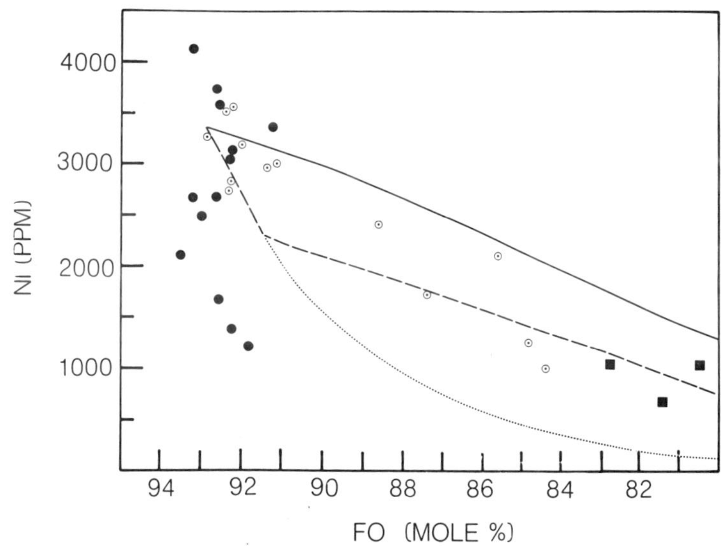

Fig. 7.8. The composition of olivines in the Dumont intrusion, northwestern Quebec. The solid circles represent olivines from the main mineralized zone that exhibit a striking Ni depletion trend that may have been enhanced by a subsolidus equilibration with the coexisting sulfide. The open circles represent olivines at some distance from the mineralized zone, either along the strike or stratigraphically above. The squares are corrected whole rock analyses of samples from the top of the ultramafic zone in which the olivine has been completely serpentinized. The solid line is the model trend from sulfide-unsaturated fractionation, the dotted line is the trend for fractional segregation of olivine and sulfide in a 60:1 ratio. The dashed line is the Dumont model, deduced on the basis of geologic evidence. (After Naldrett et al., 1984.)

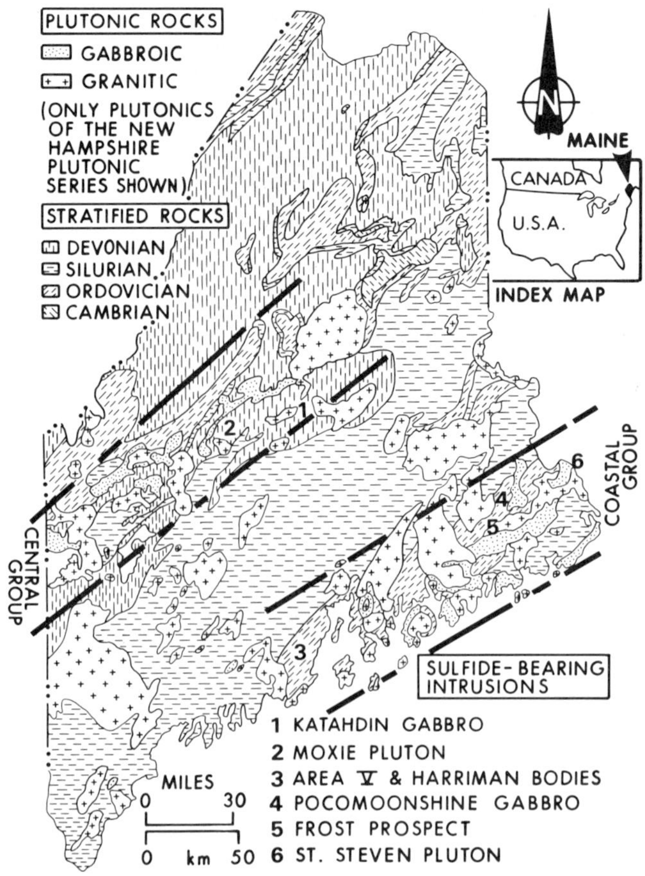

Fig. 7.9. Geology of Maine, U.S.A., showing the principal sulfide-bearing mafic/ultramafic intrusions. (After Thompson and Naldrett, 1984.)

and central parts, where the basal contact approaches a sulfide-rich formation within the country rocks (the Hildreth's Formation; Fig. 7.10). The S isotope ($\delta^{34}S = -8$ to -23) and the S/Se ratios (S/Se = 10,000 to 40,000) of the sulfides of these showings leaves little doubt that much of the S was derived from the Hildreth's Formation ($\delta^{34}S = -26$; S/Se = 27,000 to 50,000).

Simpkin and Smith (1970) compiled analyses of olivine from a wide range of intrusive and extrusive rocks, and showed that these rocks defined a reasonably tight trend as illustrated in Figure 7.11. The positive correlation between Ni and the forsterite content of olivine reflects both the amount of partial melting that gave rise to the parent magma of each body, and the subsequent fractionation of olivine from it. If a given set of olivine analyses do not fall within this field, they can be regarded as anomalous and an explanation for their composition is required.

Compositions of olivines from the Moxie body are shown superimposed on Simpkin and Smith's field. Their very steep trend and generally Ni-depleted composition is apparent. Modeling of the compositional variation of the Moxie olivines indicates the subtraction of silicates plus sulfide in a ratio of 100:1 or lower. This suggests therefore, that the magma was saturated in sulfide either during or very shortly after intrusion, and that fractional crystallization of the magma was accompanied by sulfide segregation. Whether the segregation of sulfide led to concentrations greater than the three showings described previously is not clear. What is clear is that a very large amount of Ni (equivalent to about twice that currently found in solid–solution in all of the olivine of the Moxie body) has been concentrated in sulfide not yet located.

Thompson and Naldrett (1984) used the range in olivine compositions associated with each of the sulfide prospects at Moxie to compute the range in tenor of the sulfides expected through equilibrium with the olivines. These ranges are compared with the observed tenors of each individual prospect in Table 7.1; it is seen that there is a reasonable correspondence between calculated and observed ranges. Better estimates for each showing would probably have been obtained

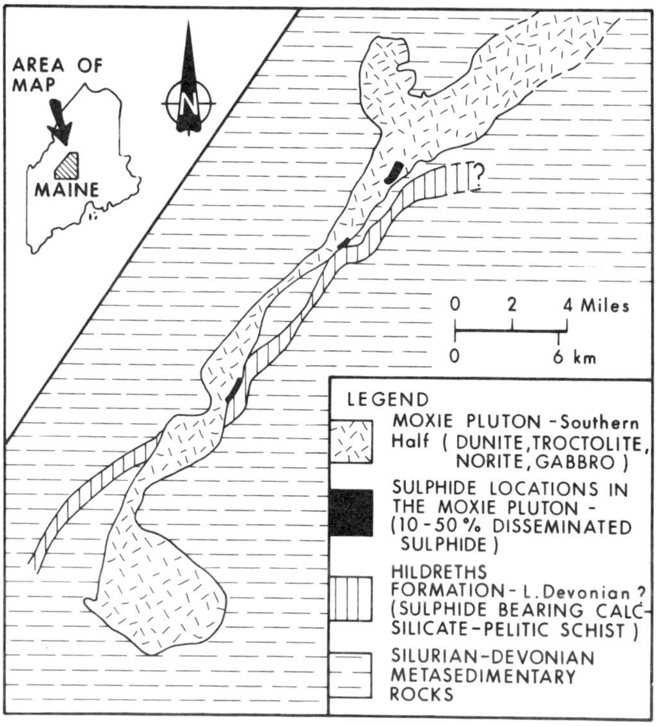

Fig. 7.10. The Moxie pluton, emphasizing the locations of bodies of disseminated sulfides and their proximity to the sulfide-bearing Hildreth's Formation. (After Thompson and Naldrett, 1984.)

if more olivines had been analyzed. Nevertheless, the Table illustrates the principle by which one can obtain a guide to the tenor of the ore to be expected from the composition of the olivine in the associated rocks.

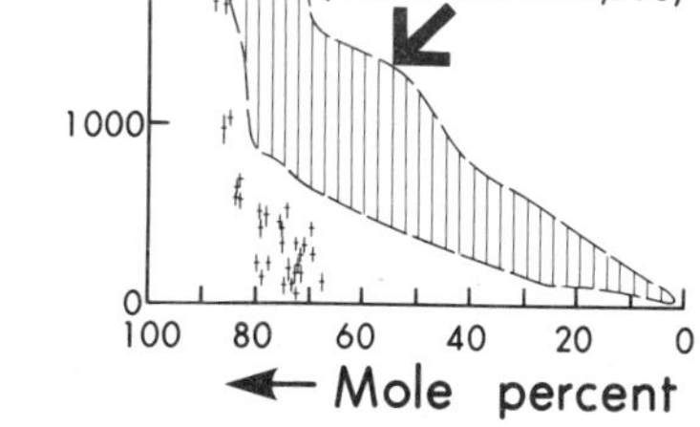

Fig. 7.11. Variation of Ni with forsterite content in olivines from the Moxie body, Maine. The Moxie olivines (crosses) are depleted in Ni in comparison with the field defined by olivines from a range of layered intrusions tabulated by Simpkin and Smith (1970). (Modified after Thompson and Naldrett, 1984.)

7.1.2.2.2.2 The Katahdin Gabbro. The Katahdin gabbro occurs (Fig. 7.12) as a small 4.5- by 1.5-km stock, 15 km south of the Moxie pluton and was also studied by Thompson and Naldrett (1984). It consists of olivine gabbro and the similarity in rock types, rock chemistry, and age leave little doubt that the stock has been formed by a magma that was very similar to that responsible for the Moxie body. The only outcrop is in the south where a deposit of in excess of 200 million tonnes of nearly massive magmatic sulfides occurs within the gabbro. The tenor of these sulfides is 0.2 wt percent Ni, 0.1 wt percent Cu, and 0.17 wt percent Co. Their isotopic composition ($\delta^{34}S$ = -11 to -22) and S/Se ratios (23,000 to 33,000) indicate that a high proportion of country-rock S has been incorporated in the deposit.

As might be expected from the preceding discussion of the Moxie pluton, Thompson and Naldrett (1984) found the extremely low Ni tenor of the sulfides at Katahdin to be matched by an extremely low Ni content in the olivine of the gabbro (Fig. 7.13). Comparison of the Ni tenor of the sulfides calculated from the olivine composition with the observed tenor also produced a reasonable match with the observed tenor (Table 7.1).

The unexposed portions of the Katahdin gabbro contain numerous electrically conductive anomalies and the question has arisen as to whether these anomalies might mark sulfides of a

TABLE 7.1. *Calculated and Observed Ni Tenors (wt %) in Sulfides in Moxie and Katahdin Plutons*

Deposit	Observed Ni in Olivine (ppm)	Calculated Ni Tenor	Observed Ni Tenor	
			Range	Mean
Black Narrows	260–700	0.8–2.1	0.8–4	1.9
Burnt Nubble	160–170	0.5–0.52	0.3–0.5	0.44
Big Indian Pond	160–320	0.5–1.0	0.4–1.3	0.84
Katahdin	70–170	0.21–0.5		0.21

higher Ni tenor than those of the known deposit. Cores from a few scattered drill holes in the unexposed area (Fig. 7.12) were investigated to determine the Ni contents of the olivines. All of the olivine proved to contain extremely little Ni and it is concluded that the magma of the Katahdin gabbro had been involved in batch equilibration with country-rock S at an extremely low R value. Consequently, the original Ni content of the magma was depleted to a very great extent, so that subsequently the magma crystallized olivine with essentially no Ni. For this reason, little hope is held out for finding a deposit of high Ni tenor within the Katahdin gabbro.

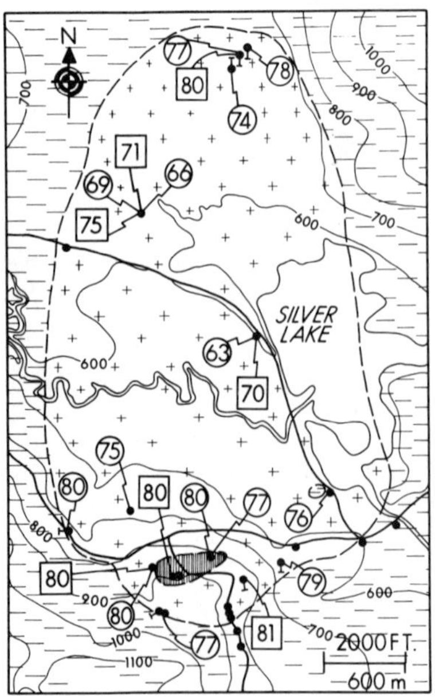

Fig. 7.12. The outline of the Katahdin stock (based largely on aeromagnetic surveys) showing the location of the sulfide body and the MgNos of olivine and orthopyroxene. (After Thompson and Naldrett, 1984.)

7.1.2.2.3 The Insizwa Complex, Transkei, Southern Africa. The Insizwa complex consists of four Karoo-age intrusions (Insizwa, Ingeli, Tonti, and Tabankulu; Fig. 7.14), each ranging in composition from picrite to quartz gabbro. Both the Insizwa and Tabankulu intrusions consist of a basal zone of gabbro, overlain by a picrite unit, overlain in turn by a troctolite and a series of gabbros (Lightfoot et al., 1984). The only significant sulfide showing is that at Waterfall Gorge, which consists of 600,000 tonnes of material grading 0.5 wt percent Ni + Cu (Scholz, 1952).

Lightfoot et al. (1984) studied the compositions of olivine in all units at both Insizwa and Tabankulu (Fig. 7.15). At Insizwa, the basal olivine gabbro is characterized by somewhat Ni-depleted olivine; the picrite by distinctly depleted olivine containing 1,100 ppm Ni (instead of the "normal" 2,500 ppm, according to Simpkin and Smith, 1970); and the upper gabbros by nearly normal olivine. The basal olivine gabbro and picrite at Tabankulu are characterized by very Ni-depleted olivine with 350 ppm Ni (instead of a "normal" 2,500 ppm) and the troctolite and upper gabbro by normal olivine.

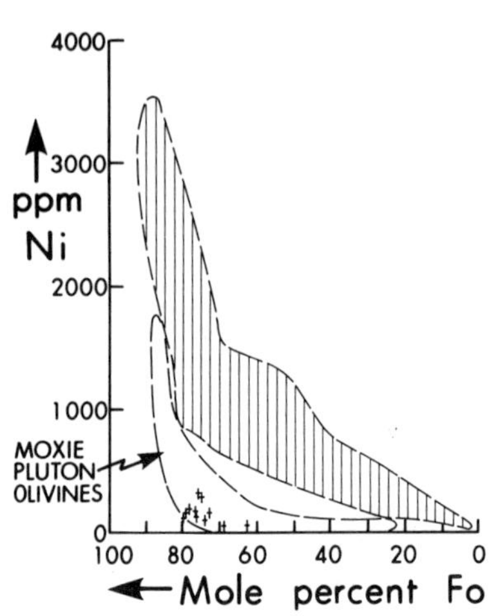

Fig. 7.13. Variations of Ni with forsterite content of olivines from the Katahdin body (crosses). These are strongly depleted in comparison with those from a variety of layered intrusions (hatched field) tabulated by Simpkin and Smith (1970). The field of Moxie olivines is also shown. (After Thompson and Naldrett, 1984.)

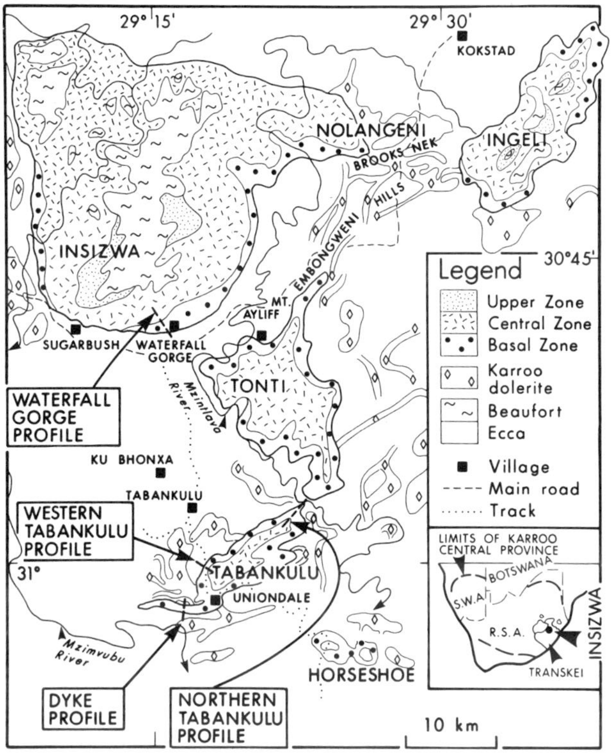

Fig. 7.14. Intrusions of the Insizwa Complex. (After Lightfoot et al., 1984.)

Lightfoot et al. (1984) interpreted the Ni-depleted basal gabbro and picrite as having crystallized from pulses of magma that had undergone extensive equilibration with sulfide, and the troctolite and upper gabbros as having formed from later pulses that had not undergone such extensive reaction. The picrite contains very little sulfide, and they concluded that it had equilibrated with the sulfides in a batch process and that the sulfides settled out before introduction of the picrite. Using the same approach that was applied at Moxie, and making a conservative estimate of the thickness of the picrite in each intrusion, they calculated that the quantities of Ni metal shown in Table 7.2 had been removed by sulfides with tenors also shown in Table 7.2. The tenor of the sulfides calculated for Insizwa matches very closely with the observed tenor of the mineralization. On the basis of what has been found so far, the sulfides at Waterfall Gorge contain a very small portion of the Ni missing from Insizwa. The location of the balance, and the location of all of the Ni missing from Tabankulu should intrigue exploration geologists for a long time to come.

7.1.3 Conclusions

Two approaches can be taken as to the use in exploration of the information included in this book. The first is qualitative; the search for the appropriate rocks, for the suitable tectonic environment where the rocks might have had the optimum opportunity to segregate Ni-bearing sulfides, and the application of traditional exploration methods. This approach will, and should, remain the explorationists prime methodology. The second approach is the quantitative approach based on the appreciation that the segregation of sulfides from bodies of magma leaves, in many cases, a mark on the magma that is reflected in the rocks that crystallize from it. The magnitude of this mark has been assessed above and has turned out to be close to that actually observed, both in chilled magma from lavas and in the olivine from intrusive bodies. The mark cannot only be used to identify favorable targets, but gives an indication of the tenor of Ni in the possible magmatic sulfide mineralization.

7.2 PGE DEPOSITS

Differences between PGE-rich deposits of magmatic sulfides and typical Ni–Cu ores include:

1. The former tend to occur as relatively weak disseminations of sulfide within silicates, rather than as massive concentrations.
2. They tend to occur at specific, clearly defined

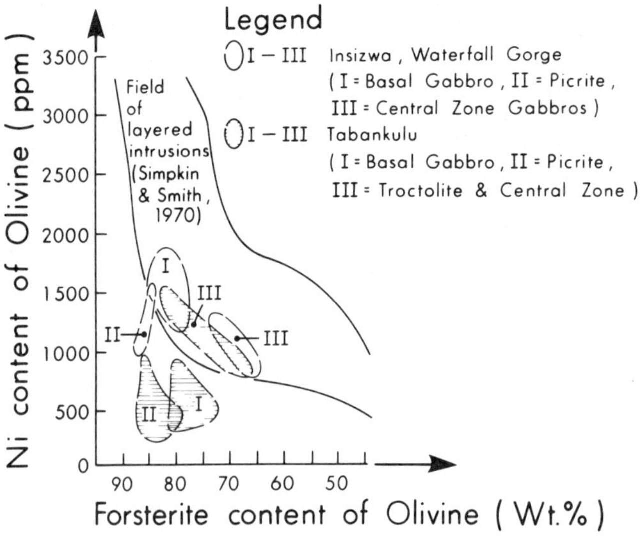

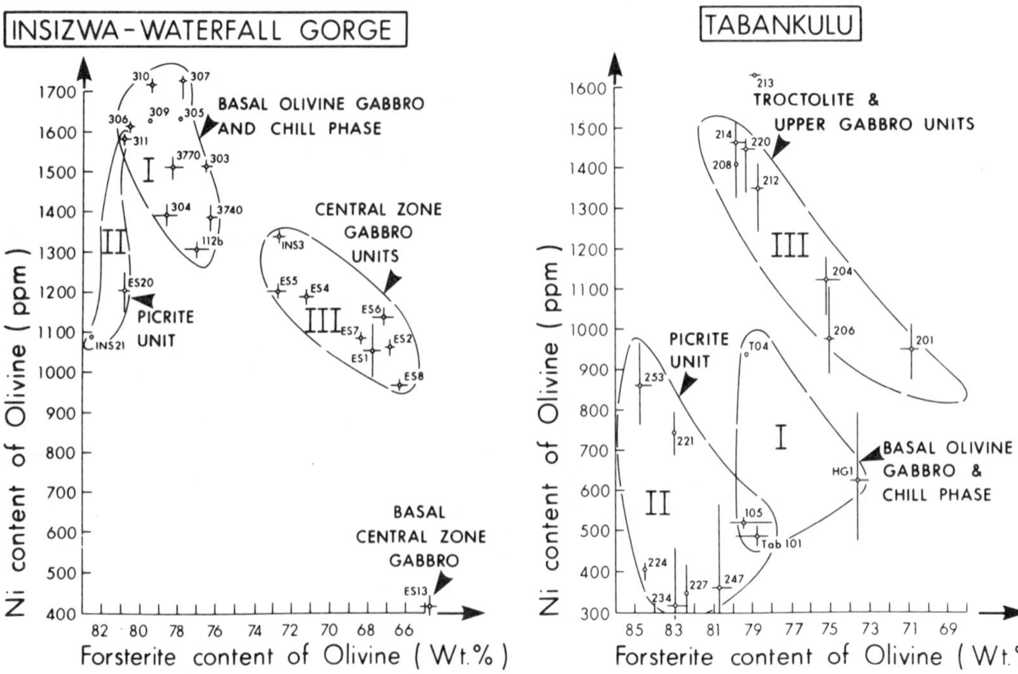

Fig. 7.15. Variation of Ni with forsterite in olivines from the Tabankulu and Insizwa intrusions. (After Lightfoot et al., 19

TABLE 7.2. *Possible Tenor and Tonnes of Sulfide at Insizwa and Tabankulu*

Rock Unit	Observed Ni in Olivine (ppm)	Expected Ni in Olivine (ppm)	Possible Sulfide: Ni Tenor (wt %)	Possible Sulfide: Millions of Tonnes
Insizwa picrite	1,100	2,500	4.8	600
Tabankulu	350	2,500	1.8	15

horizons within layered intrusions, with no particular association with the base.

3. They often show an association with chromite, either occurring as very sparse sulfides within massive chromitite, or within strata that also contain small amounts of chromite, in contrast with chromite-free strata above and below the mineralized horizons.
4. The Merensky Reef and Middle and Upper Group chromitites of the Bushveld Complex, the J–M Reef of the Stillwater Complex, and the mineralized horizons of the Penikat intrusion, all of which are distinctly enriched in PGE, occur closely associated with plagioclase-bearing rocks, within a few 100 m of the first appearance of cumulus plagioclase in their respective layered intrusions. The Main Sulfide Zone of the Great Dyke of Zimbabwe is an exception to this rule. In general, chromitite layers that occur at horizons below the incoming of cumulus plagioclase in layered intrusions are much lower in PGE than those that occur above, although the A chromitite of the Stillwater is an exception and is enriched in PGE.
5. In both the Bushveld and Stillwater Complexes, cryptic variation and the sequence of appearance and disappearance of cumulus phases in the cyclic units carrying and close to those carrying PGE mineralization suggest that the interaction of two magmas is involved in the crystallization of these units. This has been confirmed in the case of the Bushveld by isotope studies.
6. The presence of hydrous minerals, high Cl/F ratios in apatite, high Cl in biotite, fluid inclusions, graphite and pegmatoidal recrystallization suggest that the Merensky and J–M Reefs concentrated volatiles more than their surrounding rocks.

In both Chapters 5 and 6, variation in magma/sulfide ratio has been proposed as the explanation for differences in the concentration of PGE with respect to other chalcophile elements in magmatic sulfide ores. High magma/sulfide ratios can account for high PGE concentrations in the sulfides, but will not result in Ni depletion in silicates crystallizing as the sulfides segregate. This rules out one of the exploration criteria described previously for exploring Ni–Cu ores. It is possible that the depletion of PGE in silicates could be used in the same way as Ni, but as a much more sensitive indicator in view of their much higher partition coefficients. At the present time, however, the analytic problem of detecting PGE at the ppb or sub-ppb level in silicate grains remains insoluble. It is possible that bulk rock analyses may help to pinpoint levels within an intrusion at which PGE become depleted, but the large variation in background that can result from the incorporation of only small amounts of sulfide in bulk samples causes problems with this technique.

It is shown in Chapter 5 that the model by Campbell et al. (1983) of magma mixing in a turbulent plume can account for sulfide immiscibility, the achievement of a high magma/sulfide ratio, the characteristics of both the J–M and Merensky Reefs, including their high volatile contents, and the association of PGE-rich sulfides with chromite. This model requires plagioclase to have started to crystallize within a layered intrusion, and thus would direct attention to that part of the cumulus pile just above which plagioclase-bearing rocks first occur. This would have been successful in directing exploration toward the UG-2, Merensky, and J–M Reefs, and the mineralization in the Penikat intrusion. It would not have drawn attention to the Main Sulfide Zone of the Great Dyke.

Systematic study of PGE concentrations is at a less advanced stage at the time of writing than that of Ni–Cu sulfide deposits, but as more types of PGE deposits are studied and more numerous and more sophisticated genetic models are developed, it is likely that more and better exploration criteria will be developed.

REFERENCES

Abel, M. K., Buchan, R., Coats, C. J. A., and Penstone, M. E. (1979). Copper mineralization in the footwall complex, Strathcona mine, Sudbury, Ontario. *Can. Mineral.* 17:275–286.

Abraham, K. P., Davies, M. W., and Richardson, F. D. (1960). Sulfied capacities of silicate melts, Pt. I. *J. Iron & Steel Inst.* 196:309–312.

Alapieti, T. T., and Lahtinen, J. J. (1986). Stratigraphy, petrology and platinum-group element mineralization of the early Proterozoic Penikat layered intrusion, northern Finland. *Econ. Geol.* 81(5):1126–1136.

Alapieti, T. (1982). The Koillismaa layered igneous complex, Finland—Its structure, mineralogy and geochemistry, with emphasis on the distribution of chromium. *Finland Geol. Survey Bull.* 319:116 pp.

Alapieti, T., Hugg, R., and Piirainen, T. (1979). Structure, mineralogy and chemistry of the Syote section in the early Proterozoic Koillismaa layered instrusion, northeastern Finland. *Finland Geol. Surv. Bull.* 299:43 pp.

Allard, G. O. (1976). The Dore Lake Complex and its importance to Chibougamau geology and metallogeny. *Quebec. Min. Natl. Res.,*: DP 368:446 pp. and map at 1:100,000.

Allard, G. O., Duquette, G., Latulippe, M., and van de Walle, Me. (1972). Precambrian geology and mineral deposits of the Noranda-Val d'Or and Matagami-Chibougamau greenstone belts Quebec. *24th Internat. Geol. Cong.,* Montreal 1972, Field Guide to Excursions *A41* and *C41.*

Allard, G. O. (1970). The Dore Lake complex of Chibougamau, Quebec—A metamorphosed Bushveld-type layered complex. *Geol. Soc.* S.Africa Spec. Pub. 1:477–491.

Anderson, I. G. (1981). Personal communication.

Archibald, N. J., Bettenay, L. F., Binns, R. A., Groves, D. I., and Gunthorpe, R. J. (1978). The evolution of Archean greenstone terrains, Eastern Goldfields province, Western Australia. *Precambrian Res.* 6:89–91.

Arndt, N. T., and Naldrett, A. J. (1987). Komatiites in Munro Township, Ontario. In: *Geological Society of America Centennial Field Guide—Northeastern Section,* Article 69:317–322.

Arndt, N. T. (1986a). Komatiitites: A dirty window into the Archean Mantle. *Terra Cognita* 6:59–66.

Arndt, N. T. (1986b). Differentiation of komatiite flows. *J. Petrol.* 27:279–302.

Arndt, N. T., and Jenner, G. A. (1986). Crustally contaminated komatiites from Kambalda, Western Australia. *Chemical Geology* 56:229–256.

Arndt, N. T., and Nisbet, E. G. (1982). *Komatiites.* London: Allen & Unwin, 526 pp.

Arndt, N. T., Naldrett, A. J., and Pyke, D. R. (1977). Komatiitic and iron-rich tholeiitic lavas of Munro Twp., northeast Ontario. *J. Petrol.* 18:319–369.

Arndt, N. T. (1977). The partioning of nickel between olivine and ultrabasic and basic komatiite liquids. In: *Carnegie Inst. Washington Year Book, 1976.*

Ashcroft, W., and Munro, M. (1978). The structure of the eastern part of the Insch Mafic Intrusion, Aberdeenshire. *Scottish J. Geol.* 14:55–79.

Atamanov, A. V., Lugov, S. F., and Feygin, Ya. M. (1962). New results on the Loverzero massif. *Int. Geol. Review* 4:570–577.

Bailey, J. C., Melchior, L., and Sorensen, H. (1981). Introduction to the Ilimaussaq intrusion with a summary of the reported investigations. *Rapp. Gronlands Geol. Unders.* 103:5–17.

Ballhaus, C. G., and Stumpfl, E. F. (1985a). Fluid inclusions in Merensky and Bastard Reefs, Western Bushveld Complex (Abst.). *Can. Mineral.* 23:294.

Ballhaus, C. G., and Stumpfl, E. F. (1985b). Graphite, platinum and the C-O-H-S system (Abst.). *Can. Mineral.* 23:293–299.

Baragar, W. R. A., and Scoates, R. F. J. (1981). The Circum-Superior belt: A Proterozoic plate margin?. In: *Precambrian Plate Tectonics,* Kroner, A. (ed.). Amsterdam: Elsevier, pp. 297–330.

Barker, W. W. (1983). The Fe-Ni-S system (Co,Cu). *CSIRO Aust. FP Report* 26:23.

Barnes, S. J., and Naldrett, A. J. (1986). Geochemistry of the J-M Reef of the Stillwater Complex, Minneapolis Adit Area II. Silicate Mineralogy and Paragenesis. *J. Petrol.* 27:791–825.

Barnes, S. J., and Naldrett, A. J. (1985). Geochemistry of the J-M (Howland) Reef of the Stillwater Complex, Minneapolis adit area. I. Sulfide chemistry and sulfide-olivine equilibrium. *Econ. Geol.* 80:627–645.

Barnes, S. J., Coats, C. J. A., and Naldrett, A. J. (1982). Petrogenesis of a proterozoic nickel sulfide-komatiite association: The Katiniq Sill Ungava, Quebec. *Econ. Geol.* 77:413–429.

Barnes, S. J. (1979). Petrology and geochemistry of the Katiniq nickel deposit and related rocks, Ungava, northern Quebec. M.Sc. thesis, Univ. of Toronto, Toronto, Ontario.

Barnes, S.-J., Naldrett, A. J., and Gorton, M. P. (1985). The origin of the fractionation of platinum-group elements in terrestrial magmas. *Chem. Geol.* 53:303–323.

Barnes, S.-J. (1983). The fractionation of platinum group elements in Archean komatiites of the Abitibiti Greenstone Belt, Northern Ontario, Canada. Unpubl. Ph.D. thesis, Univ. of Toronto, Toronto, Ontario.

Barnes., S.-J., Gorton, M. P., and Naldrett, A. J. (1983). A comparative study of olivine and clinopyroxene flows from Alexo, Abitibi Greenstone Belt, Ontario, Canada. *Contrib. Mineral. Petrol.* 83:293–308.

Barrett, F. M., Binns, R. A., Groves, D. I., Marston, R. J., and McQueen, K. G. (1977). Structural history and metamorphic modification of Archean volcanic type nickel deposits, Yilgarn Block, Western Australia. *Econ. Geol.* 72:1195–1224.

Barrett, F. M., Groves, D. I., and Binns, R. A. (1976). Importance of metamorphic processes at the Nepean nickel deposit, Western Australia. *Inst. Min. Metal. Trans.* (sec. B.) 85:B252–B273.

Barrie, C. T. (1987). The Kamiscotia Gabbroic Complex, western Abitibi Subprovince, Ontario. Part 1. Geology and tectonic setting (Abstr.). Indaba on the tectonics of layered intrusions. University of Pretoria, Pretoria.

Barton, P. B. (1973). The system Cu-Fe-S Part I. The Cu-S and Cu-Fe-S join. *Econ. Geol.* 68:455–465.

Basaltic Volcanism Study Project. (1981). *Basaltic Volcanism on the Terrestrial Planets.* New York: Pergamon Press, 1286 pp.

Bavington, O. A. (1981). The nature of sulfidic metasediments at Kambalda and their broad relationship with associated ultramafic rocks and nickel ores. *Econ. Geol.* 76:1606–1628.

Bavington, O. A. (1979). Interflow sedimentary rocks from the Kambalda ultramafic sequence: Their geochemistry, metamorphism and genesis. Unpubl. Ph.D. thesis, Australian National Univ., Canberra.

Bazunov, E. A. (1976). Development of the main structures of the Siberian platform: History and dynamics. *Tectonophys.* 36:289–300.

Bebien, J., Ohnenstetter, D., Ohnenstetter, M., and Vegely, P. (1980). Diversity of the Greek ophiolites: Birth of oceanic basins in transcurrent systems. In: *Tethyan Ophiolites,* Ofioliti, Special issue, P. Rocci, (ed.). 2:129–177.

Beccaluva, L., Ohnenstetter, D., Ohnenstetter, M., and Paupy, A. (1984). Two magmatic series with island arc affinities within the Vuorinos ophiolite. *Contr. Min. Pet.* 85:253–271.

Bell, P. M., England, J. W., and Kullerud, G. (1964). Pentlandite: Pressure effect on breakdown. *Carnegie Inst. Washington Year Book* 63:206–207.

Besson, M., Clark, T., Joron, J. L., Ohnenstetter, D., Roger, G., Rohon, M. L., Treuil, M., and Vidal, Ph. (1986). Indices de Cu-Ni sulfures liés au magmatisme mafique et ultramafique Proterozoic dans la Fosse du Labrador (Quebec, Canada) (Abstr.) In: *Facteurs de Concentration des Matiérès Premieres Minerales,* Montpelier, Colloque PIRSEM. p. 25.

Besson, M., Boyd, R., Czamanske, G., Foose, M., Groves, D., von Gruenewaldt, G., Naldrett, A., Nilsson, G., Page, N., Papunen, H., and Peredery, W. (1979). IGCP Project #161 and a proposed classification of Ni-Cu-PGE sulfide deposits. *Can. Mineral.* 17:143–144.

Bichan, R. (1969). Origin of chromite seams in the Hartley complex of the Great Dyke, Rhodesia. *Econ. Geol.* Monograph 4:95–113.

Billington, L. G. (1986). Geological review of the Agnew nickel deposit, Western Australia. In: *Sulfide Deposits in Mafic and Ultramafic Rocks.* Buchanan, D. L., and Jones, M. L. (eds.) London: Spec. Pub. of the Inst. of Mining and Metallurgy, pp. 4–54.

Binns, R. A., and Groves, D. I. (1976). Iron-nickel partition in metamorphic olivine-sulfide assemblages from Perseverance, W. Australia. *Am. Mineral.* 61:782–787.

Boctor, N. Z., and Yoder, H. S., Jr. (1983). Partitioning of nickel between silicate and iron sulfide melts. *Carnegie Inst. Washington Year Book* 82:275–277.

Boctor, N. Z. (1982). The effect of fo_2 and fs_2, and temperature on Ni partitioning between olivine and iron sulfide melts. *Carnegie Inst. of Washington Year Book* 81:366–369.

Boctor, N. Z. (1981). Partitioning of nickel between olivine and iron monosulphide melts. *Carnegie Inst. Washington Year Book* 80:356–359.

Bog, S., and Rosenqvist, T. (1958). A thermodynamic study of the iron sulfide-oxide melts. *Meddelebse Nr 12 fra Metallurgisk Komite.* Trondheim, Norway.

Bonnichsen, W. (1974). Copper and nickel resources in the Duluth complex, northeastern Minnesota. *Minnesota Geol. Survey* Inf. Circular 10.

Bonnichsen, W. (1972). Southern part of the Duluth complex. In: *Geology of Minnesota: A Centennial Volume, Sims,* P. K. and Morey, G. B. (eds.). St. Paul: Minnesota Geol. Survey, pp. 361–387.

Bonnichsen, B. (1970). Geologic map of the Allen, Babbitt, Babbitt NE, Babbitt SE, and Babbitt SW 7 ½ minute quadrangles. *Minn. Geol. Surv.* Open File Maps.

Boone, G., McG. (1974). Acadian plate junction beneath western Maine gabbro belt (Abst.). *Geol. Assoc. Canada-Mineral. Assoc. Canada* Ann. Mtg., St. John's, Volume of Abstracts, p.11.

Borthwick, A. A., and Naldrett, A. J. (1984). Platinum group elements in Layered Intrusions. In: *Geosciences Research Programme Summary of Research 1983–1984,* Milne V. G. (ed.). Ontario Geological Survey Misc. Publ., 121, pp. 13–25.

Boudreau, A. E. (1988). Investigations of the Stillwater Complex: Part IV. The role of volatiles in the petrogenesis of the J-M Reef. *Can. Mineral.* 26:193–208.

Boudreau, A. E., Mathez, E. A., and McCallum, I. (1986). Halogen geochemistry of the Stillwater and Bushveld Complexes: Evidence for transport of the platinum-group elements by Cl-rich fluids. *J. Petrol.* 27:967–986.

Boudreau, A. E., and McCallum, I. S. (1985). Features of the Picket Pin Pt-Pd Deposit. In: *Stillwater Complex,* Czamanske, G. K., and Zienteck, M. L. (eds.). Montana Bureau of Mines Spec. Pub. 92, pp. 346–357.

Bow, C. S., Wolfgram, D., Turner, A. R., Barnes, S. J., Evans, J., Zdepski, M., and Boudreau, A. (1982). Investigations of the H. P. Reef of the Stillwater Complex, Minneapolis Adit area: Stratigraphy, structure and mineralization. *Econ. Geol.* 77:1481–1492.

Boyd, R., and Mathiesen, C. O. (1979). The nickel mineralization of the Rana mafic intrusion, Nordland, Norway. *Can. Mineral.* 17:287–298.

Boyd, R., and Munro, M. (1978). Deformation of the Belhelvie Massif, Aberdeenshire. *Scottish J. Geol.* 14:29–44.

Brett, R., and Bell, P. M. (1969). Melting relations in the Fe-rich portion of the system Fe-FeS at 30 kb pressure. *Earth and Planet. Sci.* (Lett.) 6:479–482.

Brugmann, G. E., Arndt, N. T., Hofmann, A. W., and Tobschall, H. J. (1987). Noble metal abundances in komatiite suites from Alexo, Ontario and Gorgona Island, Colombia. *Geochim. et Cosmochim. Acta* 51:2159–2169.

Brynard, H. J., de Villers, J. P. R., and Viljoen, E. A. (1976). A mineralogical investigation of the Merensky reef at the Western Platinum Mine near Marikana, South Africa. *Econ. Geol.* 71:1299–1307.

Buchanan, D. L., and Rowe, J. E. (1984). Role of contamination in the precipitation of sulfides in the Platreef of the Bushveld complex. In: *Sulfide Deposits in Mafic and Ultramafic Rocks,* Buchanan, D. L., and Jones, M. J. (eds). London:Inst. of Min. & Metal., pp. 141–146.

Buchanan, D. L., Nolan, J., Wilkinson, N., and de Villiers, J. R. R. (1983). An experimental investigation of sulfur solubility as a function of temperature in synthetic silicate melts. *Geol. Soc. S. Africa,* Spec. Publ. 7:383–391.

Buchanan, D. L., and Nolan, J. (1979). Solubility of sulfur and sulfide immiscibility in synthetic tholeiitic melts and their relevance to Bushveld Complex rocks. *Can. Mineral.* 17:483–494.

Buntin, T. J., Grandstaff, D. E., Ulmer, G. C., and Gold, D. P. (1985). A pilot study of geochemical and redox relationships between potholes and adjacent normal Merensky Reef of the Bushveld Complex. *Econ. Geol.* 80:975–987.

Burt, D. R. L., and Sheppy, N. R. (1975). Mount Keith nickel sulfide deposit. In: *Economic Geology of Australia and Papua New Guinea, I. Metals: Australasian Inst. Min. Metal,* Knight, C. L. (ed.). Monograph 5, pp. 159–168.

Cabri, L. J. (1973). New data on phase relations in the Cu-Fe-S system. *Econ. Geol.* 68:443–454.

Cameron, E. N. (1982). The Upper Critical Zone of the eastern Bushveld Complex—Precursor to the Merensky Reef. *Econ. Geol.* 77:1307–1327.

Cameron, E. N. (1980). Evolution of the Lower Critical Zone, Central Sector, Eastern Bushveld Complex and its chromite deposits. *Econ. Geol.* 75:845–871.

Cameron, E. N. (1978). The lower zone of the eastern Bushveld complex in the Olifants river trough. *J. Petrol.* 19:437–462.

Cameron, E. N., and Desborough, G. A. (1964). Origin of certain magnetite-bearing pegmatites in the eastern part of the Bushveld complex, South Africa. *Econ. Geol.* 59:197–225.

Cameron, G. H. (1988). A geochemical investigation into the origin of the Upper Critical Zone of the Eastern Bushveld Complex, South Africa. Unpubl. M.Sc. thesis, Univ. of Toronto, Toronto, Ontario.

Campbell, F. E., and Roeder, P. L. (1968). The stability of olivine and pyroxene in the Ni-Mg-Si-O system. *Am. Mineral.* 53:257–258.

Campbell, I. H., and Barnes, S. J. (1984). A model for the geochemistry of the platinum group elements in magmatic sulfide deposits. *Can. Mineral.* 22:151–160.

Campbell, I. H., Naldrett, A. J., and Barnes, S. J. (1983). A model for the origin of the platinum-rich sulfide horizons in the Bushveld and Stillwater Complexes. *J. Petrol.* 24:133–165.

Campbell, I. H., and Naldrett, A. J. (1979). The influence of silicate: Sulfide ratios on the geochemistry of magmatic sulfides. *Econ. Geol.* 74:1503–1505.

Campbell, I. H., Naldrett, A. J., and Roeder, P. L. (1979). Nickel activity in silicate liquids: Some preliminary results. *Can. Mineral.* 17:495–506.

Campbell, I. H. (1977). A study of cumulate processes and macrorythmic layering in the Jimberlana intrusion of Western Australia, Part I: The upper layered series. *J. Petrol.* 18:183–215.

Card, K. D., and Pattison, E. F. (1973). Nipissing diabase of the Southern Province. In: *Huronian Stratigraphy and Sedimentation,* Young, G. M. (ed.). Waterloo, Ontario: Geological Association of Canada Special Paper 12, pp. 7–30.

Casey, J. F., and Karson, J. A. (1981). Magma chamber profiles from the Bay of Islands ophiolite complex. *Nature* 292:295–301.

Cawthorn, R. G., Barton, J. M., Jr., and Viljoen, M. J. (1985). Interaction of floor rocks with the Platreef on Overysel, Potgietersrus, Northern Transvaal. *Econ. Geol.* 76:1480–1502.

Cawthorn, R. G., and Davies, G. (1983). Experimental data at 3 K bars pressure on parental magma to the Bushveld Complex. *Contr. Mineral. Petrol.* 83:128–135.

Cawthorne, R. G., Barton, J. M., Jr., and Viljoen, M. J. (1982). Interaction of floor rocks with the Platreef on Overysel, Potgietersrus, Northern Transvall. *Econ. Geol.* 80:1016–1026.

Celmer, R. S. (1987). The distribution of minor elements in nickel matte smelting. Unpubl. Ph.D. thesis, Univ. of Toronto, Toronto, Ontario, 341 pp.

Christie, D. M., Carmichael, I. S. E., and Langmuir, C. H. (1986). Oxidation states of mid-ocean ridge basalt glasses. *Earth Planet Sci.* (Lett.) 79:397–411.

Chuang, Y. Y., Hsieh, D.-C., and Chang, Y. A. (1985). Thermodynamics and phase relationships of transition metal sulfur systems: Part V. A re-evaluation of the Fe-S system using an associated solution model for the liquid phase. *Metall. Trans. B.* 1613:277–285.

Chauvel, C., Dupre, B., and Jenner, G. A. (1985). The Sm-Nd age of Kambalda volcanics is 500 Ma too old! *Earth Planet. Sci.* (Lett.) 74:314–324.

Church, W. R., and Riccio, L. (1977). Fractionation trends in the Bay of Islands ophiolite of Newfoundland: Polycyclic cumulate sequences in ophiolites and their classification. *Can. J. Earth Sci.* 14:1156–1165.

Chyi, L. L., and Crocket, J. H. (1976). Partition of platinum, palladium, iridium, and gold among co-existing minerals from the Deep Ore Zone, Strathcona Mine, Sudbury, Ontario. *Econ. Geol.* 71:1196–1205.

Clark, T., and Naldrett, A. J. (1972). The distribution of Fe and Ni between synthetic olivine and sulfide at 900°c. *Econ. Geol.* 67:939–952.

Coats, C. J. A., and Snajdr, P. (1984). Ore deposits of the North Range, Onaping-Levack Area, Sudbury. In: *The Geology and Ore Deposits of the Sudbury Structure,* Pye, E. J., Naldrett, A. J., and Giblin, P. E. (eds.). Ont. Geol. Surv. Spec. Vol. No. 1, pp. 327–346.

Coats, C. J. A., Quirke, T. T., Jr., Bell, C. K., Cranstone, D. A., and Campbell, F. H. A. (1972). Geology and mineral deposits of the Flin Flon, Lynn Lake and Thompson areas, Manitoba, and the Churchill-Superior boundary of the Western Precambrian shield. *Internat. Geol. Cong.,* 24th, Montreal 1972, Guidebook, Field Excursions A31-C31, pp. 1–96.

Cochrane, L. B. (1984). Ore deposits of the Copper Cliff Offset. In: *The Geology and Ore Deposits of the Sudbury Structure,* Pye, E. G., Naldrett, A. J., and Giblin, P. E. (eds.). Ont. Geol. Survey Spec.Vol. No. 1. pp. 97–138.

Coleman, R. G., and Irwin, W. P. (1977). North American ophiolites. *Oregon Dept. of Geological and Mineral Industries* Bull. 95:183 pp.

Collins, W. H. (1936). Life history of the Sudbury nickel irruptive (iii). *Roy. Soc. Canada Trans.* sec. IV, 30:29–53.

Collins, W. H. (1934). Life history of the Sudbury nickel irruptive (i). *Roy. Soc. Can. Trans.* sec. IV, 28:123–177.

Compston, W., Williams, I. S., Campbell, I. H., and Gresham, J. J. (1986). Zircon xenocrysts from the Kambalda volcanics: Age constraints and direct evidence for older continental crust beneath the Kambalda-Norseman greenstones. *Earth Planet. Sci.* (Lett.) 76:299–311.

Connolly, J. W. D., and Haughton, D. R. (1972). The volume of sulfur in glass of basaltic composition formed under conditions of low oxidation potential. *Am. Mineral.* 57:1515–1517.

Cousins, C. A. (1964). The platinum deposits of the Merensky Reef. In: *The Geology of Some Ore Deposits in Southern Africa, vol. II,* Haughton H. S. (ed.). Geol. Soc. S. Africa, Johannesburg, pp. 225–238.

Cowan, J. C. (1968). The geology of the Strathcona ore deposit, Sudbury district, Ontario. *CIM Bull.* 61:38–54.

Cowden, A., and Woolrich, P. (1987). Geochemistry of the Kambalda iron-nickel sulfides:

Implications for models of sulfide-silicate partitioning. *Can. Mineral.* 25:21–36.

Cowden, A., Donaldson, M. J., Naldrett, A. J., and Campbell, I. H. (1986). Platinum-group elements in the komatiite-hosted Fe-Ni-Cu sulfide deposits at Kambalda, Western Australia. *Econ. Geol.* 81:1226–1235.

Craig, J. R., and Scott, S. D. (1974). In: *Sulfide Phase Equilibria: Mineral. Soc. of Amer. Short Course Notes,* vol. 1, Ribbe, P. H. (ed.). Sulfide mineralogy, CS1-CS110.

Craig, J. R. (1973). Penntlandite-pyrrhotite and other low-temperature relations in the Fe-Ni-S Systems. *Am. J. Sci.* 273-A:496–510.

Craig, J. R. (1971). Violarite stability relations. *Am. Mineral.* 56:1303–1311.

Craig, J. R., and Kullerud, G. (1969). Phase relations in the Cu-Fe-Ni-S system and their application to magmatic ore deposits. *Econ. Geol. Mon.* 4:344–358.

Craig, J. R., and Naldrett, A. J. (1967). Minimum melting of nickeliferrous pyrrhotite ores. *Ann. Rep. Dir. of Geophys. Lab.,* Carnegie Inst. Washington Year Book 66.

Crocket, J. H., 1981. Geochemistry of the Platinum Group Elements. In: *Platinum Group Elements: Mineralogy, Geology and Recovery,* Cabri, L. (ed.). Can. Inst. Min. Met. Special volume.

Crocket, J. H. (1979). Platinum group elements in mafic and ultramafic rocks: A survey. *Can. Mineral.* 17 (2):391–403.

Czamanske, G. K., and Zientek, M. L. (1985). The Stillwater Complex, Montana: Geology and guide. *Mont. Bur. Mines Geol.,* Special Publication 92, 396 pp.

Czamanske, G. K., Himmelberg, G. R., and Groff, F. E. (1976). Zoned Cr, Fe-spinel from the La Perouse layered gabbro, Fairweather Range, Alaska. *Earth Planet. Sci.* (Lett.) 33:111–118.

Darken, L. S., and Gurry, R. W. (1953). *Physical Chemistry of Metals.* New York: McGraw-Hill, 535 pp.

Davis, D. W., Corfu, F., and Krogh, T. E. (1982). Zircon U-Pb ages from the Wabigoon-Manitou Lakes region, Wabigoon Subprovince, northwest Ontario. *Can. J. Earth Sci.* 19:254–266.

Davies, G., and Tredoux, M. (1985). The platinum group element and gold contents of the marginal rocks and sills of the Bushveld Complex. *Econ. Geol.* 80:838–848.

DePaolo, D. J., and Wasserburg, G. J. (1979). Neodymium isotopes in flood basalts from the Siberian platform and inferences about their mantle sources. *Proc. Natl. Acad. Sci. USA* 76:3056–3060.

de Wit, M. J., Hart, R. A., and Hart, R. J. (1987). The Jamestown Ophiolite Complex, Barberton mountain belt: A section through 3.5 Ga oceanic crust. *J. African Earth Sci.* 6:681–730.

De Young, J. H., Jr., Sutphin, D. M., Werner, A. B. T., and Foose, M. P. (1985). *International Strategic Minerals Inventory Summary Report—Nickel.* U. S. Geol. Surv. Circ. 930-D, 62 pp.

Dillon-Leitch, H. C. H., Watkinson, D. H., and Coats, C. J. A. (1986). Distribution of platinum-group elements in the Donaldson West deposit, Cape Smith belt, Quebec. *Econ. Geol.* 81:1147–1158.

Distler, V. V., Genkin, A. D., and Dyuzhikov, O. A. (1986). Sulfide petrology and genesis of copper-nickel ore deposits. In: *Geology and Metallogeny of Copper Deposits,* Friedrich, G. H., Genkin, A. D., Naldrett, A. J., Ridge, J. D., Sillitoe, R. H., and Vokes, F. M. (eds.). Society for Geology Applied to Ore Deposits Special Publication No. 4, Berlin, Heidelberg: Springer-Verlag, pp. 111–123.

Distler, V. V., and Genkin, A. D. (1980). Deposits of sulfide-copper-nickel ores of the USSR and their connection with cratonal volcanism. Proc. 5th Quat. IAGOD Symp., E. Schweigerbart'sche Verlag-freckhandburg, Stuttgart, pp. 275–295.

Distler, V. V., Malevsky, A. Yu, and Laputina, P. P. (1977). Distribution of platinoids between pyrrhotite and pentlandite in crystallization of a sulfide melt. *Geochem. Internat.* 14(6):30–40.

Donaldson, C. H. (1982). Spinifex-textured komatiites: A review of textures, compositions and layering. In: *Komatiites,* Arndt, N. T., and Nisbet, E. G. (eds.). London: Allen and Unwin, pp. 213–244.

Donaldson, M. J., and Groves, D. I. (1985). The relevance of chromite as a petrogenetic indicator in metamorphosed komatiitic rocks and associated Ni ores. *Can. Mineral.* 23:326.

Donaldson, M. J. (1981). Redistribution of ore elements during serpentinization and talc-carbonate alteration of some Archean dunites, Western Australia. *Econ. Geol.* 76:1698–1713.

Doyle, C. D., and Naldrett, A. J. (1987). The oxygen content of 'sulfide' magma and its effect on the partitioning of nickel between co-existing olivine and molten ores. *Econ. Geol.* 82:208–211.

Doyle, C. D., and Naldrett, A. J. (1986). Ideal mixing of divalent cations in mafic magma and its effect on the solution of ferrous oxide. *Geochim. et Cosmochim. Acta* 50:435–443.

Dressler, B. O. (1984). The effects of the Sudbury event and the intrusion of the Sudbury igneous complex on the footwall of the Sudbury structure. In: *The Geology and Ore Deposits of the Sudbury Structure.* Pye, E. G.,

Naldrett, A. J., and Giblin, P. E. (eds.). Ont. Geol. Survey Specl Vol. No. 1, pp. 97–138.

Duke, J. M. (1986). The Dumont nickel deposit: A genetic model for disseminated magmatic sulfide deposits of komatiitic affinity. In: *Metallogeny of Basic and Ultrabasic Rocks.* Gallagher, M. J., Ixer, R. A., Neary, C. R., and Prichard, H. M. (eds.). Special publication of Inst. Min. Metall., London, pp. 151–160.

Duke, J. M. (1985). Petrology and economic geology of the Dumont sill: An Archean intrusion of komatiitic affinity in northwestern Quebec. *Geol. Surv. Can. Econ. Geol.* Rep. No. 35.

Duke, J. M. (1979). Computer simulation of the fractionation of olivine and sulfide from mafic and ultramafic magmas. *Can. Mineral.* 17:507–514.

Duke, J. M., and Naldrett, A. J. (1978). A numerical model of the fractionation of olivine and molten sulfide from komatiite magma. *Earth Plan. Sci.* (Lett.) 39:255–266.

Dupuy, C. Dostal, J., and Leblanc, M. (1981). Geochemistry of an ophiolitic complex from New Caledonia. *Contrib. Mineral. Petrol.* 76:77–83.

Dyuzhikov, O. A., Fedorenko, V. A., Nestorovoskiy, V. A., and Dermisdovich, V. I. (1976). The new North Kharayelakh ore find and its nickel potential, Doklady. *Earth Sci. Sect* 229:123–125.

Eckstrand, O. R. (1975). The Dumont serpentinite: A model for control of opaque nickel ferrous mineral assemblages by alteration reactions in ultramafic rocks. *Econ. Geol.* 70:183–201.

Elliott, W. C., Grandstaff, D. E., Ulmer, G. C., and Buntin, T. (1982). An intrinsic oxygen fugacity study of platinum carbon associations in layered intrusions. *Econ. Geol.* 77:1493–1510.

Elthon, D., Casey, J. F., and Komor, S. (1982). Mineral chemistry of ultramafic cumulates from the North Arm Mountain massif of the Bay of Islands ophiolite: Evidence for high-pressure fractionation of oceanic basalts. *J. Geophys. Res.* 87:8717–8734.

Emeleus, C. H. (1987). The Rhum layered complex, Inner Hebrides, Scotland. In: *Origins of Igneous Layering.* Parsons, I. (ed). NATO Advanced Sci. Inst. Series, Series C, vol. 196, pp. 263–286.

Emeleus, C. H., and Upton, B. G. J. (1976). The Gardar period in southern Greenland. In: *Geology of Greenland,* Escher, A., and Watt, W. S. (eds.). Copenhagen: Gronlands Geologiske Undersogelse.

Evarts, R. C. (1977). The geology and petrology of the Del Puerto ophiolite, Diablo range, central California coast ranges. In: *North American Ophiolites.* Coleman, R. G., and Irwin, W. P. (eds.). Oregon Dept. of Geological and Mineral Industries, Bull. 95, pp. 121–140.

Farquhar, J. (1985). The geology of Western Platinum mine. In: *Some Ore Deposits in Southern Africa; Excursion Guidebook.* Vermaak, E., and von Gruenewaldt, G. (eds.), (compilers, Bushveld excursion). Third Int. Pt. Symp., Geol. Soc. of S. Africa, pp. 16–19.

Faure, G., Chaudhuri, S., and Fenton, M. (1969). Ages of the Duluth gabro complex and the Endion sill, Duluth, Minnesota. *J. Geophys. Res.* 74:720–725.

Ferguson, J. (1964). Geology of the Ilimaussaq intrusion, South Greenland, Part I. Description of map and structure. *Bull. Gronlands Geol. Unders.* 39:1–81.

Fincham, C. J. B., and Richardson, F. D. (1954). The behaviour of sulfur in silicate and aluminate melts. *Proc. Roy. Soc.* A223:40–62.

Findlay, D. C. (1969). Origin of the Tulameen ultramafic-gabbro complex, southern British Columbia. *Can. J. Earth Sci.* 6:399–426.

Fisher, D. (1979). The petrology of the Mt. Edwards nickel sulfide deposit, Widgiemooltha, Western Australia. Unpubl. Ph.D. thesis, Univ. of Toronto, 320 pp.

Fleet, M. E., and MacRae, N. D. (1987). Partition of Ni between olivine and sulfide: The effect of temperature, fO_2 and fS_2. *Contrib. Mineral. Petrol.* 95:336–342.

Fleet, M. F., and MacRae, N. D. (1983). Partition of Ni between olivine and sulfide and its application to Ni-Cu sulfide deposits. *Contr. Mineral. Petrol.* 83:75–81.

Fleet, M. E. (1977). Origin of disseminated copper-nickel sulfide ore at Frood, Sudbury, Ontario. *Econ. Geol.* 72:1449–1456.

Fleet, M. E., MacRae, N. D., and Herzberg, C. T. (1977). Partition of nickel between olivine and sulfide: A test for immiscible sulfide liquids. *Contr. Mineral. Petrol.* 65:191–197.

Foose, M., and Weiblen, P. (1986). The physical and petrologic setting and textural and compositional characteristics of sulfides from the South Kawishiwi Intrusion, Duluth Complex, Minnesota, USA. In: *Geology and Metallogeny of Copper Deposits,* Freidrich, G. H., Genkin, A. D., Naldrett, A. J., Ridge, J. D., Sillitoe, R. H., and Vokes, F. M. (eds.). Heidelberg, Berlin: Springer-Verlag, pp. 8–24.

Ford, A. B. (1970). The development of the layered series and the capping granophyre of the Dufek intrusion of Antartica. *Geol. Soc. South Africa,* Spec. Pub. 1, pp. 492–510.

Ford, A. B., and Boyd, W. W., Jr. (1968). The Dufek intrusion, a major stratiform gabbroic body in the Pensacola mountains, Antarctica. Internat. Geol. Cong., 23rd, Prague 1968, Proc. 2. pp. 213–228.

Fudali, R. F. (1965). Oxygen fugacities of basaltic

and andesitic magmas. *Geochim et Cosmochim. Acta,* 29:1063–1075.

Gaal, G. (1985). Nickel metallogeny related to tectonics. *Bull. Comm. Geol. Finland* 333:145–154.

Gain, S. B. (1985). The geologic setting of the platiniferous UG-2 chromitite layer on the farm Maandagshoek, Eastern Bushveld Complex. *Econ. Geol.* 80:925–943.

Gain, S. B., and Mostert, A. B. (1982). The geological setting of the platinoid and base metal sulfide mineralization in the Platreef of the Bushveld complex in Drenthe, north of Potgietersrus. *Econ. Geol.* 77:1395–1404.

Gain, S. B. (1980). The geology and PGE distribution in the Upper Group chromitite layers at Maandagshoek 254, K. T., eastern Bushveld Complex. *Inst. Geol. Res. on the Bushveld, Res. Rept. 22,* 24 pp.

Gaas, I. G., and Smewing, J. D. (1973). Intrusion, extrusion and metamorphism at constructive margins: Evidence from the Troodos massif, Cyprus. *Nature* 242:26–29.

Gass, I. G. (1967). The ultrabasic volcanic assemblage of the Troodos massif, Cyprus. In: *Ultramafic and Related Rocks,* Wyllie, P. J. (ed.) New York: John Wiley and Sons, pp. 121–134.

Gemuts, I., and Theron, A. (1975). The stratigraphy and mineralization of the Archean between Norseman and Coolgardie, Western Australia. In: *Economic Geology of Australia and Papua New Guinea, I. Metals,* Knight, C. L. (ed.). Australasian Inst. Min. Metal. Monograph 5, pp. 66–74.

Genkin, A. D., Distler, V. V., Gladyshev, G. D., Filiminova, A. A., Evstigneeva, T. L., Kovalenker, V. A., Laputina, I. P., Smirnov, A. V., and Grokhovskaya, T. L. (1981). Sulfide copper-nickel ores of the Noril'sk deposits. Moscow, Nauka, 234 pp. (in Russian).

Genkin, A. D., Evstigneeva, T. L., and Kovalenker, V. A. (1980). Platinum group minerals of cooper-nickel ores and some aspects of their origin. Internat. Mineral Assoc. Cong., 11th, Novosibirsk, Proc. pp. 165–171, Moscow, Nauka.

Gerassimovsky, V. I., Volkov, V. P., Kogarko, L. L., Poljakov, A. I., Saprykina, T. V., and Balshov, Yu. A. (1966). Geochemistry of the Loverzero alkaline massif. Izd. "Nauka", Moskva, pp. 1–395. (in Russian).

Glazkovsky, A. A., Gorbunov, G. I., and Sysoev, F. A. (1977). Deposits of nickel. In: *Ore Deposits of the USSR,* Vol. II, pp. 3–79. Smirnov, V. I. (ed.), (Translated into English by D.A. Brown).

Godlevsky, M. N., and Likhachev, A. P. (1986). Types and distinctive features of ore-bearing formations of copper-nickel deposits. In: *Geology and Metallogeny of Copper Deposits.* Friedrich, G. H., Genkin, A. D., Naldrett, A. J., Ridge, J. D., Sillitoe, R. H., and Vokes, F. M. (eds.). Berlin: Springer-Verlag, pp. 111–123.

Godlevsky, M. N., and Grinenko, L. N. (1963). Some data on the isotopic composition of sulfur in the sulfides of the Noril'sk deposit. *Geochem.* 1:335–341.

Godlevsky, M. N. (1959). *Traps and Ore-bearing Intrusions of the Noril'sk Region.* Moscow: Gosgeoltekhizdat, 69 pp.

Gole, M. J., Hill, R. E. T., and Barnes, S. J. (1988). Evidence for the formation of erosion channels by komatiite. *Geology* (in press).

Gole, M. J., Hill, R. E. T., and Barnes, S. J. (1986). *Komatiite Flows and Lava Lakes—Environments for Formation of Spinifex and Adcumulate Textures.* Volume of Abstracts, International Volcanological Congress, New Zealand.

Good, D. J. (1985). Petrology and geochemistry of the Bucko Lake nickel deposit, Thompson Nickel Belt, Manitoba. Unpubl. M.Sc. thesis, Univ. of Toronto, Toronto, Ontario.

Gorbachev, N. S., and Kashirceva, G. A. (1986). Fluid-magmatic differentiation of basaltic magma and equilbrium with magmatic sulfides. In: *Experiments in the Study of Important Problems in Geology.* Moscow: Nauka, pp. 98–1198.

Gorbachev, N. S., and Grinenko, L. N. (1973). The sulfur isotope ratios of the sulfides and sulfates of the Oktyabr'sk sulfide deposit, Noril'sk region, and the problem of its origin. *Geokhimiya* 8:1127–1136.

Gorbunov, G. I., Zagorodny, V. G., and Robonen, W. I. (1985a). Main features of the geological history of the Baltic shield and the epochs of ore formation. *Geol. Surv. Finland, Bull.* 333:17–41.

Gorbunov, G. I., Yakolev, Yu. N., Goncharov, Yu. V., Gorelov, V. A., and Tel'nov, V. A. (1985b). The nickel areas of the Kola peninsula. *Geol. Surv. Finland, Bull.* 333:42–122.

Gorbunov, G. I. (1968). Geology and origin of the copper-nickel sulfide ore deposits of Pechenga (Petsamo). Moscow: Nedra, 352 pp. (in Russian).

Grant, R. W., and Bite, A. (1984). Sudbury quartz diorite offset dikes: In: *The Geology and Ore Deposits of the Sudbury Structure,* Pye, E. G., Naldrett,. J., and Giblin, P.E. (eds.). Ont. Geol. Survey Spec. V. No. 1, pp. 97–138.

Graterol, M., and Naldrett, A. J. (1971). Mineralogy of the Marbridge No.3 and No.4 deposits with some comments on low temperature equilibration in the Fe-Ni-S system. *Econ. Geol.* 66:886–900.

Gray, F., Page, N. J., Carlson, C.A., Wilson, S. A., and Carlson, R. R. (1986). Platinum-group element geochemistry of zoned ultramafic suites, Klamath mountains, California and Oregon. *Econ. Geol.* 81:1252–1260.

Green, A. H., and Naldrett, A. J. (1981). Volcanic peridotite-associated nickel sulfide deposits of the Abitibi belt, Canada with special reference to Langmuir. *Econ. Geol.* 76:1503–1523.

Green, A. H. (1978). Evolution of Fe-Ni sulfide ores associated with Archean ultramafic komatiites, Langmuir Township, Ontario. Unpubl. Ph.D. thesis, Univ. of Toronto, Toronto, Ontario.

Green, D. H., Nicholls, I. A., Viljoen, M., and Viljoen, R. (1975). Experimental demonstration of the existence of peridotitic liquids in earliest Archean magmatism. *Geol.* 3:11–14.

Green, D. H. (1975). Genesis of archean peridotite magmas and constraints on Archean geothermal gradients and tectonics. *Geology* 3(1):15–18.

Green, J. C., Phinney,W. C., and Weiblen, P. W. (1966). Gabbro Lake Quadrangle, Lake County, Minnesota. *Minn. Geol. Surv.Misc.* Map M2.

Gresham, J. J. (1986). Depositional environment of volcanic peridotite-associated nickel-copper sulfide deposits with special reference to the Kambalda dome. In: *Geology and Metallogeny of Copper Deposits,* Friedrich, G. H., Genkin, A. D., Naldrett, A. J., Ridge, J. D., Sillitoe, R. H., and Vokes, F. M. (eds.). Society for Geology Applied to Mineral Deposits, Special Publication No. 4., pp. 63–90.

Gresham, J. J., and Loftus-Hills, G.D. (1981). The Geology of the Kambalda Nickel Field, Western Australia. *Econ. Geol.* 76:1373–1416.

Grinenko, L.N. (1985). Sources of sulfur of the nickeliferous and barren gabbro-dolerite intrusions of the northwest Siberian platform. *Internat. Geology Rev.* pp. 695–708.

Grinenko, L. N. (1967). The sulfur-isotope ratios of sulfides of some copper-nickel deposits and prospects of the Siberian platform. In: *The Petrology of the Trap-rocks of the Siberian Platform, Nedra Press, Leningrad.* pp. 221–229.

Grout, F. F. (1918). The lopolith; an igneous form exemplified by the Duluth abbro. *Am. Jour. Sci.* 4th ser., 46:516–522.

Groves, D. I., Korkiakoski, E. A., McNaughton, N. J., Lesher, C. M., and Cowden, A. (1986). Field evidence for thermal erosion by komatiites at Kambalda and the genesis of nickel ores. *Nature* 319:136–139.

Groves, D. I., Lesher, C. M., and Gee, R. D. (1984). Tectonic setting of sulfide nickel deposits of the Western Australian shield. In: *Sulfide Deposits in Mafic and Ultramafic Rocks,* Buchanan. D.L., Jones, M. J. (eds.). Special Publ. of Inst. Min. Metall., pp. 1–13.

Groves, D. I., and Lesher, C. M. (1982). Regional geology and nickel deposits of the Norseman-Wiluna belt, Western Australia. In: *Field Guide to IGCP Projects 161 and 91.* Publication No. 7, Geology Department and University Extensions, The University of Western Australia.

Groves, D. I., and Keays, R. R. (1979). Mobilization of ore forming elements during alteration of dunites Mt. Keith-Betheno, Western Australia. *Can. Mineral.* 17:373–390.

Groves, D. I., Barrett, F. M., and McQueen, K. G. (1979). The relative roles of magmatic segregation, volcanic exhalation and regional metamorphism in the generation of volcanic-associated nickel ores of Western Australia. *Can. Mineral.* 17:319–336.

Groves, D. I., Barrett, F. M., and McQueen, K. G. (1977). Spinel phases associates with metamorphosed volcanic-type iron-nickel sulfide ores from Western Australia. *Econ. Geol.* 72:1224–1244.

Groves, D. I., Barrett, F. M., Binns, R. A., Marston, R. J., and McQueen, K. G. (1976). A possible volcanic exhalative origin for lenticular nickel sulfide deposits of volcanic association with special reference to those in Western Australia. Discussion. *Can. J. Earth Sci.* 13:1646–1650.

Gupta, V. K., Grant, F. S., and Card, K. D. (1986). Gravity and Magnetic Characteristics of the Sudbury Structure. In: *The Geology and Ores Deposits of the Sudbury Structure,* Pye, E. G., Naldrett, A. J., and Giblin, P. E. (eds.). Ont. Geol. Surv. Spec. Vol. No.1, pp. 381–410.

Hall, A. L. (1932). The Bushveld Igneous Complex of the central Transvaal. *South Africa Geol. Survey* Mem. 28, 560 pp.

Hallberg, J. A., and Williams, D. A. C. (1972). Archean mafic and ultramafic rock associations in the Eastern Goldfields region, Western Australia. *Earth Planet. Sci.* (Lett.) 15:191–200.

Hamilton, J. (1977). Sr isotope and trace element studies of the Great Dyke and Bushveld mafic phases and their relation to early Proterozoic magma genesis in Southern Africa. *J. Petrol.* 18:24–52.

Hamlyn, P. R., and Keays, R. R. (1986). Sulfur saturation and second-stage melts. Application to the Bushveld platinum metal deposits. *Econ. Geol.* 81:1431–1445.

Hamlyn, P. R., and Keays, R. R. (1985). Precious metals in magnesian low-Ti lavas: Implications for metallogenesis and sulfur saturation in primary magmas. *Geochim. et Cosmochim. Acta* 49:1797–1811.

Hammerbeck, E. C. I. (1984). Aspects of nickel metallogeny of Southern Africa. In: *Sulfide Deposits in Mafic and Ultramafic Rocks,*

Buchanan, D. L., and Jones, M. J. (eds.). London: Special Publication, Inst. Min. and Metal., pp. 135–140.

Harmer, R. E., and Sharpe, M. R. (1985). Field relations and strontium isotope systematics of the marginal rocks of the Eastern Bushveld Complex. *Econ. Geol.* 80:813–837.

Haughton, D. R., Roeder, P. L., and Skinner, B. J. (1974). Solubility of sulfur in mafic magmas. *Econ. Geol.* 69:451–467.

Hawley, J. E. (1965). Upside-down zoning at Frood, Sudbury, Ontario. *Econ. Geol.* 60:529–575.

Helz, R. T. (1985). Compositions of fine-grained mafic rocks from sills and dikes associated with the Stillwater Complex. In: *Stillwater Complex,* Czamanske, G. K., and Zientek, M. L. (eds.). Montana Bureau of Mines and Geology Spec. Pub. No. 92, pp. 97–117.

Hess, H. H. (1960). Stillwater igneous complex, Montana. *Geol. Soc. Am.* Mem. 80, 230 pp.

Hiemstra, S. A. (1986). The distribution of chalcophile and platinum-group elements in the UG-2 chromitite layer of the Bushveld Complex. *Econ. Geol.* 81(5):1080–1087.

Hiemstra, S. A. (1985). The distribution of some platinum group elements in the UG-2 chromitite layer of the Bushveld Complex. *Econ. Geol.* 80:944–957.

Hill, R. E. T., Gole, M. J., and Thompson, J. F. H. (1985). Characteristics of centres of Archean komatiitic volcanism, exemplified by lithologies in the Agnew area, Yilgarn block, Western Australia. *Can. Mineral.* 23:327.

Hill, R. E. T. (1984). Experimental study of phase relations at 600°C in a portion of the Fe-Ni-Cu-S system and its application to natural sulfide assemblages. In: *Sulfide Deposits in Mafic and Ultramafic Rocks,* Buchanan, D. L., and Jones, M. J. (eds.). London, Spec. Publ. of the Inst. of Mining and Metallurgy, pp. 14–21.

Hiller, J. E., and Probsthain, K. (1956). Thermische und rontgenographische Untersuchungen an Kupferkres: *Z. Kristallogo.* 108:108–129.

Hilty, D. C., and Crafts, W. (1952). Liquidus surface of the Fe-S-O system. *J. Metals* 4:1307–1312.

Himmelberg, G. R., and Ford, A. B. (1976). Pyroxenes of the Dufek intrusion, Antarctica. *J. Petrol.* 17:219–243.

Hoffman, E. L., Naldrett, A. J., Alcock, R. A., and Hancock, R. G. V. (1979). The noble metal content of ore in the Levack West and Little Stobie mines, Ontario. *Can. Mineral.* 17:437–452.

Hopson, C. A., Coleman, R. G., Gregory, R. T., Pallister, J. S., and Bailey, E. H. (1981). Geologic section through the Samail ophiolite and associated rocks along a Muscat-Ibra transect, southeastern Oman mountains. *J. Geophys. Res.* 86:2527–2544.

Hriskevich, M. E. (1968). Petrology of the Nipissing Diabase sill of the Cobalt area, Ontario, Canada. *Geol. Soc. Am. Bull.* 79:1387–1404.

Hunter, D. R. (1976). Some enigmas of the Bushveld complex. *Econ. Geol.* 71:229–248.

Huppert, H. E., and Sparks, R. S. J. (1985a). Komatiites I: Eruption and flow. *J. Petrology* 26:694–725.

Huppert, H. E., and Sparks, R. S. J. (1985b). Cooling and contamination of mafic and ultramafic magmas during ascent through continental crust. *Earth and Planet. Sci.* (Lett.) 74:371–386.

Huppert, H. E., Sparks, R. S. J., and Whitehead, J. A. (1984). *Plumes: Magma Chamber Replenishment by Light Inputs,* Proc. of Conf. on Open Magmatic Systems, Dungan, M. A. Grove, T. L., and Hildreth, W. (eds.). Inst. for the Study of Earth and Man and Dept. of Geological Sci. of Southern Methodist Univ. pp. 83–84.

Huppert, H. E., and Sparks, R. S. J. (1980). The fluid dynamics of a basic magma chamber replenished by hot, dense ultrabasic magma. *Contr. Mineral. Petrol.* 75:279–289.

Hynes, A., and Francis, D. (1982a). Komatiite basalts of the Cape Smith foldbelt, New Quebec, Canada. In: *Komatiites,* Arndt, N. T., and Nesbit, E. G. (eds.). London: Allen & Unwin, pp. 159–170.

Hynes, A., and Francis, D. (1982b). A transect of the early Proterozoic Cape Smith foldbelt, New Quebec. *Tectonophysics* 88:23–59.

Huang, W. -L., and Williams, R. J. (1980). Melting relations of portions of the system Fe-S-Si-O to 32 kb with implication to the nature of the mantle core boundary (Abstr.). Lunar & Planet. Sc. XI, Lunar and Planetary Inst., Houston, p. 486–488.

Irvine, T. N., Keith, D. W., and Todd, S. G. (1983). The J-M platinum palladium reef: II. Origin by double-diffusive convective magma mixing and implications for the Bushveld Complex. *Econ. Geol.* 78:1287–1334.

Irvine, T. N. (1980). Magmatic infiltration metasomatism, double diffusive fractional crystallization and adcumulus growth in the Muskox intrusion and other layered intrusions. In: *Physics of Magmatic Processes,* Hargraves, R. B. (ed.). Princeton, NJ: Princeton University Press, pp. 325–383.

Irvine, T. N. (1977). Origin of chromite layers in the Muscox intrusion and other stratiform intrusions; A new interpretation. *Geology* 5:273–277.

Irvine, T. N. (1975). Crystallization sequences of the Muskox intrusion and other layered intrusions—II. Origin of chromitite layers and similar deposits of other magmatic ores. *Geochim. et Cosmochim. Acta* 39:991–1020.

Irvine, T. N. (1974). Petrology of the Duke Island ultramafic complex, southeastern Alaska. *Geol. Soc. Am.* Mem. 138, 240 pp.

Irvine, T. N. (1970). Crystallization sequences in the Muskox intrusion and other layered intrusions: I. Olivine-pyroxene-plagioclase relations. *Geol. Soc. South Africa* Special Publication 1:441–447.

Irvine, T. N., and Smith, C. H. (1967). The ultramafic rocks of the Muskox intrusion, N.W.T., Canada. In: *Ultramafic and Related Rocks,* Wyllie, P. J. (ed.). New York: John Wiley and Sons, pp. 38–49.

Jackson, E. D., Green, H. W., and Moores, E. M. (1975). The Vuorinos ophiolite, Greece: Cyclic units of lineated cumulates overlying harzburgite tectonite. *Geol. Soc. Am. Bull.* 86:390–398.

Jackson, E. D. (1970). the cyclic unit in layered intrusions, a comparison of repetitive stratigraphy in the ultramafic parts of the Stillwater, Muskox, Great Dyke and Bushveld Complex (with discussion). In: *Symposium on the Bushveld Igneous Complex and Other Layered Intrusions,* Visser, D. J. L., and von Gruenewaldt, G. (eds.). Geol. Soc. of Africa Spec. Pub. No. 1, pp. 391–424.

Jackson, E. D. (1961). Primary textures and mineral associations in the ultramafic zone of the Stillwater complex. *U.S. Geol. Surv. Prof. Paper* 358, 106 pp.

Johan, S., and Auge, T. (1986). Ophiolitic mantle sequences and their evolution: Mineral chemistry constraints. In: *Metallogeny of Basic and Ultrabasic Rocks,* Gallagher, M. J., Ixer, R. A., Neary, C. R., and Pritchard, H. M. (eds.) London: Spec. Pub. of Inst. of Mining & Metallurgy, pp. 305–318.

Johan, S., and Watkinson, D. H. (1985). Significance of a fluid phase in platinum-group element concentration: Evidence from the Critical Zone, Bushveld Complex (Abstr.). *Can. Mineral.* 23:305–306.

Jones, W. R., Peoples, J. W., and Howland, A. L. (1960). Igneous and tectonic structures of the Stillwater complex, Montana, *U.S. Geol. Surv. Bull.* 1071-H, pp. 281–335.

Juteau, T., Nicolas, A., Dubessy, J., Fruchard, J. C., and Bouchez, J. L. (1977). Structural relationships in the Antalya ophiolite complex, Turkey: possible model for an ocean ridge. *Geol. Soc. Am. Bull.* 88:1740–1748.

Keays, R. R., Nickel, E. H., Groves, D. I., and McGoldrick, P. J. (1982). Iridium and palladium as discriminants of volcanic-exhalative, hydrothermal and magmatic nickel sulfide mineralization. *Econ. Geol.* 77:1535–1547.

Keays, R. R., and Campbell, I. H. (1981). Precious metals in the Jimberlana Intrusion, Western Australia: Implications for genesis of platiniferous ores in layered intrusions. *Econ. Geol.* 76:1118–1141.

Keays, R. R., Ross, J. R., and Woolrich, P. (1981). Precious metals in volcanic peridotite-associated nickel sulfide deposits in Western Australia. II. Distribution within ores and host-rocks at Kambalda. *Econ. Geol.* 76:1645–1674.

Keays, R. R., and Davison, R. M. (1976). Palladium, iridium and gold in the ores and host rocks of nickel sulfide deposits in Western Australia. *Econ. Geol.* 71:1214–1228.

Keays, R. R., and Crocket, J. H. (1970). A study of precious metals in the Sudbury Nickel Irruptive ores. *Econ. Geol.* 65:438–450.

Kozlov, E. K. (1973). *Natural Series of Nickel Bearing Intrusive Rocks and Their Metallogeny in the Kola Peninsula.* Nauka: Leningrad, 288 pp. (in Russian).

Krogh, T. E., Davis, D. W., and Corfu, F. (1984). Precise U-Pb Zircon and Baddeleyite ages for the Sudbury Area. In: *The Geology and Ore Deposits of the Sudbury Structure,* Pye, E. G., Naldrett, A. J., and Giblin, P. E. (eds.). Ont. Geol. Survey Spec. Vol. No. 1, pp. 431–447.

Kruger, F. J., and Marsh, J. S. (1985). The mineralogy, petrology and origin of the Merensky cyclic unit in the Western Bushveld Complex. *Econ. Geol.* 80:958–974.

Kruger, F. J., and Marsh, J. S. (1982). Significance of Sr^{87}/Sr^{86} ratios in the Merensky cyclic unit of the Bushveld Complex. *Nature* 298:53–55.

Kucha, H. (1982). Platinum-group metals in the Zechstein Copper Deposits, Poland. *Econ. Geol.* 77:1578–1591.

Kullerud, G., Yund, R. A., and Moh, G. (1969). Phase relations in the Fe-Ni-S, Cu-Fe-S and Cu-Ni-S systems. *Econ. Geol.* Monograph 4, pp. 323–343.

Kullerud, G. (1967). Sulfide studies. In: *Researches in Geochemistry,* 2, New York: John Wiley & Sons.

Kullerud, G. (1963). Thermal stability of pentlandite. *Can. Mineral.* 7:353–366.

Kullerud, G., and Yund, R. A. (1962). The Ni-S system and related minerals. *J. Petrology* 3:126–175.

Kullerud, G., and Yoder, H. S. (1959). Pyrite stability relations in the Fe-S system. *Econ. Geol.* 54:533–572.

Kullerud, G. (1953). The FeS–ZnS system: A geological thermometer. *Norsk Geol. Tidsskr.* 32:61–147.

Kuo, H. -Y., and Crocket, J. H. (1979). Rare earth elements in the Sudbury nickel irruptive: Comparison with layered gabbros and implications for nickel irruptive petrogenesis. *Econ. Geol.* 79:590–605.

Kuznetzov, B. G., Sokolova, M. A., Palkina, K. K., and Popova, Z. V. (1966). The system Co-S. *Isv. Akad. Nauk SSSR, Neorgan Materialy* 1:675.

Lauder, W. R. (1970). Origin of the Merensky reef. *Nature* 227:365–366.

Lee, C. A. (1983). Trace and platinum-group element geochemistry and the development of the Merensky unit of the Western Bushveld Complex. *Mineralium Deposita* 18:173–190.

Lehmann, J., and Roux, J. (1986). Experimental and theoretical study of (Fe^{2+},Mg) (Al $Fe^{3+})_2O_4$ spinels: Activity-composition relationships, miscibility gaps, vacancy contents. *Geochim. et Cosmochim. Acta* 50:1765–1783.

Lenz, J. G., Conard, R. S., and Warner, J. (1978). The liquidus surface and tie-lines in the iron-nickel-sulfur-system between 1473 and 1673°K. *Metall. Trans. B* 9B:459–462.

Leroy, L. W. (1985). Troctolite-anorthosite Zone 1 and the J-M Reef: Frog Pond Adit to the Graham Creek Area. In: *Stillwater Complex,* Czamanske, G. K., and Zientek, M. L. (eds.). Montana Bureau of Mines and Geology Spec. Pub. No. 92, pp. 325–333.

Lesher, C. M., and Groves, D. I. (1986). Controls on the formation of komatiite-associated nickel-copper sulfide deposits. In: *Geology and Metallogeny of Copper Deposits,* Friedrich, G. H., Genkin, A. D., Naldrett, A. J., Ridge, J. D., Sillitoe, R. H., and Vokes, F. M. (eds.). Society for Geology Applied to Mineral Deposits, Spec. Pub. No. 4, pp. 43–62.

Lesher, C. M., Arndt, N. T., and Groves, D. I. (1984). Genesis of komatiite-associated nickel sulfide deposits at Kambalda, Western Australia; A distal volcanic model. In: *Sulfide Deposits in Mafic and Ultramafic Rocks.* Buchanan, D. L., and Jones, M. J. (eds.). London: Inst. of Mining and Metallurgy, pp. 70–80.

Lesher, C. M. (1983). Localization and genesis of komatiite-associated Fe-Ni-Cu sulfide mineralization at Kambalda, Western Australia. Unpubl. Ph.D. thesis, Univ. of Western Australia.

Lesher, C. M., Lee, R. F., Groves, D. I., Bickle, M. J., and Donaldson, M. J. (1981). Geochemistry of komatiites from Kambalda, Western Australia: 1. Chalcophile element depletion—A consequence of sulfide liquid separation from komatiite magmas. *Econ. Geol.* 76:1714–1728.

Liebenberg, L. (1970). The sulfides in the layered sequence of the Bushveld Igneous Complex. *Geol. Soc. S.Africa* Spec. Pub. No. 1, pp. 108–207.

Lightfoot, P. C., Conrod, D., Naldrett, A. J., and Evensen, N. M. (1987). Grant 230, Petrogenetic, chemical, isotopic and economic potential studies of the Nipissing Diabase. In: Geoscience Res. Grant Program Summary of Res., Ont. Geol. Surv. Misc. Paper 136, pp. 4–26.

Lightfoot, P. C., Conrod, D., Naldrett, A. J., and Evensen, N. M. (1986). Grant 230, Petrogenetic, chemical isotopic and economic potential studies of the Nipissing Diabase. In: Geoscience Res. Grant Program Summary of Res., Ont. Geol. Surv. Misc. Paper 130, pp. 87–107.

Lightfoot, P. C., Naldrett, A. J., and Hawkesworth, C. J. (1984). The geology and geochemistry of the Waterfall Gorge section of the Insizwa complex with particular reference to the origin of nickel sulfide deposits. *Econ. Geol.* 79:1857–1879.

Lightfoot, P. C., and Naldrett, A. J. (1983). The geology of the Tabankulu section of the Insizwa complex, Transkei, Southern Africa. *Trans. Geol. Soc. S. Africa* 86 (3):169–187.

Listerud, W. H., and Meinekke, D. G. (1977). Minnesota Dept. of Natural Resources, Div. of Minerals, Minerals Exploration Section, Report 93.

Lusk, J. (1976). A possible volcanic-exhalative origin for lenticular nickel sulfide deposits of volcanic association with reference to Western Australia. *Can. J. Earth Sci.* 13:451–458.

Mackovicky, M., Mackovicky, E., and Rose-Hansen, J. (1986). Experimental studies on the solubility and distribution of platinum group elements in base-metal sulfides in platinum deposits. In: *Metallogeny of Basic and Ultrabasic rocks,* Gallagher, M. J., Ixer, R. A., Neary, C. R., and Prichard, H. M., (eds.). Inst. of Mining and Metallurgy, Spec. Publ. pp. 415–426.

MacLean, W. H. (1978). Personal communication.

MacLean, W. H. (1969). Liquidus phase relations in the FeS-FeO-Fe_3O_4-SiO_2 systems and their application in geology. *Econ. Geol.* 64:865–884.

Mainwaring, P. R., and Naldrett, A. J. (1977). Country-rock assimilation and the genesis of Cu-Ni sulfides in the Waterhen intrusion, Duluth Complex, Minnesota. *Econ. Geol.* 72:1269–1284.

Maksimov, Ye. M., and Rudkevich, M. Ya. (1971). Quantitative evaluation of the dynamics of Mesozoic and Cenozoic vertical movements over the West Siberian plate. *Geotect.* 1-6:245–248.

Malich, N. S., and Tuganova, Ye. V. (1976). Distribution patterns of mineral resources in the Siberian plaform cover. *Internat. Geol. Rev.* 18:417–424.

Malpas, J. (1981). Petrology and tectonic significance of Newfoundland ophiolites, with examples from the Bay of Islands. In: *North American Ophiolites,* Coleman, R. G., and

Irwin, W. P. (eds.). Oregon Dept. of Geological and Mineral Industries Bull. 95, pp. 13–24.

Malpas, J. (1979). The dynamothermal aureole of the Bay of Islands Ophiolite suite. *Can. J. Earth Sci.* 16:2086–2101.

Mann, E. L., Lipin, B. R., Page, N. J., Foose, M. P., and Loferski, P. J. (1985). Guide to the Stillwater Complex. In: *Stillwater Complex,* Czamanske, G. K., and Zientek, M. L. (eds.). Montana Bureau of Mines Spec. Pub. No. 92. pp. 231–246.

Mann, E. L., and Lin, C.-P. (1985). Geology of the West Fork adit. In: *Stillwater Complex,* Czamanske, G. K., and Zientek, M. L. (eds.). Montana Bureau of Mines and Geology Spec. Pub. No. 92, pp. 325–333.

Marston, R. J., Groves, D. I., Hudson, D. R., and Ross, J. R. (1981). Nickel sulfide deposits in Western Australia: A review. *Econ. Geol.* 69:1330–1363.

Marston, R. J., and Kay, B. D. (1980). The distribution, petrology and genesis of nickel ores at the Juan complex, Kambalda, Western Australia. *Econ. Geol.* 75:546–565.

Maske, S. (1966). The petrography of the Ingeli mountain range. *Univ. Stellenbosch Annals,* ser. A, 41(1):109.

Mathez, E. A., Boudreau, A. E., and McCallum, I. S. (1985). Apatite and biotite from the Bushveld and Stillwater Comlexes and the nature of hydrothermal activity. *Can. Mineral* 23:308.

Mathez, E. A. (1976). Sulfur solubility and magmatic sulfides in submarine basalt glass. *J. Geophys. Res.* 81:4269–4276.

Matlock, W. F., and Watowich, S. N. (1980). Geology and sulfide mineralization of the Duluth Compex–Virginia formation contact, Minnamax deposit, Minnesota (Abstr.). *Geol. Soc. Am.* Abstracts with Programs, pp. 477–478.

McCallum, I. S., Raedeke, L. D., and Mathez, E. A. (1980). Investigations in the Stillwater Complex. Part I. Stratigraphy and structure of the banded zone. In: *The Jackson Volume,* Irving, A. J., and Dungan, M. A. (eds.) Am. J. Sci. 280-A, pp. 59–87.

McCullough, M. R., and Wasserburg, G. J. (1978). Sm-Nd and Rb-Sr chronology of continental crust formation. *Science* 200:1003–1011.

McLaren, C. H., and de Villiers, J. P. R. (1982). The platinum-group chemistry and mineralogy of the UG-2 chromitite layers of the Bushveld Complex. *Econ. Geol.* 77:1348–1386.

McQueen, K. G. (1981). Volcanic-associated nickel deposits from around the Widgiemooltha dome, Western Australia. *Econ. Geol.* 76:1417–1443.

McRae, N. D. (1969). Ultramafic intrusions of the Abitibi area, Ontario. *Can. J. Earth Sci.* 6:281–304.

Merwin, H. E., and Lombard, R. H. (1937). The system Cu-Fe-S. *Econ. Geol.* 32:203–284.

Miller, L. J., and Smith, M. E. (1975). Sherlock Bay nickel-copper deposit. In: *Economic Geology of Australia and Papuna, New Guinea,* Knight, C. L. (ed.). Aust. Inst. of Min. & Metal. Monograph 5, pp. 168–174.

Misra, K. C., and Fleet, M. E. (1974). Chemical composition and stability of violarite. *Econ. Geol.* 69:391–403.

Misra, K. C., and Fleet, M. E. (1973). The chemical compositions of synthetic and natural pentlandite assemblages. *Econ. Geol.* 68:518–539.

Molyneaux, T. G. (1974). A geological investigation of the Bushveld Complex in Sekhuneland and a part of the Steelpoort Valley. *Geol. Soc. South Africa Trans.* 77:329–338.

Moores, E. M., and Vine, F. J. (1971). The Troodos massif, Cyprus, and other ophiolites as oceanic crust; Evaluation and implications. *Roy. Soc. London Philos. Trans.* ser. A, 268:443–466.

Moores, E. M. (1969). Petrology and structure of the Vourinos ophiolitic complex of northern Greece. *Geol. Soc. Am.* Spec. Paper 118:1–74.

Morrison, D. A., Maczuga, D. E., Phinney, W. C., and Ashwal, L. D. (1986). Stratigraphy and petrology of the Mulcahy Lake layered gabbro: An Archean intrusion in the Wabigoon Subprovince, Ontario. *J. Petrol.* 27:303–342.

Muir, T. L., and Peredery, W. V. (1984). The Onaping formation. In: *The Geology and Ore Deposits of the Sudbury Structure,* Pye, E. G., Naldrett, A. J., and Giblin, P. E. (eds.). Ont. Geol. Survey Spec. Vol. no. 1, pp. 139–210.

Muir, T. L. (1984). The Sudbury structure: Consideration of models of an endogenic origin. In: *The Geology and Ore Deposits of the Sudbury Structure,* Pye, E. G., Naldrett, A. J., and Giblin, P. E. (eds.). Ontario Geological Survey Spec. Vol. No. 1, pp. 449–490.

Mukaiyama, H., and Izawa, E. (1970). Phase relations in the Cu-Fe-S system: The copper-deficient part. In: *Volcanism and Ore Genesis,* Tatsumi, T. (ed.). Tokyo: Univ. Tokyo Press, pp. 339–355.

Nagamori, M., and Kameda, M. (1965). Equilibria between Fe-S-O system melts and CO-CO_2-SO_2 gas mixtures at 1200° C. *Trans. Japan Inst. Met.* 6:21–30.

Naldrett, A. J., and von Gruenewaldt, G. (1989). The association of PGE with chromitite in layered intrusions and ophiolite complexes. *Econ. Geol.* (in press).

Naldrett, A. J., and Lehmann, J. (1988). Spinel non-stoichiometry as the explanation for Ni-, Cu-, and PGE-enriched sulphides in chromitites. In: *Geoplatinum '87.* London: Elsevier, pp. 93–110.

Naldrett, A. J., Cameron, G., von Gruenewaldt, G., and Sharpe, M. R. (1987). The formation of stratiform PGE deposits in layered intrusions. In: *Origins of Igneous Layering,* Parsons, I. (ed.). NATO Advanced Sc. Inst. Series, Series C., Vol. 196, pp. 313–397, Reidel, Dudrecht.

Naldrett, A. J., and Barnes, S.-J. (1986). The behavior of platinum group elements during fractional crystallization and partial melting with special reference to the composition of magmatic sulfide ores. *Fortschritte der Mineralogie* 64:113–133.

Naldrett, A. J., Gasparrini, E. C., Barnes, S. J., von Gruenewaldt, G., and Sharpe, M. R. (1986a). The upper critical zone of the Bushveld Complex and a model for the origin of Merensky-type ores. *Econ. Geol.* 81:1105–1118.

Naldrett, A. J., Rao, B. V., and Evensen, N. M. (1986b). Contamination at Sudbury and its role in ore formation. In: *Metallogeny of Basic and Ultrabasic Rocks,* Gallagher, M. J., Ixer, R. A., Neary, C. R., and Pritchard, H. M. (eds.). London: Spec. Pub. of Inst. of Mining & Metallurgy, pp. 75–92.

Naldrett, A. J. (1986). New developments in Sudbury geology: A model for the Complex and its ores. In: *Geology and Metallogeny of Copper Deposits,* Friedrich, G. Genkin, A., Naldrett, A. J., Ridge, J. D., Sillitoe, R. H., and Vokes, F. M. (eds.). Berlin, Heidelberg, New York, Tokyo: Springer, pp. 91–110.

Naldrett, A. J., Gasparrini, C., Barnes, S. J., Sharpe, M. R., and von Gruenewaldt, G. (1984). The petrology of the upper part of the critical zone of the Bushveld Complex and its bearing on the origin of the Merensky Reef. Proc. 27th Int. Geol. Congress, Moscow, 9, pp. 373–389.

Naldrett, A. J. (1984). Summary, discussion and synthesis. In: *The Geology and Ore Deposits of the Sudbury Structure,* Pye, E. G., Naldrett, A. J., and Giblin, P. E. (eds.). Ont. Geol. Surv. Spec. Vol. No. 1, p. 533.

Naldrett, A. J., and Campbell, I. H. (1982). Physical and chemical constraints on genetic models for komatiite related Ni-Sulfide deposits. In: *Komatiites,* Arndt, N. T., and Nisbet, E. C. (eds.). London: Allen and Unwin.

Naldrett, A. J., Innes, D. G., Sowa, J., and Gorton, M. (1982). Compositional variation within and between 5 Subury ore deposits. *Econ. Geol.* 77:1519–1534.

Naldrett, A. J. (1981a). Nickel sulfide deposits: Classification, composition and genesis. *Econ. Geol.* 75th Anniv. Vol. pp. 628–685.

Naldrett, A. J. (1981b). Pt group element deposits. In: *Platinum Group Elements: Mineralogy, Geology, Geochemistry.* Cabri, L. C. (ed.). Can. Inst. Min. Met. Spec. Vol. 23, pp. 197–232.

Naldrett, A. J., and Smith, I. E. M. (1981). Mafic and ultramafic volcanism during the Archean. In: *Basaltic Volcanism in the Terrestrial Planets: Basaltic Volcanism, Study Project.* New York: Pergamon Press, pp. 5–28.

Naldrett, A. J., and Duke, J. M. (1980). Platinum metals in magmatic sulfide ores. *Science* 208:1417–1428.

Naldrett, A. J., and MacDonald, A. J. (1980). Tectonic settings of some Ni-Cu sulfide ores: Their importance in genesis and exploration. *Geol. Assoc. Can.* Spec. Paper 20, pp. 633–657.

Naldrett, A. J., Hoffman, E. L., Green, A. H., Chou, C. L., Naldrett, S. R., and Alcock, R. A. (1979). The composition of Ni-sulfide ores with particular reference to their content of PGE and Au. *Can. Mineral.* 17(2):403–416.

Naldrett, A. J. (1979). Partitioning of Fe, Co, Ni and Cu between sulfide liquid and basaltic melts and the composition of Ni-Cu sulfide deposits—A reply and further discussion. *Econ. Geol.* 74:1520–1528.

Naldrett, A. J., and Turner, A. R. (1977). The geology and petrogenesis of a greenstone belt and related nickel sulfide mineralization at Yakabindie, Western Australia. *Precamb. Res.* 5:43–103.

Naldrett, A. J., and Goodwin, A. M. (1977). Volcanic rocks of the Blake River group, Abitibi greenstone belt, Ontario, and their sulfur content. *Can. Jour. Earth Sci.* 14:539–550.

Naldrett, A. J., and Cabri, L. J. (1976). Ultramafic and related mafic rocks: Their classification and genesis with special reference to the concentration of nickel sulfides and platinum-group elements. *Econ. Geol.* 71:1131–1158.

Naldrett, A. J. (1973). Nickel sulfide deposits—Their classification and genesis with special emphasis on deposits of volcanic association. *Trans. Can. Inst. Min. Met.* 76:183–201.

Naldrett, A. J., Hewins, R. H., and Greenman, L. (1972). The main irruptive and the sub-layer at Sudbury, Ontario. Proc. 24th Int. Geol. Congress (Section 4), pp. 206–214.

Naldrett, A. J., and Gasparrini, E. L. (1971). Archean nickel sulfide deposits in Canada: Their classification, geological setting and genesis with some suggestions as to exploration. *Geol. Soc. Australia,* Spec. Pub. No. 3, pp. 201–226.

Naldrett, A. J., Bray, J. G., Gasparrini, E. L., Podolsky, T., and Rucklidge, J. C. (1970). Cryptic variation and the petrology of the Sudbury Nickel Irruptive. *Econ. Geol.* 65:122–155.

Naldrett, A. J. (1969). A portion of the Fe-S-O

and its application to sulfide ore magmas. *J. Petrol.* 10:171–201.

Naldrett, A. J., and Mason, G. D. (1968). Contrasting Archean ultramafic igneous bodies in Dundonald and Clergue townships, Ontario. *Can. J. Earth Sci.* 5:111–143.

Naldrett, A. J., and Richardson, S. W. (1967). Effect of water on the melting of pyrrhotite-magnetite assemblages. Ann. Rep. Dir. Geophys. Lab., Carnegie Inst. of Washington Year Book, 66, pp. 429–431.

Naldrett, A. J., and Kullerud, G. (1967). A study of the Strathcona mine and its bearing on the origin of the nickel-copper ores of the Sudbury district, Ontario. *J. Petrol.* 8:453–531.

Naldrett, A. J., Craig, J. R., and Kullerud, G. (1976). The central portion of the Fe-Ni-S system and its bearing on pentlandite solution in iron-nickel sulfide ores. *Econ. Geol.* 62:826–847.

Nesbitt, R. W. (1971). Skeletal crystal forms in the ultramafic rocks of the Yilgarn block, Western Australia: Evidence for an Archean ultramafic liquid. *Geol. Soc. Australia,* Spec. Pub. No. 3, pp. 331–350.

Page, N. J., Zientek, M. L., Lipin, B. R., Raedeke, L. D., Wooden, J. L., Turner, A. R., Loferski, P. J., Foose, M. P., Moring, B. C., and Ryan, M. P. (1985). Geology of the Stillwater Complex exposed in the Mountain View area and on the west side of the Stillwater Complex. In: *Stillwater Complex,* Czamanske, G. K., and Zientek, M. L. (eds.). Montana Bureau of Mines and Geology, Spec. Pub. No. 92, pp. 147–210.

Page, N. J. (1977). Stillwater complex, Montana: Rock succession, metamorphism and structure of the complex and adjacent rocks. U.S. Geol. Surv. Prof. Paper 999.

Pallister, J. S., and Hopson, C. A. (1981). Samail ophiolite plutonic suite: Field relations, phase variation; cryptic variation and layering, and a model of a spreading ridge magma chamber. *J. Geophys. Res.* 86:2593–2644.

Pallister, J. S., and Knight, R. J. (1981). Rare-earth element geochemistry of the Samail ophiolite near Ibra, Oman. *J. Geophys. Res.* 86:2673–2697.

Palme, H., Suess, J. H. E., and Zeh, H. D. (1981). Abundances of the elements in the solar system. In: *Landolt-Bornstein 82 (Astronomy and Astrophysics),* Schaifers, K., and Voight, H. H. (eds.). Berlin, Heidelberg, New York: Springer, pp. 257–273.

Papunen, H., and Vorma, A. (1985). Nickel deposits in Finland, a review. *Geol. Surv. Finland Bull.* 333:123–144.

Papunen, H., and Gorbunov, G. I. (1985). Nickel-copper deposits of the Baltic Shield and Scandinavian Caledonides. *Finland Geol. Surv. Bull.* 333:394.

Parks, J., and Hill, R. E. T. (1986). Phase compositions and cryptic variation in a 2.2-km section of the Windimurra layered gabbroic intrusion, Yilgarn block, Western Australia—A comparison with the Stillwater complex. *Econ. Geol.* 81:1196–1202.

Parsons, I. (1981). The Klokken gabbro-syenite complex, South Greenland: Quantitative interpretation of mineral chemistry. *J. Petrology* 22:233–260.

Parsons, I. (1979). The Klokken gabbro-syenite complex, South Greenland: Cryptic variation and origin of inversely graded layering. *J. Petrology* 20:653–694.

Paterson, H. L., Donaldson, M. J., Smith, R. N., Lenard, M. F., Gresham, J. J., Boyack, D. J., and Keays, R. R. (1984). Nickeliferous sediments and sediment-associated nickel ores at Kambalda, Western Australia. In: *Sulfide Deposits in Mafic and Ultramafic Rocks,* Buchanan, D. L., and Jones, M. J. (eds.). London: Inst. of Min. & Metal, pp. 81–94.

Pattison, E. F. (1979). The Sudbury sub-layer: Its characteristics and relationships with the main mass of the Sudbury Irruptive. *Can. Mineral.* 17(2):257–274.

Pauly, H. (1969). White cast iron with cohenite, schreibersite, and sulfides from tertiary basalts on Disko, Greenland. *Geol. Soc. Denmark Bull.* 19(1):30.

Peredery, W. V., and Morrison, G. G. (1984). Discussion of the origin of the Sudbury structure. In: *Geology and Ore Deposits of the Sudbury Structure,* Pye, E. G., Naldrett, A. J., and Giblin, P. E. (eds.). Ont. Geol. Surv. Spec. vol. 1, pp. 491–512.

Peredery, W. V., and Geological Staff, (1982). Geology and nickel sulfide deposits of the Thompson belt, Manitoba. In: *Robinson Volume,* Hutchinson, R. W. (ed.). Geol. Assoc. Can. Spec. Paper No. 25, pp. 165–209.

Peredery, W. V. (1979). Relationships of ultramafic amphibolites to metavolcanic rocks and serpentinites in the Thompson belt, Manitoba. *Can. Mineral.* 17(2):187–200.

Peredery, W. V., and Naldrett, A. J. (1975). Petrology of the Upper Irruptive rocks, Sudbury, Ontario, Canada. *Econ. Geol.* 70:164–175.

Phillips, B., and Muan, A. (1960). Stability relations of iron oxides: Phase equilibria in the system Fe_3O_4–Fe_2O_3 at oxygen pressures up to 45 atmospheres. *J. Phys. Chem.* 64:1451.

Phinney, W. C. (1969). The Duluth Complex in the Gabbro Lake quadrangle, Minnesota. *Minn. Geol. Surv. Rep. Invest.* 9:20.

Phinney, W. C. (1972). Northwestern part of the Duluth Complex. In: *Geology of Minnesota: A Centennial Volume.* Minn. Geol. Surv., St. Paul, Minnesota, pp. 333–345.

Platt, J. P., Allchurch, P. D., and Rutland, R. W.

R. (1978). Archean tectonics in the Agnew supracrustal belt, Western Australia. *Precambrian Res.* 7:3–30.

Podmore, F., and Wilson, A. H. (1987). A reappraisal of the structure, geology and emplacement of the Great Dyke, Zimbabwe. In: *Mafic Dyke Swarms,* Halls, H. C., and Fahrig, W. F. (eds.). Geol. Assoc. Canada, Spec. Paper 34, pp. 317–330.

Prendergast, M. D. (1988). The geology and economic potential of the PGE rich main sulfide zone of the Great Dyke, Zimbabwe. In: *Geoplatinum '87,* Pritchard, H., Potts, P. J., Bowles, J. F. W., and Cribb, S. J. (eds.). Proc. of the Symp. GeoPlatinum '87, Open University. London: Elsevier, pp. 281–302.

Prichard, H. M., Neary, C. R., and Potts, P. J. (1986). Platinum group minerals in the Shetland Ophiolite. In: *Metallogeny of Basic and Ultrabasic Rocks,* Gallagher, M. J., Ixer, R. A., Neary, C. R., and Prichard, H. M. (eds.). London: Special Publication of Inst. Min. Metall.

Prinzhofer, A., Nicolas, A., Cassard, D., Moutte, J., Leblanc, M., Paris, J. P., and Rabinivitch, M. (1980). Structures in the New Caledonia peridotites-gabbros: Implications for oceanic mantle and crust. *Tectonophysics* 69:85–112.

Pye, E., Naldrett, A. J., and Giblin, P. The geology and ore deposits of the Sudbury structure. *Ont. Geol. Surv.* Spec. Vol. No. 1, 603 pp.

Pyke, D. R., Naldrett, A. J., and Eckstrand, O. R. (1973). Archean ultramafic flows in Munro Township, Ontario. *Geol Soc. Am. Bull.* 84:955–978.

Pyke, D. R., and Middleton, R. S. (1970). Distribution and characteristics of the sulfide ores of the Timmins area. Ont. Div. Mines, MP 41, 14 pp. Accompanied by map, scale 1 inch to 2 miles.

Quirke, T. T. Jr., Cranstone, D. A., Bell, C. K., and Coats, C. J. A. (1970). Geology of the Moak-Setting Lakes area, Manitoba (Manitoba Nickel Belt). *Geol. Assoc. Canada* 23rd Ann. Mtg. Winnipeg, Guidebook, Field trip No. 1.

Rae, D. R. (1975). Inclusions in the sublayer from Strathcona mine, Sudbury and their significance. Unpubl. M.Sc. thesis, Univ. of Toronto, Toronto, Ontario.

Raedeke, L. D., and Vian, R. W. (1986). A three-dimensional view of mineralization in the Stillwater J-M Reef. *Econ. Geol.* 81:1187–1195.

Raedeke, L. D., and McCallum, I. S. (1984). Investigations in the Stillwater Complex, Part II. Petrology and petrogenesis of the Ultramafic series. *J. Petrol.* 25:395–420.

Rajamani, V., Naldrett, A. J. (1978). Partitioning of Fe, Co, Ni, and Cu between sulfide liquid and basaltic melts and the composition of Ni-Cu sulfide deposits. *Econ. Geol.* 73:82–93.

Rao, B. V., Naldrett, A. J., and Evensen, N. M. (1984). Grant 146. Crustal contamination in the sublayer, Sudbury Igneous Complex: A combined trace element and strontium isotope study. In: *Geoscience Research Programme, Summary of Research 1983–1984,* Milne, V. C., (ed.). Ontario Geolog. Survey Misc. Paper 121, pp. 128–146.

Rao, B. V., Naldrett, A. J., Evensen, N. M., and Dressler, B. O. (1983). Grant 146. Contamination and genesis of the Sudbury Ores. In: *Geoscience Research Programme, Summary of Research 1982–1983,* Milne, V. C. (ed.). Ontario Geolog. Survey Misc. Paper 113, pp. 139–152.

Rao, B. V., and Ripley, E. M. (1983). Petrochemical studies of the Dunka Road Cu-Ni deposit. *Econ. Geol.* 78:1222–1238.

Ridler, R. H. (1966). Petrographic study of a Crow Lake ultrabasic sill, Keewatin volcanic belt, Northwestern Ontario. Unpubl. M. A. Sc. thesis, Univ. of Toronto, Toronto, Ontario.

Ripley, E. M. (1986). Applications of stable isotope studies to problems of magmatic sulfide ore genesis with special reference to the Duluth Complex, Minnesota. In: *Geology and Metallogeny of Copper Deposits,* Friedrich, G. H., Genkin, A. D., Naldrett, A. J., Ridge, J. D., Sillitoe, R. H., and Vokes, F. M. (eds.). Society for Geology Applied to Ore Deposits, Spec. Publ. No. 4, Berlin, Heidelberg: Springer-Verlag, pp. 25–42.

Ripley, E. M. (1981). Sulfur isotopic abundances of the Dunka Road Cu-Ni deposit, Duluth Complex, Minnesota. *Econ. Geol.* 76:619–620.

Robertson, A. H. F., and Woodcock, N. H. (1980). Tectonic setting of the Troodos massif in the eastern Mediterranean. In: *Ophiolites: International Ophiolite Symposium, Cyprus,* Panayiotou, A. (ed.). Geology, 11, pp. 400–404.

Robins, B. (1982). Geology and petrology of the Rognsund intrusion, Finnmark, northern Norway. *Norg. Geol. Unders.* 371:1–55.

Robins, B., and Gardner, P. M. (1974). Synorogenic layered basic intrusions in Seiland province, Finnmark. *Norg. Geol. Unders.* 312:91–130.

Roeder, P. L. (1974). Activity of iron and olivine solubility in basaltic liquids. *Earth Planet Sci.* (Lett.) 23:397–410.

Roeder, P. L., and Emslie, R. F. (1970). Olivine-liquid equilibrium. *Contr. Mineral. Petrol.* 29:275–289.

Rosenqvist, T. (1954). A thermodynamic study of the iron, cobalt and nickel sulfides. *J. Iron & Steel Inst.* 176:37–57.

Ross, J. R., and Keays, R. R. (1979). Precious metals in volcanic-type nickel sulfide deposits in Western Australia. Part 1: Relationship with the composition of the ores and their host rocks. *Can. Mineral.* 17(2):417–436.

Ross, J. R., and Hopkins, G. M. F. (1975). The nickel sulfide deposits of Kambalda, Western Australia. In: *Economic Geology of Australia and Papua-New Guinea,* Knight, C. (ed.). Australian Inst. Min. Metall. Monograph 5, 1, Metals, pp. 100–121.

Ross, J. V., Mercier, J. C., Ave-Lallement, H. G., Carter, N. L., and Zimmerman, J. (1980). The Vuorinos ophiolite complex, Greece: The tectonite suite. *Tectonophysics* 70:63–83.

Ruckmick, J. C., and Noble, J. A. (1959). Origin of the ultramafic complex at Union Bay southeastern Alaska. *Bull. Geol. Soc. Am.* 70:981–1018.

Ruttan, G. D. (1955). Geology of Lynn Lake, Canada. *Inst. Min. Metall. Bull.* 48:339–348.

Ryshenko, B., and Kennedy, G. C. (1973). The effect of pressure on the eutectic in the system Fe-FeS. *Am. J. Sci.* 273:803–810.

Sage, R. P., and Troop, W. R. (1974). Operation Winisk, Big Trout Lake Area; Ontario Div. of Mines, Preliminary Map P. 860, scale 1 inch to 1 mile.

Schiffries, C. M. (1982). The petrogenesis of a platiniferous dunite pipe in the Bushveld complex: Infiltration metasomatism by a chloride solution. *Econ. Geol.* 77:1439–1453.

Schlegel, H., and Schiller, A. (1952). Die Schmelzund Kristallisations gleischgeinwichte in System Cu-Fe-S und ihre Bedentung for Kupfergewinnung. *Freiberg. Forsch. Sec.* Be., 2:1–32.

Schmincke, H. -U., Rautenschlein, M., Robinson, P. T., and Mehegen, J. T. (1983). Troodos extrusive series of Cyprus: A comparison with oceanic crust. *Geology* 11:405–409.

Schmulian, M. L. (1984). Windarra nickel deposits, Western Australia. In: *Sulfide Deposits in Mafic and Ultramafic rocks,* Buchanan, D. L., and Jones, M. J. (eds.). London: Institut Min. Metall. Spec. Publ. pp. 95–102.

Scholtz, D. L. (1952). Fourth bi-annual report on the Insizwa mine. Unpubl. company report, Nickel Corp. of Africa Ltd., 9 pp.

Scholtz, D. L. (1936). The magmatic nickeliferous ore deposits of East Griqualand and Pondoland. *Trans Geo. Soc. S. Afr.* 39:81–210.

Schwellnus, J. S. I., Hiemstra, S. A., and Gasparrini, E. (1976). The Merensky reef at the Atok platinum mine and its environs. *Econ. Geol.* 71:249–260.

Scoates, R. F. J., and Eckstrand, O. R. (1986). Platinum-group elements in the upper central layered zone of the Fox River sill, Northeastern Manitoba. *Econ. Geol.* 81:1137–1146.

Scoates, R. F. J. (1984). A preliminary stratigraphic examination of the ultramafic zone of the Bird River Sill. *Geol. Surv. Canada Rep.* GS-12A, pp. 70–83.

Scoates, R. F. J. (1983). Preliminary stratigraphic examination of the ultramafic zone of the Bird River sill. *Manitoba Mineral Res. Div. Rept. Field Activities,* pp. 70.–83.

Scott, R. W. (1980). The geology and petrology of a portion of the Bell River Complex in Bourbaux Township, Quebec. Unpubl. M.Sc. thesis, Univ. of Toronto, Toronto, Ontario.

Scott, S. D., Naldrett, A. J., and Gasparrini, E. (1974). Regular solution model for the $Fe_{1-x}S - Ni_{1-x}S$ (Mss) solid solution (Absr.). *Internat. Mineral. Assoc.* 9th Gen. Meeting, Berlin and Regensburg, p. 172.

Scribbins, B., Rae, D. R., and Naldrett, A. J. (1984). Mafic and ultramafic inclusions in the sublayer of the Sudbury Igneous Complex. *Can. Mineral.* 22(1):67–75.

Seccombe, P. K., Groves, D. I., Bickle, M. J., and Donaldson, M. J. (1981). Sulfide paragenesis and sulfur mobility in Fe-Ni-Cu sulfide ores at Lunnon and Juan main shoots, Kambalda: textural and sulfur isotopic evidence. *Econ. Geol.* 76:1675–1685.

Seccombe, P. K., Groves, D. I., Binns, R. A., and Smith, J. W. (1977). A sulfur isotope study to test a genetic model for Fe-Ni sulfide mineralization at Mt. Windarra, Western Australia. In: Stable isotopes in Earth Sciences. *New Zealand Dept. of Scientific and Industrial Research Bull.* 220:187–202.

Segerstrom, K., and Carlson, R. R. (1982). Geological map of the banded upper zone of the Stillwater Complex and adjacent rocks, Stillwater, Sweet Grass and Park Counties, Montana. US Geol. Survey Map 1-1383, 2 sheets, scale 1:24,000.

Sharma, R. C., and Chang, Y. A. (1980). Thermodynamics and phase relationsips of transition metal-sulfur systems: IV. Thermodynamic properties of Ni-S liquid. Phase and calculation of Ni-S phase diagrams. *Metall. Trans.* B. 11B:139–146.

Sharp, J. I. (1965). Field relations of Matagami sulfide masses bearing on their disposition in time and space. *Trans. Can. Inst. Min. Met.* 68:265–278.

Sharp, W. E. (1969). Melting curves of sphalerite, galena and pyrrhotite and the decomposition of pyrite between 30 and 65 kilobars. *J. Geophys. Res.* 74:1645–1652.

Sharpe, M. R. (1985). Strontium isotopic evidence for preserved density stratification from the main zone of the Bushveld Complex, South Africa. *Nature* 316:119–126.

Sharpe, M. R. (1981). The chronology of magma in fluxes to the eastern compartment of the

Bushveld Complex as exemplified by its marginal border groups. *J. Geol. Soc. London* 138:206–307.

Shewman, R. W., and Clark, L. A. (1970). Pentlandite phase relations in the Fe-Ni-S system and notes on the monosulfide solid solution. *Can. J. Earth Sci.* 7:67–85.

Shima, H., and Naldrett, A. J. (1975). Solubility of sulfur in an ultramafic melt and the relevance of the system Fe-S-O. *Econ. Geol.* 70:960–967.

Shimazaki, H., and MacLean, W. H. (1976). An experimental study on the partition of zinc and lead between silicate and sulfide liquids. *Mineral. Deposita* II:125–132.

Simkin, T., and Smith, J. V. (1970). Minor element distribution in olivine. *J. Geol.* 78:304–325.

Simonen, A. (1980). The Precambrian in Finland. *Bull. Comm. Geol. Finlande* 168:58.

Sims, P. K., and Morey, G. B. (1972). Geology of Minnesota: A centennial volume. *Minn. Geol. Survey,* St. Paul, Minn. p. 632.

Skinner, B. J., Luce, F. D., Dill, J. A., Ellis, D. E., Hagan, H. A., Lewis, D. M., Odell, D. A., Sverjensky, D. A., and Williams, N. (1976). Phase relations in the ternary portion of the system Pt-Pd-Fe-As-S. *Econ. Geol.* 71:1469–1475.

Slaught, W. H. (1951). A petrographic study of the Copper Cliff offset in the Sudbury District; Unpubl. M.Sc. thesis, McGill University.

Smirnov, M. F. (1966). *The Noril'sk Nickeliferous Intrusions and Their Sulfide Ores.* Moscow: Nedra press.

Smirnov, V. I. (1977). Ore deposits of the USSR, Vol. III. (in English). San Francisco: Pitman, 492 pp.

Smith, F. G. (1963). *Physical Geochemistry.* Reading: Addison-Wesley, pp. 304–307.

Soderholm, K., and Inkinen, O. (1982). The Tornio layered intrusion—A recently discovered intrusion with chromite horizons in northern Finland. *Geol. Soc. Finland Bull.* 54:15–24.

Souch, B. E., Podolsky, T., and Geological Staff of the International Nickel Co. of Canada Ltd. (1969). The sulfide ores of Sudbury: Their particular relation to a distinctive inclusion-bearing facies of the nickel irruptive. *Econ. Geol..* Monograph 4, pp. 252–261.

Sparks, R. S. J., Huppert, H. E., and Turner, J. S. (1984). The fluid dynamics of evolving magma chambers. *Phil. Trans. Roy. Soc.* A.310:511–534.

Stockman, H. W., and Hlava, P. F. (1984). Platinum-group minerals in alpine chromitites from southwestern Oregon. *Econ. Geol.* 79:491–508.

Stolz, G. W., and Nesbitt, R. W. (1981). The Komatiite Nickel Sulfide Association at Scotia: A petrochemical investigation of the ore environment. *Econ. Geol.* 76:1480–1502.

Stumpfl, E. F. (1986). Distribution, transport and concentration of platinum group elements. In: *Metallogeny of Basic and Ultrabasic Rocks,* Gallagher, M. J., Ixer, R. A., Neary, C. R., and Prichard, H. M., (eds.). Spec. Publ., Inst. Mining and Metallurgy, pp. 379–394.

Stumpfl, E. F. (1974). The genesis of platinum deposits: Further thoughts. *Minerals. Sci. Eng.* 6:120–141.

Stumpfl, E. F. (1962). Some aspects of the genesis of Pt deposits. *Econ. Geol.* 57:619–622.

Sturt, B. A., Pringle, I. R., and Ramsay, D. M. (1978). The Finnmarkian phase of the Caledonian orogeny. *J. Geol. Soc. London* 135:597–610.

Tamrazyan, G. P. (1971). Siberian continental drift. *Tectonophysics* 11:433–460.

Taylor, S. R. (1974). Abundances of chemical elements in the continental crust: A new table. *Geochim. et Cosmochim. Acta* 28:1273–1285.

Thompson, J. F. H., Barnes, S. J., and Duke, J. M. (1984). The distribution of nickel and iron between olivine and magmatic sulfides in some natural assemblages. *Can. Mineral.* 22:55–66.

Thompson, J. F. H., and Naldrett, A. J. (1984). Sulfide–silicate reactions as a guide to Ni-Cu-Co mineralization in central Main. In: *Sulfide Deposits in Mafic and Ultramafic Rocks,* Buchanan, D. L., and Jones, M. J. (eds.). London: Inst. Min. Metall. Special Publ., pp. 103–113.

Todd, S. G., Keith, D. W., LeRoy, L. W., Schissel, D. J., Mann, E. L., and Irvine, T. N. (1982). The J-M platinum-palladium reef of the Stillwater Complex, Montana. I. Stratigraphy and Petrology. *Econ. Geol.* 77:1454–1480.

Trueman, D. (1971). Petrological, structural and magmatic studies of a layered basic intrusion, Bird River sill, Manitoba. Unpubl. M.Sc. thesis, Univ. Manitoba, 67 pp.

Turner, A. R., Wolfgram, D., and Barnes, S. J. (1985). Geology of the Stillwater County Sector of the J-M Reef, including the Minneapolis adit. In: *Stillwater Complex,* Czamanske, G. C., and Zientek, M. L. (eds.). Montana Bureau of Mines and Geology Spec. Pub. No. 92, pp. 210–230.

Turner, J. S., Huppert, H. E., and Sparks, R. S. J. (1986). Komatiites II: Experimental and theoretical investigations of post-emplacement cooling and crystallization. *J. Petrol.* 27:397–438.

Ueno, T. (1981). A study on the system Cu-Fe-S—Phase equilibrium in the central portion. Unpubl. Ph.D. thesis, Tohoku University, 189 pp.

Usselman, T. M., Hodge, D. S., Naldrett, A. J.,

and Campbell, I. H. (1979). Physical constraints on the characteristics of nickel sulfide ore in ultramafic lavas. *Can. Mineralogist* 17:(2):361–372.

Usselman, T. M. (1975). *Am. J. Sci.* 275:278–290.

van der Merwe, M. J. (1976). The layered sequence of the Potgietersrus limb of the Bushveld complex. *Econ. Geol.* 71:1337–1351.

Van Zyl, J. P. (1970). The petrology of the Merensky reef and associated rocks in Swartklip 988, Rustenburg. In: *Symposium on the Bushveld and Igneous Complex and other Layered Intrusions,* Visser, D. J. L., and von Gruenewaldt, G. (eds.). Geol. Soc. South Africa Spec. Publ. 1, pp. 242–265.

Vermaak, C. F. (1976a). The nickel pipes of Vlakfontein and vicinity, Western Transvaal. *Econ. Geol.* 71:261–286.

Vermaak, C. F. (1976b). The Merensky reef—Thoughts on its environment and genesis. *Econ. Geol.* 71:1270–1298.

Viljoen, M. J., Bernasconi, A., van Coller, N., Kinlock, E., and Viljoen, R. P. (1976). The geology of the Shangani nickel deposit. *Econ. Geol.* 71:76–95.

Viljoen, M. J., and Viljoen, R. P. (1969). Evidence of the existence of a mobile extrusive peridotitic lava from the Komati formation of the Onverwacht group. *Geol. Soc. S. Africa* (Spec. Pub.) 2:87–113.

Viljoen, R. P., and Viljoen, M. J. (1970). The geology and the geochemistry of the layered ultramafic bodies of the Kaapmuiden area, Barberton Moutainland. Symposium on the Bushveld and other layered intrusions. *Geol. Soc. S. Africa* (Spec. Pub.) pp. 661–688.

Vlasov, K. A., Kuz'menko, M. V., and Es'kova, E. M. (1959). *The Loverzero Alkaline Massif.* (in Russian) Izd. Acad. Sci., Moskva, pp. 1–624. (English translation: Oliver and Boyd, Edinburgh, 1966).

Von Gruenewaldt, G., Hatton, C. J., Merkle, R. K. W., and Gain, S. B. (1986). Platinum-group element-chromitite associations in the Bushveld Complex. *Econ. Geol.* 81:1067–1079.

Von Gruenewaldt, G., Sharpe, M. R., and Hatton, C. J. (1985). The Bushveld complex: Introduction and review. *Econ. Geol.* 80:803–812.

Von Gruenewaldt, G. (1979). A review of some recent concepts of the Bushveld complex with particular reference to the sulfide mineralization. *Can. Mineral.* 17(2):233–256.

Von Gruenewaldt, G. (1977). The mineral resources of the Bushveld complex. *Mineral. Sci. Eng.* 9:83–95.

Von Gruenewaldt, G. (1973). The main and upper zones of the Bushveld complex in the Roossenekal area, Eastern Transvaal. *Geol. Soc. S. Afr. Trans.* 76:207–227.

Waal de, S. A., (1977). Carbon dioxide and water from metamorphic reactions as agents for sulfide and spinel precipitation in mafic magmas. *Geol. Soc. S. Africa Trans.* 80:193–197.

Wadsworth, W. J. (1970). The Aberdeenshire layered intrusion of north-east Scotland. *Geol. Soc. S. Africa* (Spec. Pub.) 1:565–575.

Wager, L. R., and Brown, G. M. (1968). *Layered Igneous Rocks.* London: Oliver and Boyd.

Wager, R. E., Podolsky, T., Alcock, R. A., Weiblen, P. W., and Phinney, W. C. (1969). A comparison of the Cu-Ni deposits of the Sudbury and Duluth basins. Am. Inst. Mining Engineers 52nd Ann. Mtg., Minnesota Sect., 30th Ann. Mining Symposium, Proc. pp. 95–96.

Wagner, P. A. (1929). *The Platinum Deposits and Mines of South Africa.* Cape Town: C. Struik (Pty) Ltd., 356 pp.

Wandke, A., and Hoffman, R. (1924). A study of the Sudbury ore deposits. *Econ. Geol.* 19:169–204.

Weiblen, P. W., and Morey, G. B. (1980). A summary of the stratigraphy, petrology and structure of the Duluth complex. *Am. J. Sci.* 280-A:88–133.

Weiblen, P. W., and Morey, G. B. (1975). The Duluth complex—A petrographic and tectonic summary. 36th Ann. Minn. Min. Symp. Dept. of Conferences and Continuing Ed., Univ. of Minnesota, pp. 72–95.

Wendlandt, R. F. (1982). Sulfide saturation of basalt and andesite melts at high pressures and temperatures. *Am. Mineral.* 67:877–885.

Wendlandt, R. F., and Huebner, J. S. (1979). Melting relations of portions of the system Fe-S-O at high pressure and applications to the compositions of earth's core (Absr.). Lunar Planetary Sc. X, Lunar & Planetary Inst., pp. 1329–1331.

Williams, D. A. C. (1979). The association of some nickel sulfide deposits with komatiite volcanism in Rhodesia. *Can. Mineral.* 17:337–350.

Wilson, A. H., Naldrett, A. J., and Tredoux, M. Distribution and controls of platinum group element and base element mineralization in the Darwendale subchamber of the Great Dyke, Zimbabwe. *Geology* (in press).

Wilson, A. H., and Prendergast, M. D. The Great Dyke of Zimbabwe. I: Tectonic setting, stratigraphy, petrology, structure, emplacement and crystallization: In: *5th Magmatic Sulfide Field Conference, Harare, Zimbabwe,* Prendergast, M. D., and Jones, M. (eds.). London: Inst. of Mining & Metall., (in press).

Wilson, A. H., and Prendergast, M. D. (1987). The Great Dyke of Zimbabwe—An overview. In: *Abstracts and Field Guide to 5th Magmatic Sulfides Field Conference, IGCP Project 161, Harare, Zimbabwe,* Campbell, A. (ed.). pp. 23–25.

Wilson, A. H. (1982). The geology of the Great "Dyke," Zimbabwe: The ultramafic rocks. *J. Petrology,* 42:109–124.

Wilson, A. H., and Wilson J. F. (1981). The Great Dyke. In: *Precambrian of the Southern Hemisphere,* Hunter, D. R. (ed.). Amsterdam: Elsevier, pp. 572–578.

Wilson, H. D. B., Kilburn, L. C., Graham, A. R., and Rambal, K. (1969). Geochemistry of some Canadian nickeliferous ultrabasic intrusions. *Econ. Geol. Mon.* 4:294–309.

Wilson, H. D. B., and Anderson, D. T. (1959). The composition of Canadian sulfide ore deposits. *Trans. Can. Inst. Min. Met.* 62:327–339.

Wilson, J. R. (1987). Petrology of the marginal and lower layered series rocks of the Hyllingen series, Fongen-Hyllingen complex, Norway. In: *Origins of Igneous Layering,* Parsons, I. (ed.). NATO Advanced Research Series, D. Reidel, Dordrecht (in press).

Wisser, D. J. L., and von Gruenewaldt, G. (1970). Symposium on the Bushveld and other layered complexes. *Geol. Soc. S. Africa* (Spec. Publ.) 1:763.

Wolfe, W. (1970). Distribution of Cu, Ni, Co and S in mafic intrusive rocks of the Kamiscotia-Whitesides area, district of Cochrane. Ont. Dept. of Mines Misc. Paper 44, 28 pp.

Woodall, R., and Travis, G. A. (1969). The Kambalda nickel deposits, Western Australia. Commonwealth Min. Metal. Cong., 9th, Pub., Paper 26, 17 pp.

Woolrich, P., Cowden, A., and Giorgetta, N. E. (1981). The chemical and mineralogical variations in the nickel mineralization associated with the Kambalda dome, Western Australia. *Econ. Geol.* 76:1629–1644.

Worst, B. G. (1960). The Great Dyke of Southern Rhodesia. *Southern Rhodesia Geol. Surv. Bull.* 47.

Wyllie, P. J. (ed). (1967). *Petrography and Petrology. Ultramafic and Related Rocks.* New York: John Wiley and Sons.

Yakolev, Yu. N., and Yakolava, A. K. (1986). On the role of metamorphism in the formation of nickel-copper sulfide deposits in the Kola peninsula. In: *Geology and Metallogeny of Copper Deposits,* Friedrich, G. H., Genkin, A. D., Naldrett, A. J., Ridge, J. D., Sillitoe, R. H., and Vokes, F. M. (eds.). Berlin, Heidelberg: Springer-Verlag, pp. 135–150.

Yund, R. A., and Kullerud, G. (1966). Thermal stability of assemblages in the Cu-Fe-S system. *J. Petrol.* 7:454–488.

Zientek, M. L., Foose, M. P., and Mei, L. (1986). Palladium, platinum and rhodium contents of rocks near the lower margin of the Stillwater complex, Montana. *Econ. Geol.* 81:1169–1170.

Zientek, M. L., Czamanske, G. K., and Irvine, T. N. (1985). Stratigraphy and nomenclature for the Stillwater Complex. In: *Stillwater Complex,* Czamanske, G. K., and Zientek, M. L. (eds.). Montana Bureau of Mines and Geology Spec. Pub. No. 92, pp. 21–32.

Zolotuchin, V. V., and Vasilev, V. R. (1967). *Particular Characteristics of Formation of Some Trap Intrusives of the South East Siberian Platform.* Moscow: Nauka, 231 pp.

Zurbrigg, H. F. (1963). Thompson mine geology. *Can. Inst. Min. Met. Trans.* 66:227–236.

INDEX